Introduction to
Functional
Equations

Introduction to
Functional
Equations

Prasanna K. Sahoo

Palaniappan Kannappan

CRC Press
Taylor & Francis Group
Boca Raton   London   New York

CRC Press is an imprint of the
Taylor & Francis Group an **informa** business
A CHAPMAN & HALL BOOK

Chapman & Hall/CRC
Taylor & Francis Group
6000 Broken Sound Parkway NW, Suite 300
Boca Raton, FL 33487-2742

First issued in paperback 2017

© 2011 by Taylor and Francis Group, LLC
Chapman & Hall/CRC is an imprint of Taylor & Francis Group, an Informa business

No claim to original U.S. Government works

ISBN-13: 978-1-4398-4111-2 (hbk)
ISBN-13: 978-1-138-11455-5 (pbk)

Library of Congress Cataloging-in-Publication Data

Sahoo, Prasanna,
    Introduction to functional equations / Prasanna K. Sahoo, Palaniappan Kannappan.
        p. cm.
    Includes bibliographical references and index.
    ISBN 978-1-4398-4111-2 (hardback)
    1. Functional equations. I. Kannappan, Pl. (Palaniappan) II. Title.

QA431.S15 2011
515'.75--dc22   2010045164

Visit the Taylor & Francis Web site at
http://www.taylorandfrancis.com

and the CRC Press Web site at
http://www.crcpress.com

Dedication

Dedicated by

Prasanna Sahoo
to his wife Sadhna and son Amit

and

Palaniappan Kannappan
to his wife Renganayaki and his grandchildren

Contents

Preface

The subject of functional equations forms a modern branch of mathematics. The origin of functional equations came about the same time as the modern definition of function. From 1747 to 1750, J. d'Alembert published three papers. These three papers were the first on functional equations. The first significant growth of the discipline of functional equations was stimulated by the problem of the parallelogram law of forces (for a history see Aczél (1966)). In 1769, d'Alembert reduced this problem to finding solutions of the functional equation $f(x + y) + f(x - y) = 2f(x)f(y)$. Many celebrated mathematicians including N.H. Abel, J. Bolyai, A.L. Cauchy, J. d'Alembert, L. Euler, M. Fréchet, C.F. Gauss, J.L.W.V. Jensen, A.N. Kolmogorov, N.I. Lobačevskii, J.V. Pexider, and S.D. Poisson have studied functional equations because of their apparent simplicity and harmonic nature.

Although the modern study of functional equations originated more than 260 years ago, a significant growth of this discipline occurred during the last sixty years. In 1900, David Hilbert suggested in connection with his fifth problem that, while the theory of differential equations provides elegant and powerful techniques for solving functional equations, the differentiability assumptions are not inherently required. Motivated by Hilbert's suggestion many researchers have treated various functional equations without any (or with only mild) regularity assumption. This effort has given rise to the modern theory of functional equations. The comprehensive books by S. Pincherle (1906, 1912); E. Picard (1928); G. Hardy, J.E. Littlewood and G. Polya (1934); M. Ghermanescu (1960); J. Aczél (1966); and M. Kuczma (1968) also advanced considerably the discipline of functional equations. Recent books by A.N. Sarkovskii and G.P. Reljuch (1974); J. Aczél and Z. Daróczy (1975); J. Dhombres (1979); M. Kuczma (1985); J. Aczél (1987); J. Smital (1988); J. Aczél and J. Dhombres (1989); M. Kuczma, B. Choczewski, and R. Ger (1990); B. Ramachandran and K.-S. Lau (1991); L. Székelyhidi (1991); E. Castillo and M.R. Ruiz-Cobo (1992); C.R. Rao and D.N. Shanbhag (1994); B.R. Ebanks, P.K. Sahoo and W. Sander (1998); P.K. Sahoo and T. Riedel (1998); D.H. Hyers, G. Isac and Th.M. Rassias (1998); S.-M. Jung (2001); S. Czerwik (2002); I. Risteski and V. Covachev (2002); and Pl. Kannappan (2009) have contributed immensely to the further advancement of this discipline.

This book grew out of a set of classnotes by the first author who taught functional equations as a graduate level introductory course at the University of Louisville. Our goal in writing this book is to communicate the mathematical ideas of this subject to the reader and provide the reader with an elementary exposition of the discipline. All functions appearing in the functional equations treated in this book are real or complex valued. We did not cover any functional equation where the unknown functions take on values on algebraic structures such as groups, rings or fields. The reason for this is to make the presentation as accessible as possible to students from a variety of disciplines. However, at the end of each chapter we have included a section to point out various developments of the main equations treated in that chapter. In addition, we discuss functional equations in abstract domains like semigroups, groups, or Banach spaces. The innovation of solving functional equations lies in finding the right tricks for a particular equation. We have tried to be generous with explanations. Perhaps there will be places where we belabor the obvious. Each chapter (except Chapters 5 and 6) ends with a set of exercises and some of these problems are adapted from Kuczma (1964), Stamate (1971), and Makarov et al. (1991).

We now give a brief description of the contents. Chapters 1 through 17 deal with a wide variety of functional equations and Chapters 18 through 24 deal with stability of some of the functional equations considered in the earlier chapters. Chapter 1 gives an account of additive functions. In this chapter, we treat the additive Cauchy functional equation and show that continuous or locally integrable additive functions are linear. We further explore the behavior of discontinuous additive functions and show that they display a very strange behavior: their graphs are dense in the plane. To this end, we briefly discuss the Hamel basis and its use for constructing discontinuous additive functions. A discussion of complex additive functions is also provided in this chapter. This chapter ends with a set of concluding remarks where we point out some developments and some open problems related to the additive Cauchy functional equation.

In Chapter 2, the remaining three Cauchy functional equations are treated. Chapter 3 presents all four Cauchy functional equations in several variables. In this chapter, we show that every additive function in n variables is a sum of n different additive functions in one variable. Similarly, we show that every multiplicative function in n variables is a product of n distinct multiplicative functions in one variable. Analogous results for exponential functions and logarithmic functions of n variables are also provided in this chapter. Chapter 4 deals with the problem of extending the additive Cauchy functional equation from a smaller domain to a larger domain that contains the smaller domain.

Chapter 5 examines some applications of Cauchy functional equations. In this chapter, Cauchy functional equations are used for deriving formulas for the area of a rectangle, laws of logarithm, simple and compound interest rates, and radioactive disintegration. In this chapter, using Cauchy functional equations, we characterize the geometric probability distribution, the discrete normal probability distribution, and the normal probability distribution.

In Chapter 6, some more applications of Cauchy functional equations are given. Suppose $f_k(n) = 1^k + 2^k + \cdots + n^k$, where n is a positive integer and k is a nonnegative integer. Then $f_k(n)$ denotes the sum of the kth power of the first n natural numbers. Finding formulas for $f_k(n)$ has interested mathematicians for more than 300 years since the time of James Bernoulli (1658–1705). In this chapter, using functional equations, we give formulas for $f_k(n)$ for $k = 1, 2, 3$, and for arbitrary k we suggest a functional relationship for finding a formula. Chapter 6 also contains formulas for the number of possible pairs among n things, cardinality of a power set, and sum of certain finite series. These formulas are all obtained using Cauchy functional equations.

Chapter 7 deals with the Jensen functional equation which arises from the mid-point convexity condition. In this chapter, we derive the solution of this equation and show that every continuous Jensen function is affine. We also give continuous solution of the Jensen functional equation on a closed and bounded interval. In Chapter 8, pexiderized versions of the Cauchy functional equations as well as the Jensen functional equation are studied.

In Chapter 9, we study biadditive functions, quadratic functions, and the quadratic functional equation. First we give the solution of the quadratic functional equation assuming the unknown function to be continuous. Then we present the solution without assuming any regularity condition on the unknown function. Finally, we treat the pexiderized version of the quadratic functional equation.

Chapter 10 examines the solution of the d'Alembert functional equation. In this chapter, we show that every continuous nontrivial solution $f : \mathbb{R} \to \mathbb{R}$ of the d'Alembert functional equation is either $f(x) = \cosh(\alpha x)$ or $f(x) = \cos(\beta x)$, where α and β are arbitrary constants. Furthermore, we show that every nontrivial solution $f : \mathbb{R} \to \mathbb{C}$ of the d'Alembert functional equation is of the form $f(x) = \frac{1}{2}[E(x) + E(x)^{-1}]$, where $E : \mathbb{R} \to \mathbb{C}^\star$ is an exponential function. Section 4 of this chapter presents a characterization of the cosine function by means of a functional equation due to Van Vleck (1910).

In Chapter 11 we continue with the study of functional equations related to various trigonometric functions. In Section 2 of this chapter we determine the general solution of a cosine-sine functional equation $f(x - y) = f(x)f(y) + g(x)g(y)$. In Section 3, we determine the general solution of a sine-cosine functional equation $f(x + y) = f(x)g(y) + g(x)f(y)$. In Section 4, we present the general solution of a sine functional equation $f(x + y)f(x - y) = f(x)^2 - f(y)^2$. Here, we also present the general solution of the functional $f(x+y)g(x-y) = f(x)\,g(x) - f(y)\,g(y)$. Section 5 of this chapter deals with a sine functional inequality. In Section 6, we present an elementary functional equation due to Butler (2003) and its solution due to M.Th. Rassias (2004).

Chapter 12 focuses on a functional equation of Pompeiu (1946), namely $f(x + y + xy) = f(x) + f(y) + f(x)f(y)$. We first present the general solution of this functional equation and then determine the general solution of a generalized Pompeiu functional equation. Chapter 13 deals with the Hosszú functional equation and a generalization of it. In Chapter 14, first we present the continuous solution of the Davison functional equation. Then we find the general solution of this functional equation without any regularity assumption. Chapter 15 examines the Abel functional equation. In this chapter, the general solution of the Abel functional equation is determined without any regularity assumption on the unknown functions.

Chapter 16 deals with some functional equations that arise from the mean value theorem of differential calculus. Functional equations of this type were originated by Pompeiu (1930), but the actual study of this type of functional equations was started by Aczél (1985), Haruki (1979) and Kuczma (1991b). In this chapter, we study a mean value type functional equation and several of its generalizations. In Chapter 17, we examine four functional equations, namely, $f(pr, qs) + f(ps, qr) = (r + s)f(p, q) + (p + q)f(r, s)$; $f(pr, qs) + f(ps, qr) = f(p, q)\,f(r, s)$; $f_1(pr, qs) + f_2(ps, qr) = g(p, q) + h(r, s)$; and $f_1(pr, qs) + f_2(ps, qr) = (r+s)g(p, q) + (p+q)h(r, s)$ that arise in the characterization of distance measures. In this chapter, we determine the general solution of these functional equations on open unit interval $(0, 1)$. In the last section of this chapter we point out several functional equations whose solutions are not presently known.

In 1940 S.M. Ulam (see [142]) posed the following problem: If we replace a given functional equation by a functional inequality, then under what conditions can we say that the solutions of the inequality are close to the solutions of the equation. For example, given a group $(G_1, \cdot)$, a metric group $(G_2, \star)$ with a metric $d(\cdot, \cdot)$ and a positive number ε, the Ulam question is: Does there exist a real number $\delta > 0$ such that if the map $f : G_1 \to G_2$ satisfies $d(f(x \cdot y), f(x) \star f(y)) < \delta$ for all $x, y \in G_1$, then a homomorphism $T : G_1 \to G_2$ exists with $d(f(x), T(x)) < \varepsilon$ for all $x, y \in G_1$? The first affirmative answer to this question was given by D. H. Hyers (1941). Hyers result initiated much of the present research in the stability theory of functional equations. Chapters 18 through 24 present stability of several functional equations studied in the earlier chapters.

In Chapter 18, the Hyers-Ulam stability of the additive Cauchy functional equation is treated. In Section 3, a stability result due to Hyers (1941) is presented by considering the Cauchy difference $(x, y) \mapsto f(x + y) - f(x) - f(y)$ to be bounded. In Section 4, Hyers' theorem is generalized by allowing the Cauchy difference to be unbounded. In this section, the contributions of Aoki (1950) and Rassias (1978) are presented with their proofs. We have also included works of Gajda (1991) and Rassias and Šemrl (1992).

Chapter 19 deals with the Hyers-Ulam stability of the exponential as well as the multiplicative Cauchy functional equations. The notion of superstability is introduced in this chapter. Ger type stability of the exponential and the multiplicative Cauchy functional equations is also investigated. In Section 4, we point out various developments concerning the superstability of exponential and multiplicative functional equations.

The stability of the d'Alembert functional equation and the sine functional equations are the main topics of Chapter 20. In Section 2, we consider the Hyers-Ulam stability of the d'Alembert functional equation and show that if a function $f : \mathbb{R} \to \mathbb{C}$ satisfies the inequality $|f(x+y) + f(x-y) - 2f(x)f(y)| \leq \delta$ for all $x, y \in \mathbb{R}$ and for some $\delta > 0$, then either f is bounded or f is a solution of the d'Alembert functional equation. In Section 3, we treat the stability of sine functional equation, namely, $f(x+y)f(x-y) = f(x)^2 - f(y)^2$. In this section, we prove that any unbounded function $f : \mathbb{R} \to \mathbb{C}$ satisfying the functional inequality $|f(x+y)f(x-y) - f(x)^2 + f(y)^2| \leq \delta$ for all $x, y \in \mathbb{R}$ and for some $\delta > 0$ has to be a solution of the sine functional equation.

Chapter 21 explores the stability of the quadratic functional equation. The stability of the pexiderized quadratic functional equation as well as the Drygas functional equation are also presented. Chapter 22 treats the Hyers-Ulam stability of Davison's functional equation. Here the generalized stability of the Davison functional equation is also considered.

In Chapter 23, we deal with the stability of the Hosszú functional equation. We also consider the stability of a generalization of this functional equation. In Chapter 24 the Hyers-Ulam stability of the Abel functional equation is treated.

This book would not have been possible without the support and encouragement from many of our colleagues. In particular the first author is very thankful to Robert Powers, Thomas Riedel, and Themistocles M. Rassias for their constant encouragement from the very beginning to the end. He would also like to thank many of his students who read the draft copy of this book and made many suggestions for improvements. The University of Louisville has provided the first author with support in the form of sabbatical leave while this book was in its early stage of completion and for this he is very thankful. Perhaps the one person besides us who is most responsible for this book is the senior editor, Sunil Nair of Taylor & Francis Group, who encouraged and published this book. Thanks, Sunil. Our special thanks also go to Judith Simon, project editor; Sarah Morris, editorial assistant; Jessica Vakili, project coordinator; Carole Gustafson, proofreader; and Kevin Craig, cover designer at the Taylor & Francis Group for their commitment to excellence in all aspects of the production of the book. In the book, we have used results from many researchers and many of our collaborators, and we have made honest efforts to pay credit to everyone whose results we have used. If we have missed anyone, we apologize. Finally, this project was an intellectual pursuit under many constraints, including our busy teaching schedules and many other scholastic endeavors. So all errata and suggestions for improvements will be welcomed gratefully by the authors.

References to the bibliography are made as follows: a work is cited using the last name of the author and the year of publication shown between brackets. For instance, Abel (1823) refers to the work of Abel published in 1823. Two different works of Abel that were published in the same year 1823 are cited as Abel (1823a) and Abel (1823b) to differentiate between them by listing the year as 1823a and 1823b. Thus, Abel (1823a) refers to the first item whereas Abel (1823b) refers to the second item in the bibliography listed under Abel for the year 1823.

This book was typeset by the first author in LaTeX, a macro package written by Leslie Lamport for Donald Knuth's TeX typesetting package. The bibliography and index were compiled using BibTeX and MakeIndex, respectively.

Prasanna Sahoo
Palaniappan Kannappan

Chapter 1

Additive Cauchy Functional Equation

1.1 Introduction

The study of additive functions dates back to A.M. Legendre who first attempted to determine the solution of the Cauchy functional equation

$$f(x + y) = f(x) + f(y)$$

for all $x, y \in \mathbb{R}$. The systematic study of the additive Cauchy functional equation was initiated by A.L. Cauchy in his book *Cours d'Analyse* in 1821. Additive functions are the solutions of this additive Cauchy functional equation. This chapter gives an account of additive functions. First, we explain what a functional equation is. Then we treat the additive Cauchy functional equation and show that continuous or locally integrable additive functions are linear. We further explore the behavior of nonlinear discontinuous additive functions and show that they display a very strange behavior: their graphs are dense in the plane. To this end, we briefly discuss the Hamel basis and its use for constructing discontinuous additive functions. We also examine under what other criteria the solution of the Cauchy functional equation is linear. A discussion of complex additive functions is also provided in this chapter. The chapter ends with a set of concluding remarks where we have pointed out some developments and some open problems related to the additive Cauchy functional equation.

Kuczma (1985) gives an excellent exposition on additive functions. Additive functions have also found places in the books of Aczél (1966, 1987), Aczél and Dhombres (1989) and Smital (1988). The general solutions of many functional equations of two or more variables can be expressed in terms of additive, multiplicative, logarithmic or exponential functions. Some of the material in this chapter is adapted from Aczél (1965) and Wilansky (1967).

1

1.2   Functional Equations

An equation involving an unknown function and one or more of its derivatives is called a differential equation. Examples of differential equations are

$$f'(x) + mx = 5$$

and

$$f''(x) + f'(x) + \sin(x) = 0.$$

Differential equations are well studied. Equations involving integrals of an unknown function are called integral equations. Some examples of integral equations are

$$f(x) = e^x - \int_0^x e^{x-t} f(t)\, dt,$$

$$f(x) = \sin(x) + \int_o^1 [\, 1 - x\cos(xt)\,]\, f(t)\, dt$$

and

$$f(x) = \int_0^x [\, tf^2(t) - 1\,]\, dt.$$

As with differential equations, there is a well-studied theory of integral equations.

Functional equations are equations in which the unknowns are functions. Some examples of functional equations are

$$f(x + y) = f(x) + f(y),$$

$$f(x + y) = f(x)f(y),$$

$$f(xy) = f(x)f(y),$$

$$f(xy) = f(x) + f(y),$$

$$f(x + y) = f(x)g(y) + g(x)f(y),$$

$$f(x + y) + f(x - y) = 2f(x)f(y),$$

$$f(x + y) + f(x - y) = 2f(x) + 2f(y),$$

$$f(x + y) = f(x) + f(y) + f(x)f(y),$$

$$f(x + y) = g(xy) + h(x - y),$$

$$f(x) - f(y) = (x - y)h(x + y),$$

$$f(pr, qs) + f(ps, qr) = 2f(p, q) + 2f(r, s),$$
$$g(f(x)) = g(x) + \beta,$$
$$g(f(x)) = \alpha g(x), \quad \alpha \neq 1$$

and

$$f(t) = f(2t) + f(2t - 1).$$

The field of functional equations includes differential equations, difference equations and iterations, and integral equations. In this book we will not cover these topics. Functional equations is a field of mathematics which is over 260 years old. More than 5000 papers have been published in this area.

Functional equations appeared in the literature around the same time as the modern theory of functions. In 1747 and 1750, d'Alembert published three papers that were the first on functional equations (see Aczél (1966)). Functional equations were studied by d'Alembert (1747), Euler (1768), Poisson (1804), Cauchy (1821), Abel (1823), Darboux (1875) and many others. Hilbert (1902) suggested in connection with his 5th problem, that, while the theory of differential equations provides elegant and powerful techniques for solving functional equations, the differentiability assumptions are not inherently required. Motivated by Hilbert's suggestions many researchers in functional equations have treated various functional equations without any (or with mild) regularity assumptions. This effort has given rise to the modern theory of functional equations. The theory of functional equations forms a modern mathematical discipline, which has developed very rapidly in the last six decades.

To solve a functional equation means to find all functions that satisfy the functional equation. In order to obtain a solution, the functions must often be restricted to a specific nature (such as analytic, bounded, continuous, convex, differentiable, measurable or monotonic).

1.3 Solution of Additive Cauchy Functional Equation

In this section, we introduce the additive Cauchy functional equation and determine its regular solution.

Let $f : \mathbb{R} \to \mathbb{R}$, where $\mathbb{R}$ is the set of real numbers, be a function satisfying the functional equation

$$f(x + y) = f(x) + f(y) \tag{1.1}$$

for all x, $y \in \mathbb{R}$. This functional equation is known as the additive Cauchy functional equation. The functional equation (1.1) was first treated by A.M. Legendre (1791) and C.F. Gauss (1809) but A.L. Cauchy (1821) first found its general continuous solution. The equation (1.1) has a privileged position in mathematics. It is encountered in almost all mathematical disciplines.

Definition 1.1. *A function $f : \mathbb{R} \to \mathbb{R}$ is said to be an additive function if it satisfies the additive Cauchy functional equation*

$$f(x + y) = f(x) + f(y)$$

for all $x, y \in \mathbb{R}$.

Definition 1.2. *A function $f : \mathbb{R} \to \mathbb{R}$ is called a linear function if and only if it is of the form*

$$f(x) = c\,x \qquad (\forall\, x \in \mathbb{R}),$$

where c is an arbitrary constant.

The graph of a linear function $f(x) = cx$ is a non-vertical line that passes through the origin and hence it is called linear. The linear functions satisfy the Cauchy functional equations. The question arises, are there any other functions that satisfy the Cauchy functional equation?

We begin by showing that the only continuous solutions of the Cauchy functional equation are those which are linear. This was the result proved by Cauchy in 1821.

Theorem 1.1. *Let $f : \mathbb{R} \to \mathbb{R}$ be a continuous function satisfying the additive Cauchy functional equation (1.1). Then f is linear; that is, $f(x) = cx$ where c is an arbitrary constant.*

Proof. First, let us fix x and then integrate both sides of (1.1) with respect to the variable y to get

$$f(x) = \int_0^1 f(x)dy$$

$$= \int_0^1 [f(x + y) - f(y)]\, dy$$

$$= \int_x^{1+x} f(u)du - \int_0^1 f(y)dy, \qquad \text{where} \quad u = x + y.$$

Since f is continuous, by using the Fundamental Theorem of Calculus, we get

$$f'(x) = f(1 + x) - f(x). \tag{1.2}$$

The additivity of f yields

$$f(1 + x) = f(1) + f(x). \tag{1.3}$$

Substituting (1.3) into (1.2), we obtain

$$f'(x) = c,$$

where $c = f(1)$. Solving the above first order differential equation, we obtain

$$f(x) = cx + d, \tag{1.4}$$

where d is an arbitrary constant. Letting (1.4) into the functional equation (1.1), we see that $d = 2d$ and thus d must be zero. Therefore, from (1.4) we see that f is linear as asserted by the theorem. The proof of the theorem is now complete. □

Notice that in Theorem 1.1, we use the continuity of f to conclude that f is also integrable. The integrability of f forced the solution f of the additive Cauchy equation to be linear. Thus every integrable solution of the additive Cauchy equation is also linear.

Definition 1.3. *A function $f : \mathbb{R} \to \mathbb{R}$ is said to be locally integrable if and only if it is integrable over every finite interval.*

It is known that every locally integrable solution of the additive Cauchy equation is also linear. We give as a short proof of this using an argument provided by Shapiro (1973). Assume f is a locally integrable solution of the additive Cauchy equation. Hence $f(x + y) = f(x) + f(y)$ holds for all x and y in $\mathbb{R}$. From this and using the local integrability of f, we get

$$
\begin{aligned}
y\,f(x) &= \int_0^y f(x)dz \\
&= \int_0^y [f(x + z) - f(z)]\,dz \\
&= \int_x^{x+y} f(u)du - \int_0^y f(z)dz \\
&= \int_0^{x+y} f(u)du - \int_0^x f(u)du - \int_0^y f(u)du.
\end{aligned}
$$

The right side of the above equality is invariant under the interchange of x and y. Hence it follows that

$$y\,f(x) = x\,f(y)$$

for all $x, y \in \mathbb{R}$. Therefore, for $x \neq 0$, we obtain

$$\frac{f(x)}{x} = c,$$

where c is an arbitrary constant. This implies that $f(x) = cx$ for all $x \in \mathbb{R} \setminus \{0\}$. Letting $x = 0$ and $y = 0$ in (1.1), we get $f(0) = 0$. Together with this and the above, we conclude that f is a linear function in $\mathbb{R}$.

Although the proof of Theorem 1.1 is brief and involves only calculus, this proof is not very instructive. We will present now a different proof which will help us to understand the behavior of the solution of the additive Cauchy equation a bit more. We begin with the following definition.

Definition 1.4. *A function $f : \mathbb{R} \to \mathbb{R}$ is said to be rationally homogeneous if and only if*

$$f(rx) = r\, f(x) \tag{1.5}$$

for all $x \in \mathbb{R}$ and all rational numbers r.

The following theorem shows that any solution of the additive Cauchy equation is rationally homogeneous.

Theorem 1.2. *Let $f : \mathbb{R} \to \mathbb{R}$ be a solution of the additive Cauchy equation. Then f is rationally homogeneous. Moreover, f is linear on the set of rational numbers $\mathbb{Q}$.*

Proof. Letting $x = 0 = y$ in (1.1) see that $f(0) = f(0) + f(0)$ and hence

$$f(0) = 0. \tag{1.6}$$

Substituting $y = -x$ in (1.1) and then using (1.6), we see that f is an odd function in $\mathbb{R}$, that is

$$f(-x) = -f(x) \tag{1.7}$$

for all $x \in \mathbb{R}$. Thus, so far, we have shown that a solution of the additive Cauchy equation is zero at the origin and it is an odd function. Next, we will show that a solution of the additive Cauchy equation is rationally homogeneous. For any x,

$$f(2x) = f(x + x) = f(x) + f(x) = 2\, f(x).$$

Hence

$$f(3x) = f(2x + x) = f(2x) + f(x) = 2f(x) + f(x) = 3\, f(x);$$

so in general (using induction)

$$f(nx) = n f(x) \tag{1.8}$$

for all positive integers n. If n is a negative integer, then $-n$ is a positive integer and by (1.8) and (1.7), we get

$$\begin{aligned}
f(nx) &= f(-(-n)x) \\
&= -f(-nx) \\
&= -(-n) f(x) \\
&= n f(x).
\end{aligned}$$

Thus, we have shown $f(nx) = n f(x)$ for all integers n and all $x \in \mathbb{R}$. Next, let r be an arbitrary rational number. Hence, we have

$$r = \frac{k}{\ell}$$

where k is an integer and ℓ is a natural number. Further, $kx = \ell(rx)$. Using the integer homogeneity of f, we obtain

$$k f(x) = f(kx) = f(\ell(rx)) = \ell f(rx);$$

that is,

$$f(rx) = \frac{k}{\ell} f(x) = r f(x).$$

Thus, f is rationally homogeneous. Further, letting $x = 1$ in the above equation and defining $c = f(1)$, we see that

$$f(r) = c r$$

for all rational numbers $r \in \mathbb{Q}$. Hence, f is linear on the set of rational numbers and the proof is now complete. □

Now we present the second proof of Theorem 1.1.

Proof. Let f be a continuous solution of the additive Cauchy equation. For any real number x there exists a sequence $\{r_n\}$ of rational numbers with $r_n \to x$. Since f satisfies the additive Cauchy equation, by Theorem 1.2, f is linear on the set of rational numbers. That is,

$$f(r_n) = c r_n$$

for all n. Now using the continuity of f, we get

$$f(x) = f\left(\lim_{n \to \infty} r_n\right)$$
$$= \lim_{n \to \infty} f(r_n)$$
$$= \lim_{n \to \infty} c r_n$$
$$= c x$$

and the proof is now complete. □

The following theorem is due to Darboux (1875).

Theorem 1.3. *Let f be a solution of the additive Cauchy functional equation (1.1). If f is continuous at a point, then it is continuous everywhere.*

Proof. Let f be continuous at t and let x be any arbitrary point. Hence, we have $\lim_{y \to t} f(y) = f(t)$. Next, we show that f is continuous at x. Consider

$$\lim_{y \to x} f(y) = \lim_{y \to x} f(y - x + x - t + t)$$
$$= \lim_{y \to x} [f(y - x + t) + f(x - t)]$$
$$= \lim_{y \to x} f(y - x + t) + \lim_{y \to x} f(x - t)$$
$$= f(t) + f(x - t)$$
$$= f(t) + f(x) - f(t)$$
$$= f(x).$$

This proves that f is continuous at x and the arbitrariness of x implies f is continuous everywhere. The proof is complete. □

The following theorem is obvious from Theorem 1.1 and Theorem 1.3.

Theorem 1.4. *Let f be a solution of the additive Cauchy functional equation (1.1). If f is continuous at a point, then f is linear; that is, $f(x) = c x$ for all $x \in \mathbb{R}$.*

1.4 Discontinuous Solution of Additive Cauchy Equation

In the previous section, we showed that a continuous solution of the additive Cauchy equation is linear. In other words continuous additive functions are linear. Even if we relax the continuity condition to continuity at a point, still additive functions are linear. For many years the existence of discontinuous additive functions was an open problem. Mathematicians could neither prove that every additive function is continuous nor exhibit an example of a discontinuous additive function. It was the German mathematician G. Hamel in 1905 who first succeeded in proving that there exist discontinuous additive functions.

Now we begin our exploration on the non-linear solution of the additive Cauchy equation. First, we show that the non-linear solution of the additive Cauchy equation displays a very strange behavior.

Definition 1.5. *The graph of a function $f : \mathbb{R} \to \mathbb{R}$ is the set*

$$G = \{(x, y) \mid x \in \mathbb{R}, \quad y = f(x)\}.$$

It is easy to note that the graph G of a function $f : \mathbb{R} \to \mathbb{R}$ is subset of the plane $\mathbb{R}^2$. The proof of our next theorem is similar to one found in Aczél (1987).

Theorem 1.5. *The graph of every non-linear solution $f : \mathbb{R} \to \mathbb{R}$ of the additive Cauchy equation is everywhere dense in the plane $\mathbb{R}^2$.*

Proof. The graph G of f is given by

$$G = \{(x, y) \mid x \in \mathbb{R}, \ y = f(x)\}.$$

Choose a nonzero x_1 in $\mathbb{R}$. Since f is a non-linear solution of the additive Cauchy equation, for any constant m, there exists a nonzero real number x_2 such that

$$\frac{f(x_1)}{x_1} \neq \frac{f(x_2)}{x_2};$$

otherwise writing $c = \frac{f(x_1)}{x_1}$ and letting $x_1 = x$, we will have $f(x) = cx$ for all $x \neq 0$, and since $f(0) = 0$ this implies that f is linear contrary to our assumption that f is non-linear. This implies that

$$\begin{vmatrix} x_1 & f(x_1) \\ x_2 & f(x_2) \end{vmatrix} \neq 0,$$

so that the vectors $\mathbf{v}_1 = (x_1, f(x_1))$ and $\mathbf{v}_2 = (x_2, f(x_2))$ are linearly independent and thus span the whole plane $\mathbb{R}^2$. This means that for any vector $\mathbf{v} = (x, f(x))$ there exist real numbers r_1 and r_2 such that

$$\mathbf{v} = r_1\mathbf{v}_1 + r_2\mathbf{v}_2.$$

If we permit only rational numbers ρ_1, ρ_2, then by their appropriate choice, we can get with $\rho_1\mathbf{v}_1 + \rho_2\mathbf{v}_2$ arbitrarily close to any given plane vector $\mathbf{v}$ (since the rational numbers $\mathbb{Q}$ are dense in reals $\mathbb{R}$ and hence $\mathbb{Q}^2$ is dense in $\mathbb{R}^2$). Now,

$$\begin{aligned}
\rho_1\mathbf{v}_1 + \rho_2\mathbf{v}_2 &= \rho_1(x_1, f(x_1)) + \rho_2(x_2, f(x_2)) \\
&= (\rho_1 x_1 + \rho_2 x_2, \ \rho_1 f(x_1) + \rho_2 f(x_2)) \\
&= (\rho_1 x_1 + \rho_2 x_2, \ f(\rho_1 x_1 + \rho_2 x_2)).
\end{aligned}$$

Thus, the set

$$\widehat{G} = \{(x, y) \,|\, x = \rho_1 x_1 + \rho_2 x_2, \ y = f(\rho_1 x_1 + \rho_2 x_2), \ \rho_1, \rho_2 \in \mathbb{Q}\}$$

is everywhere dense in $\mathbb{R}^2$. Since

$$\widehat{G} \subset G,$$

the graph G of our non-linear additive function f is also dense in $\mathbb{R}^2$. The proof of the theorem is now complete. □

The graph of an additive continuous function is a straight line that passes through the origin. The graph of a non-linear additive function is dense in the plane. Next, we introduce the concept of Hamel basis to construct a discontinuous additive function.

Let us consider the set

$$S = \{s \in \mathbb{R} \,|\, s = u + v\sqrt{2} + w\sqrt{3}, \ u, v, w \in \mathbb{Q}\}$$

whose elements are rational linear combination of $1, \sqrt{2}, \sqrt{3}$. Further, this rational combination is unique. That is, if an element $s \in S$ has two different rational linear combinations, for instance,

$$s = u + v\sqrt{2} + w\sqrt{3} = u' + v'\sqrt{2} + w'\sqrt{3},$$

then $u = u'$, $v = v'$ and $w = w'$. To prove this we note that this assumption implies that

$$(u - u') + (v - v')\sqrt{2} + (w - w')\sqrt{3} = 0.$$

Letting $a = (u - u')$, $b = (v - v')$ and $c = (w - w')$, we see that the above expression reduces to

$$a + b\sqrt{2} + c\sqrt{3} = 0.$$

Next, we show that $a = 0 = b = c$. The above expression yields

$$b\sqrt{2} + c\sqrt{3} = -a,$$

and squaring both sides, we have

$$2bc\sqrt{6} = a^2 - 2b^2 - 3c^2.$$

This implies that b or c is zero; otherwise, we may divide both sides by $2bc$ and get

$$\sqrt{6} = \frac{a^2 - 2b^2 - 3c^2}{2bc}$$

contradicting the fact that $\sqrt{6}$ is an irrational number. If $b = 0$, then we have $a + c\sqrt{3} = 0$; this implies that $c = 0$ (else $\sqrt{3} = -\frac{a}{c}$ is a rational number contrary to the fact that $\sqrt{3}$ is an irrational number). Similarly if $c = 0$, we obtain that $b = 0$. Thus both b and c are zero. Hence it follows immediately that $a = 0$.

If we call

$$B = \left\{ 1,\ \sqrt{2},\ \sqrt{3} \right\},$$

then every element of S is a *unique* rational linear combination of the elements of B. This set B is called a Hamel basis for the set S. Formally, a Hamel basis is defined as follows.

Definition 1.6. *Let S be a set of real numbers and let B be a subset of S. Then B is called a Hamel basis for S if every member of S is a unique (finite) rational linear combination of B.*

If the set S is the set of reals $\mathbb{R}$, then using the axiom of choice it can be shown that a Hamel basis B for $\mathbb{R}$ exists. The proof of this is beyond the scope of this book.

There is a close connection between additive functions and Hamel bases. To exhibit an additive function it is sufficient to give its values on a Hamel basis, and these values can be assigned arbitrarily. This is the content of the next two theorems.

Theorem 1.6. *Let B be a Hamel basis for $\mathbb{R}$. If two additive functions have the same value at each member of B, then they are equal.*

Proof. Let f_1 and f_2 be two additive functions having the same value at each member of B. Then $f_1 - f_2$ is additive. Let us write $f = f_1 - f_2$. Let x be any real number. Then there are numbers b_1, b_2, ..., b_n in B and rational numbers r_1, r_2, ..., r_n such that

$$x = r_1 b_1 + r_2 b_2 + \cdots + r_n b_n.$$

Hence

$$
\begin{aligned}
f_1(x) - f_2(x) &= f(x) \\
&= f(r_1 b_1 + r_2 b_2 + \cdots + r_n b_n) \\
&= f(r_1 b_1) + f(r_2 b_2) + \cdots + f(r_n b_n) \\
&= r_1 f(b_1) + r_2 f(b_2) + \cdots + r_n f(b_n) \\
&= r_1 [f_1(b_1) - f_2(b_1)] + r_2 [f_1(b_2) - f_2(b_2)] \\
&\quad + \cdots + r_n [f_1(b_n) - f_2(b_n)] \\
&= 0.
\end{aligned}
$$

Thus, we have $f_1 = f_2$ and the proof is complete. $\qquad\square$

Theorem 1.7. *Let B be a Hamel basis for $\mathbb{R}$. Let $g : B \to \mathbb{R}$ be an arbitrary function defined on B. Then there exists an additive function $f : \mathbb{R} \to \mathbb{R}$ such that $f(b) = g(b)$ for each $b \in B$.*

Proof. For each real number x there can be found $b_1, b_2, ..., b_n$ in B and rational numbers $r_1, r_2, ..., r_n$ with

$$x = r_1 b_1 + r_2 b_2 + \cdots + r_n b_n.$$

We define $f(x)$ to be

$$r_1 g(b_1) + r_2 g(b_2) + \cdots + r_n g(b_n).$$

This defines $f(x)$ for all x. This definition is unambiguous since for each x, the choice of $b_1, b_2, ..., b_n, r_1, r_2, ..., r_n$ is unique, except for the order in which b_i and r_i are selected. For each b in B, we have $f(b) = g(b)$ by definition of f. Next, we show that f is additive on the reals. Let x and y be any two real numbers. Then

$$
\begin{aligned}
x &= r_1 a_1 + r_2 a_2 + \cdots + r_n a_n \\
y &= s_1 b_1 + s_2 b_2 + \cdots + s_m b_m,
\end{aligned}
$$

where $r_1, r_2, ..., r_n$, $s_1, s_2, ..., s_m$ are rational numbers and $a_1, a_2, ..., a_n$, b_1, $b_2, ..., b_m$ are members of the Hamel basis B. The two sets $\{a_1, a_2, ..., a_n\}$

and $\{b_1, b_2, ..., b_m\}$ may have some members in common. Let the union of these two sets be $\{c_1, c_2, ..., c_l\}$. Then $l \leq m + n$, and

$$x = u_1 c_1 + u_2 c_2 + \cdots + u_l c_l$$
$$y = v_1 c_1 + v_2 c_2 + \cdots + v_l c_l,$$

where $u_1, u_2, ..., u_l, v_1, v_2, ..., v_l$ are rational numbers, several of which may be zero. Now

$$x + y = (u_1 + v_1)c_1 + (u_2 + v_2)c_2 + \cdots + (u_l + v_l)c_l$$

and

$$\begin{aligned}
f(x+y) &= f((u_1 + v_1)c_1 + (u_2 + v_2)c_2 + \cdots + (u_l + v_l)c_l) \\
&= (u_1 + v_1)\, g(c_1) + (u_2 + v_2)\, g(c_2) + \cdots + (u_l + v_l)\, g(c_l) \\
&= [(u_1 g(c_1) + u_2 g(c_2) + \cdots + u_l g(c_l)] \\
&\quad + [(v_1 g(c_1) + v_2 g(c_2) + \cdots + v_l g(c_l)] \\
&= f(x) + f(y).
\end{aligned}$$

Hence f is additive on the set of real numbers $\mathbb{R}$ and the proof of the theorem is now complete. □

With the help of a Hamel basis, next we construct a non-linear additive function. Let B be a Hamel basis for the set of real numbers $\mathbb{R}$. Let $b \in B$ be any element of B. Define

$$g(x) = \begin{cases} 0 & \text{if } x \in B \setminus \{b\} \\ 1 & \text{if } x = b. \end{cases}$$

By Theorem 1.7, there exists an additive function $f : \mathbb{R} \to \mathbb{R}$ with $f(x) = g(x)$ for each $x \in B$. Note that this f cannot be linear since for $x \in B$ and $x \neq b$, we have

$$0 = \frac{f(x)}{x} \neq \frac{f(b)}{b}.$$

Therefore f is a non-linear additive function.

We end this section with the following remark.

Remark 1.1. *No concrete example of a Hamel basis for $\mathbb{R}$ is known; we only know that it exists. The graph of a discontinuous additive function on $\mathbb{R}$ is not easy to draw as the set $\{f(x) \mid x \in \mathbb{R}\}$ is dense in $\mathbb{R}$.*

1.5   Other Criteria for Linearity

We have seen that the graph of a non-linear additive function f is dense in the plane. That is, every circle contains a point (x, y) such that $y = f(x)$. We have also seen that an additive function f becomes linear when one imposes continuity on f. One can weaken this continuity condition to continuity at a point and still get f to be linear. In this section, we present some other mild regularity conditions that force an additive function to be linear.

Theorem 1.8. *If a real additive function f is either bounded from one side or monotonic, then it is linear.*

Proof. Suppose f is not linear. Then by Theorem 1.5, the graph of f is dense in the plane. Since f is bounded from the above, for some constant M the additive function f satisfies

$$f(x) \le M, \qquad x \in \mathbb{R},$$

and the graph of f avoids the set $A = \{\, (x, y) \in \mathbb{R}^2 \,|\, y = f(x) > M \,\}$. Therefore it cannot be dense on the plane which is a contradiction. Hence contrary to our assumption, f is linear. The rest of the theorem can be established in a similar manner. Now the proof is complete. $\square$

Remark 1.2. *Note that since f is bounded on $\mathbb{R}$ and f is linear, therefore $f(x) = 0$ for all $x \in \mathbb{R}$. To see this, suppose x_o is a number such that $f(x_o) \ne 0$. By an easy induction, we have $f(n x_o) = n f(x_o)$ for all $n \in \mathbb{N}$. We can make $|n f(x_o)|$ as large as we please by increasing n, which would contradict the boundedness of $f(x)$. Therefore $f(x) = 0$ for all $x \in \mathbb{R}$.*

In the next theorem we do not assume that f is bounded on $\mathbb{R}$; rather we assume f is bounded on a closed interval $[a, b]$ for $a, b \in \mathbb{R}$.

The proof of the following theorem is based on Young (1958).

Theorem 1.9. *If a real additive function f is bounded on an interval $[a, b]$, then it is linear; that is, there exists a constant c such that $f(x) = cx$ for all $x \in \mathbb{R}$.*

Proof. Suppose $f : \mathbb{R} \to \mathbb{R}$ is an additive function and bounded on an interval $[a, b]$. We show first that $f(x)$ is bounded on the interval $[0, b-a]$. Since $f(x)$ is bounded on $[a, b]$, there exists a positive number M such that

$$|f(y)| < M$$

for all $y \in [a, b]$. If $x \in [0, b - a]$, then $x + a \in [a, b]$, so that from

$$f(x) = f(x + a) - f(a)$$

we obtain

$$|f(x)| < M + f(a).$$

If we call $\alpha = b - a$, then $f(x)$ is bounded on $[0, \alpha]$. Suppose $m = \frac{f(\alpha)}{\alpha}$ and let $\phi(x) = f(x) - cx$. Then ϕ satisfies

$$\begin{aligned}
\phi(x + y) &= f(x + y) - c(x + y) \\
&= f(x) + f(y) - cx - cy \\
&= f(x) - cx + f(y) - cy \\
&= \phi(x) + \phi(y),
\end{aligned}$$

and we have $\phi(\alpha) = f(\alpha) - c\alpha = 0$. It follows that $\phi(x)$ is periodic of period α, for

$$\phi(x + a) = \phi(x) + \phi(\alpha) = \phi(x)$$

for all $x \in \mathbb{R}$. Further, as the difference of two functions bounded on $[0, \alpha]$, the function $\phi(x)$ is bounded on $[0, \alpha]$. Since $\phi(x)$ is periodic of period α, it is bounded on $\mathbb{R}$. Thus $\phi(x)$ is an additive function which is bounded on $\mathbb{R}$. Hence by previous remark, $\phi(x) = 0$ for all $x \in \mathbb{R}$, or $f(x) = cx$. This completes the proof of the theorem. $\square$

Definition 1.7. *A function f is said to be multiplicative if and only if $f(xy) = f(x)f(y)$ for all numbers x and y.*

Theorem 1.10. *If an additive function f is also multiplicative, then it is linear.*

Proof. For any positive number x,

$$f(x) = f\left(\sqrt{x} \cdot \sqrt{x}\right) = f\left(\sqrt{x}\right) \cdot f\left(\sqrt{x}\right) = \left[f\left(\sqrt{x}\right)\right]^2 \geq 0.$$

Therefore f is bounded from below, and hence by Theorem 1.8, we see that f is linear. This completes the proof. $\square$

Actually, if f is a nonzero function, then $f(x) = x$. Indeed, since $f(x) = cx$,

$$cxy = cx \, cy;$$

that is, $c^2 = c$. Then $f(x) = 0$ or $f(x) = x$.

Remark 1.3. *From what we have studied so far, continuous, monotonic or measurable solutions* $f : \mathbb{R} \to \mathbb{R}$ *of the additive Cauchy functional equation are always of the form* $f(x) = cx$, *where* c *is an arbitrary real constant. Thus they are analytic. We have also seen that the additive Cauchy functional equation has non-regular solutions. This is due to the fact that the general solution of the additive Cauchy functional equation can be prescribed arbitrarily on a fixed Hamel base and can be extended to* $\mathbb{R}$ *in a unique way. Therefore we have the following alternative: The solutions of the additive Cauchy functional equation are very regular (in this case, analytic) or very irregular (in this case, nowhere continuous, nonmonotonic on any proper interval, nonmeasurable). A similar alternative can be stated for many functional equations. Typically one assumes weak regularity properties of the unknown functions (such as measurability, Baire property, monotonicity, continuity) and using the functional equation derives higher order regularity properties. Results of this kind are called regularity theorems for functional equations and Jarai (2005) provides an excellent account of such results.*

1.6   Additive Functions on the Complex Plane

In this section, we present some results concerning additive complex-valued functions on the complex plane. The formal definition of complex numbers was given by William Hamilton.

Definition 1.8. *The complex number system* $\mathbb{C}$ *is the set of ordered pairs of real numbers* (x, y) *with addition and multiplication defined by*

$$(x, y) + (u, v) = (x + u, y + v)$$
$$(x, y)(u, v) = (xu - yv, xv + yu)$$

for all $x, y, u, v \in \mathbb{R}$.

Thinking of a real number as either x or $(x, 0)$ and letting i denote the purely imaginary number $(0, 1)$, we can rewrite the following expression

$$(x, y) = (x, 0) + (0, 1)(y, 0)$$

as

$$(x, y) = x + iy.$$

If we denote the left side of the above representation by z, then we have $z = x + iy$. The real number x is called the *real part* of z and is denoted by $Re\, z$. Similarly, the real number y is called the *imaginary part* of z and is denoted by $Im\, z$. If z is a complex number of the form $x + iy$, then the complex number $x - iy$ is called the *conjugate* of z and is denoted by $\bar{z}$.

An arbitrary function $f : \mathbb{C} \to \mathbb{C}$ can be written as

$$f(z) = f_1(z) + i\, f_2(z), \tag{1.9}$$

where $f_1 : \mathbb{C} \to \mathbb{R}$ and $f_2 : \mathbb{C} \to \mathbb{R}$ are given by

$$f_1(z) = Re\, f(z) \qquad \text{and} \qquad f_2(z) = Im\, f(z). \tag{1.10}$$

If f is additive, then by (1.9) and (1.10) we have

$$
\begin{aligned}
f_1(z_1 + z_2) &= Re\, f(z_1 + z_2) \\
&= Re\, [\, f(z_1) + f(z_2)\,] \\
&= Re\, f(z_1) + Re\, f(z_2) = f_1(z_1) + f_1(z_2),
\end{aligned}
$$

and

$$
\begin{aligned}
f_2(z_1 + z_2) &= Im\, f(z_1 + z_2) \\
&= Im\, [\, f(z_1) + f(z_2)\,] \\
&= Im\, f(z_1) + Im\, f(z_2) = f_2(z_1) + f_2(z_2).
\end{aligned}
$$

Theorem 1.11. *If $f : \mathbb{C} \to \mathbb{C}$ is additive, then there exist additive functions $f_{kj} : \mathbb{R} \to \mathbb{R}$ $(k, j = 1, 2)$ such that*

$$f(z) = f_{11}\, (Re\, z) + f_{12}\, (Im\, z) + i\, f_{21}\, (Re\, z) + i\, f_{22}\, (Im\, z)\,.$$

Proof. By (1.9), we obtain

$$f(z) = f_1(z) + i f_2(z),$$

where $f_1 : \mathbb{C} \to \mathbb{R}$ and $f_2 : \mathbb{C} \to \mathbb{R}$ are real-valued functions on the complex plane. Since f is an additive function, f_1 and f_2 are also additive functions. Since the functions f_1 and f_2 can be considered as functions from $\mathbb{R}^2$ into $\mathbb{R}$, applying Theorem 3.2, we have the asserted result. $\square$

Our next theorem concerns the form of complex-valued continuous additive functions on the complex plane.

Theorem 1.12. *If $f : \mathbb{C} \to \mathbb{C}$ is a continuous additive function, then there exist complex constants c_1 and c_2 such that*

$$f(z) = c_1\, z + c_2\, \bar{z}, \tag{1.11}$$

where $\bar{z}$ denotes the complex conjugate of z.

Proof. Since f is additive, by Theorem 1.11, we get

$$f(z) = f_{11}\,(Re\,z) + f_{12}\,(Im\,z) + i\,f_{21}\,(Re\,z) + i\,f_{22}\,(Im\,z),$$

where $f_{kj} : \mathbb{R} \to \mathbb{R}$ $(k, j = 1, 2)$ are real-valued additive functions on the reals. The continuity of f implies the continuity of each function f_{kj} and hence

$$f_{kj}(x) = c_{kj}\,x,$$

where c_{kj} $(k, j = 1, 2)$ are real constants. Thus, using the form of $f(z)$ and the form of f_{kj}, we get

$$
\begin{aligned}
f(z) &= c_{11}\,Re\,z + c_{12}Im\,z + i\,c_{21}Re\,z + i\,c_{22}Im\,z \\
&= (\,c_{11} + ic_{21}\,)Re\,z + (\,c_{12} + ic_{22}\,)Im\,z \\
&= a\,Re\,z + b\,Im\,z \qquad \text{where} \quad a = c_{11} + ic_{21},\ \ b = c_{12} + ic_{22} \\
&= a\,Re\,z - i\,(bi)\,Im\,z \\
&= \frac{a + bi}{2}Re\,z + \frac{a - bi}{2}Re\,z - \frac{a + bi}{2}i\,Im\,z + \frac{a - bi}{2}i\,Im\,z \\
&= \frac{a - bi}{2}Re\,z + \frac{a - bi}{2}i\,Im\,z + \frac{a + bi}{2}Re\,z - \frac{a + bi}{2}i\,Im\,z \\
&= \frac{a - bi}{2}\,(\,Re\,z + i\,Im\,z\,) + \frac{a + bi}{2}\,(\,Re\,z - i\,Im\,z\,) \\
&= \frac{a - bi}{2}\,z + \frac{a + bi}{2}\,\bar{z} \\
&= c_1\,z + c_2\,\bar{z},
\end{aligned}
$$

where $c_1 = \frac{a-bi}{2}$ and $c_2 = \frac{a+bi}{2}$ are complex constants. This completes the proof of the theorem. □

Note that unlike the real-valued continuous additive functions on the reals, the complex-valued continuous additive functions on the complex plane are not linear. The linearity can be restored if one assumes a stronger regularity condition such as analyticity or differentiability instead of continuity.

Definition 1.9. *A function $f : \mathbb{C} \to \mathbb{C}$ is said to be analytic if and only if f is differentiable on $\mathbb{C}$.*

Theorem 1.13. *If $f : \mathbb{C} \to \mathbb{C}$ is an analytic additive function, then there exists complex constant c such that*

$$f(z) = c\,z;$$

that is, f is linear.

Proof. Since f is analytic, it is differentiable. Differentiating

$$f(z_1 + z_2) = f(z_1) + f(z_2) \tag{1.12}$$

with respect to z_1, we get

$$f'(z_1 + z_2) = f'(z_1)$$

for all z_1 and z_2 in $\mathbb{C}$. Hence, letting $z_1 = 0$ and $z_2 = z$, we get

$$f'(z) = c,$$

where $c = f'(0)$ is a complex constant. From the above, we see that

$$f(z) = cz + b,$$

where b is a complex constant. Inserting this form of $f(z)$ into (1.12), we obtain $b = 0$ and hence the asserted solution follows. This completes the proof of the theorem. □

We close this section with the following remark.

Remark 1.4. *It is surprising that Theorem 1.10 fails for complex-valued functions on the complex plane. It is well known that there is a discontinuous automorphism of the complex plane (see Kamke (1927)). An automorphism is a map that is one-to-one, onto, additive and multiplicative on $\mathbb{C}$.*

1.7 Concluding Remarks

Cauchy (1821) proved that every continuous additive function $f : \mathbb{R} \to \mathbb{R}$ is linear, that is, of the form $f(x) = m\,x$, where m is an arbitrary constant. Darboux (1875) showed that a real additive function which is continuous at a point is linear. It is also known that every locally integrable real additive function is linear. Young (1958) demonstrated that if a real additive function is bounded on a closed interval, then it is linear. The proof of these results we have seen in this chapter. Banach (1920) and also Sirerpiński (1920) proved that every Lebesgue measurable additive function is also linear. Ostrowski (1929) and again Kestelman (1948) showed that if a real additive function is bounded from above or below on a set of positive measure, then it is linear. Answering a question of Erdös (1960), Jurkat (1965) proved that if $f(x)$ is a real valued function defined for almost all real x and f satisfies $f(x+y) = f(x) + f(y)$ for almost all pairs (x, y) in the senses of plane (Lebesgue) measure, then there exists a real valued function $F(x)$ which coincides with $f(x)$ for almost all x in the sense of linear (Lebesgue) measure. From this result one can show that if f satisfies $f(x+y) = f(x) + f(y)$ for almost all pairs (x, y) and is also measurable or only bounded from below on a set of positive measure, then $f(x) = m x$ for almost all x in $\mathbb{R}$. The above mentioned Erdös problem was also solved by de Bruijn (1967). I. Halperin asked about finding all additive real functions f which are of the form $f(x^{-1}) = x^{-2} f(x)$ for all nonzero reals. Jurkat (1965) proved that every real additive function $f(x)$ that satisfies $f(x^{-1}) = x^{-2} f(x)$ for all $x \in \mathbb{R} \setminus \{0\}$ is linear.

For many years the existence of discontinuous additive functions was an open problem. Mathematicians could neither prove that every additive function is continuous nor exhibit an example of a discontinuous additive function. It was the German mathematician G. Hamel in 1905 who first succeeded in proving that there exist discontinuous additive functions. Hamel (1905) proved that if $f(x)$ is an additive function not of the form $f(x) = m x$, then the graph of f has a point in every neighborhood of every point in the plane $\mathbb{R}^2$. The proof of Hamel uses the axiom of choice. Hewitt and Zuckerman (1969) gave a simple proof of this result avoiding the axiom of choice. Hamel constructed a discontinuous additive function using a Hamel basis for $\mathbb{R}$. Jones (1942) demonstrated the existence of a discontinuous additive function whose graph is connected in the topological sense.

The additive Cauchy functional equation $f(x+y) = f(x) + f(y)$ has a natural generalization: $f(x+y) = F(f(x), f(y))$ where $F(u, v)$ is a known function. Aczél has proved that $f(x+y) = F(f(x), f(y))$ has a continuous and strictly monotonic solution if and only if the function $F(x, y)$ is continuous and strictly monotonic with respect to each variable and satisfies the condition $F(F(x, y), z) = F(x, F(y, z))$ (see Aczél (1966)). The above equation can be furthered generalized to $f(ax + by + c) = F(f(x), f(y))$, where $F(u, v)$ is a known function and a, b, c are real constants such that $ab \neq 0$.

We conclude this section by giving an example of a simple looking functional equations whose general solution is not known. This and some additional problems were posed in Sahoo (1995). The first problem is the following: Find all functions $f : (0, 1) \to \mathbb{R}$ satisfying the functional equation

$$f(xy) + f(x(1-y)) + f(y(1-x)) + f((1-x)(1-y)) = 0 \qquad (1.13)$$

for all $x, y \in (0, 1)$. This problem was stated as an open problem in Ebanks, Sahoo and Sander (1990). It should be noted that if $f(x) = 4A(x) - A(1)$, where A is an additive function on the reals, then it satisfies the functional equation (1.13). If f is assumed to be continuous (or measurable), then Daróczy and Jarai (1979) have shown that $f(x) = 4ax - a$, where a is an arbitrary constant. Recently, Maksa (1993) posed the following problem at the Thirtieth International Symposium on Functional Equations: Find all functions $f : [0, 1] \to \mathbb{R}$ satisfying the functional equation

$$(1 - x - y)f(xy) = xf(y(1 - x)) + yf(x(1 - y)) \qquad (1.14)$$

for all $x, y \in [0, 1]$. One can easily show that if f is a solution of (1.14), then f is skew symmetric about $\frac{1}{2}$, that is, $f(x) = -f(1 - x)$, and $f(0) = 0$. Further, it is easy to note that Maksa's equation (1.14) implies equation (1.13). To see this, replace x by $1 - x$ in (1.14) and add the resulting equation to (1.14) to obtain

$$y\left[f(xy) + f(x(1 - y)) + f(y(1 - x)) + f((1 - x)(1 - y)) \right] = 0$$

for all $x, y \in (0, 1]$. Since $f(0) = 0$, the above equation yields (1.13) for all $x, y \in [0, 1]$. Thus, the general solution of (1.13) will provide the general solution of (1.14). Utilizing the solution of the equation (1.13) given by Daróczy and Jarai (1979), it is easy to show that if f is continuous (or measurable or almost open), then all solutions of (1.14) are of the form $f(x) = 0$ for all $x \in \mathbb{R}$.

Now at the first sight the above functional equations, (1.13) seems harmless – it looks as though anyone could solve it, but nobody has succeeded in finding all solutions of this equation.

1.8 Exercises

1. Show that any monotonic function $f : \mathbb{R} \to \mathbb{R}$ satisfies the additive Cauchy functional equation if and only if $f(x) = cx$ for every $x \in \mathbb{R}$, where c is a non-zero real constant.

2. If $f : \mathbb{R} \to \mathbb{R}$ is a solution of the additive Cauchy functional equation, then show that f is either everywhere or nowhere zero.

3. Determine all twice differentiable solutions $f : (0, 1) \to \mathbb{R}$ of the fundamental equation of information

$$f(1 - x) + (1 - x)f\left(\frac{y}{1 - x}\right) = f(1 - y) + (1 - y)f\left(\frac{x}{1 - y}\right)$$

for all $x, y \in (0, 1)$.

4. Determine all differentiable solutions $f : \mathbb{R} \to \mathbb{R}$ of the functional equation

$$f(x + y - xy) + f(xy) = f(x) + f(y)$$

for all $x, y \in \mathbb{R}$.

5. Using induction, determine all functions $f : \mathbb{Q} \to \mathbb{Q}$ that satisfy the functional equation

$$f(xy) + f(x + y) = f(x) f(y) + 1 \qquad \forall x, y \in \mathbb{Q}.$$

6. Determine all functions $f : \mathbb{R} \to \mathbb{R}$ that satisfy the functional equation

$$f(x + y) + f(z) = f(x) + f(y + z)$$

for all $x, y, z \in \mathbb{R}$.

7. Determine all functions $f : \mathbb{R} \to \mathbb{R}$ that satisfy the functional equation

$$f(x + y) = f(x) + f(y) + xy$$

for all $x, y \in \mathbb{R}$.

8. Let $f : \mathbb{R} \to \mathbb{R}$ be a solution of the additive Cauchy functional equation such that $f(\mathbb{R}) = \mathbb{R}$. Prove that f either is one-to-one or has the intermediate value property.

9. Let $f : \mathbb{R} \to \mathbb{R}$ be a solution of the additive Cauchy functional equation which is not one-to-one. Prove that for every $y \in f(\mathbb{R})$, the set $f^{-1}(y)$ is dense in $\mathbb{R}$.

10. Determine all functions $f : \mathbb{R} \to \mathbb{R}$ that satisfy the Lobačevskii functional equation

$$f(x + y) f(x - y) = f(x)^2 \qquad \forall x, y \in \mathbb{R}.$$

11. By reducing the functional equation

$$f\left(\frac{x + y}{2}\right)^2 = f(x) f(y) \qquad \forall x, y \in \mathbb{R}$$

to the additive Cauchy functional equation, find its general continuous solution.

12. Let $\mathbb{R}_0 = \mathbb{R} \setminus \{0\}$. Suppose f and $g \neq 0$ are two solutions of the additive Cauchy functional equation. Suppose there exists $n \in \mathbb{Z}$ with $n \neq 0, -1$ and a continuous function $\psi : \mathbb{R}_0 \to \mathbb{R}$ such that

$$g(x) = \psi(x) f\left(x^{-n}\right)$$

for all $x \in \mathbb{R}_0$. Then prove that $\psi(x) = c\, x^{n+1}$, where c is a real constant.

13. Let $f : \mathbb{R} \to \mathbb{R}$ be a solution of the additive Cauchy functional equation satisfying the condition

$$f(x) = x^2 f(1/x) \qquad \forall\, x \in \mathbb{R} \setminus \{0\}.$$

Then show that $f(x) = cx$, where c is an arbitrary constant.

14. Let $n \in \mathbb{N}$ (the set of natural numbers) with $n > 1$. Determine all functions $f : \mathbb{R} \to \mathbb{R}$ that satisfy the functional equation

$$f(x + y^n) = f(x) + f(y)^n$$

for all $x, y \in \mathbb{R}$.

15. Let α and a be real numbers with $a > 0$. Find all functions that satisfy the functional equation

$$f(x + y) = f(x) + f(y) + \alpha\,(1 - a^x)(1 - a^y)$$

for all $x, y \in \mathbb{R}$.

16. Determine all functions $f : \mathbb{R} \to \mathbb{R}$ that satisfy the functional equation

$$f(ax + by + c) = \alpha f(x) + \beta f(y) + \gamma \qquad \forall\, x, y \in \mathbb{R},$$

where $a, b, c, \alpha, \beta, \gamma$ are apriori chosen real numbers satisfying $ab\alpha\beta \neq 0$.

17. Let $f : \mathbb{R} \to \mathbb{R}$ be a solution of the functional equation

$$|f(x + y)| = |f(x)| + |f(y)| \qquad \forall\, x, y \in \mathbb{R}.$$

Then show that f is an additive function; that is, f satisfies the additive Cauchy functional equation.

Chapter 2

Remaining Cauchy Functional Equations

2.1 Introduction

In Chapter 5 of his book *Cours d'Analyse*, A. L. Cauchy (1821) also studied three other functional equations, namely,

$$f(x + y) = f(x)f(y), \tag{2.1}$$

$$f(xy) = f(x) + f(y) \tag{2.2}$$

and

$$f(xy) = f(x)f(y) \tag{2.3}$$

besides the additive Cauchy functional equation

$$f(x + y) = f(x) + f(y) \tag{2.4}$$

for all $x, y \in \mathbb{R}$. This chapter is devoted to solving these three Cauchy functional equations. The general solution of each of these functional equations is determined in terms of the additive function. Finally, using the general solution, the continuous solution is provided for each of these functional equations.

2.2 Solution of the Exponential Cauchy Equation

In this section, we determine the general solution of the exponential Cauchy functional equation (2.1) without assuming any regularity condition such as continuity, boundedness or differentiability on the unknown function f.

Theorem 2.1. *If the functional equation* (2.1), *that is,*

$$f(x + y) = f(x)f(y),$$

holds for all real numbers x *and* y, *then the general solution of* (2.1) *is given by*

$$f(x) = e^{A(x)} \quad \text{and} \quad f(x) = 0 \quad \forall x \in \mathbb{R}, \tag{2.5}$$

where $A : \mathbb{R} \to \mathbb{R}$ *is an additive function and* e *is the Napierian base of logarithm.*

Proof. It is easy to see that $f(x) = 0$ for all $x \in \mathbb{R}$ is a solution of (2.1). Hence from now on we suppose that $f(x)$ is not identically zero. We claim that $f(x)$ is nowhere zero. Suppose not. Then there exists a y_o such that $f(y_o) = 0$. From (2.1), we get

$$f(y) = f((y - y_o) + y_o)$$
$$= f(y - y_o) f(y_o) = 0$$

for all $y \in \mathbb{R}$. This is a contradiction to our assumption that $f(x)$ is not identically zero. Hence $f(x)$ is nowhere zero.

Letting $x = \frac{t}{2} = y$ in (2.1), we see that

$$f(t) = f\left(\frac{t}{2}\right)^2$$

for all $t \in \mathbb{R}$. Hence $f(x)$ is a strictly positive. Now taking natural logarithm of both sides of (2.1), we obtain

$$\ln f(x + y) = \ln f(x) + \ln f(y).$$

Defining $A : \mathbb{R} \to \mathbb{R}$ by $A(x) = \ln f(x)$, we have

$$A(x + y) = A(x) + A(y). \tag{2.6}$$

Hence we have the asserted solution $f(x) = e^{A(x)}$ and the proof is now complete. $\square$

The following corollary is obvious from the above theorem.

Corollary 2.1. *If the functional equation* (2.1), *that is,* $f(x + y) = f(x)f(y)$, *holds for all real numbers* x *and* y, *then the general continuous solution of* (2.1) *is given by*

$$f(x) = e^{cx} \quad \text{and} \quad f(x) = 0 \quad \forall x \in \mathbb{R}, \tag{2.7}$$

where c *is an arbitrary real constant.*

The following definition will be useful in later chapters.

Definition 2.1. *A function* $f : \mathbb{R} \to \mathbb{R}$ *is called a (real-valued) real exponential function if it satisfies* $f(x+y) = f(x)\,f(y)$ *for all* $x, y \in \mathbb{R}$.

Let n be a positive integer. Suppose the functional equation

$$f(x + y + nxy) = f(x)\,f(y) \tag{2.8}$$

holds for all reals $x > -\frac{1}{n}$ and all $y > -\frac{1}{n}$. When $n \to 0$, the functional equation (2.8) reduces to the exponential Cauchy functional equation. This equation was studied by Thielman (1949).

Theorem 2.2. *Every solution f of the functional equation (2.8) holding for all reals $x > -\frac{1}{n}$ and all $y > -\frac{1}{n}$ is of the form*

$$f(x) = 0 \quad or \quad f(x) = e^{A(\ln(1+nx))}, \tag{2.9}$$

where $A : \mathbb{R} \to \mathbb{R}$ *is an additive function.*

Proof. We write the functional equation (2.8) as

$$f\left(\frac{(1+nx)(1+ny)-1}{n}\right) = f(x)\,f(y). \tag{2.10}$$

Next we define $1 + nx = e^u$ and $1 + ny = e^v$ so that $u = \ln(1 + nx)$ and $v = \ln(1 + ny)$. Now rewriting (2.10), we obtain

$$f\left(\frac{e^{u+v}-1}{n}\right) = f\left(\frac{e^u-1}{n}\right) f\left(\frac{e^v-1}{n}\right) \tag{2.11}$$

for all $u, v \in \mathbb{R}$. Letting

$$\phi(u) = f\left(\frac{e^u-1}{n}\right) \tag{2.12}$$

in (2.11), we have

$$\phi(u + v) = \phi(u)\,\phi(v) \tag{2.13}$$

for all $u, v \in \mathbb{R}$. Hence by Theorem 2.1, we have

$$\phi(x) = e^{A(x)} \quad or \quad \phi(x) = 0 \quad \forall x \in \mathbb{R}, \tag{2.14}$$

where $A : \mathbb{R} \to \mathbb{R}$ is an additive function and e is the Napierian base of logarithm. Therefore from (2.12) and (2.14), we obtain

$$f(x) = 0 \quad or \quad f(x) = e^{A(\ln(1+nx))},$$

where $A : \mathbb{R} \to \mathbb{R}$ is an additive function. The proof of the theorem is now complete. $\square$

The following corollary is obvious.

Corollary 2.2. *Every continuous solution f of the functional equation (2.8) holding for all reals $x > -\frac{1}{n}$ and all $y > -\frac{1}{n}$ is of the form*

$$f(x) = 0 \quad \text{or} \quad f(x) = (1 + nx)^k, \qquad (2.15)$$

where k is an arbitrary constant.

2.3   Solution of the Logarithmic Cauchy Equation

Now we consider the second Cauchy functional equation (2.2). This functional equation is known as the logarithmic Cauchy equation.

Theorem 2.3. *If the functional equation (2.2), that is,*

$$f(xy) = f(x) + f(y),$$

holds for all $x, y \in \mathbb{R} \setminus \{0\}$, then the general solution of (2.2) is given by

$$f(x) = A(\ln |x|) \quad \forall x \in \mathbb{R} \setminus \{0\}, \qquad (2.16)$$

where A is an additive function.

Proof. First we substitute $x = t$ and $y = t$ in (2.2) to get

$$f(t^2) = 2f(t).$$

Similarly, letting $x = -t$ and $y = -t$ in (2.2), we have

$$f(t^2) = 2f(-t).$$

Hence we see that

$$f(t) = f(-t) \quad \forall t \in \mathbb{R} \setminus \{0\}. \qquad (2.17)$$

Next, suppose the functional equation (2.2) holds for all $x > 0$ and $y > 0$. Let

$$x = e^s \quad \text{and} \quad y = e^t \qquad (2.18)$$

so that

$$s = \ln x \quad \text{and} \quad t = \ln y. \qquad (2.19)$$

Note that $s, t \in \mathbb{R}$ since $x, y \in \mathbb{R}_+$ where $\mathbb{R}_+ = \{x \in \mathbb{R} \,|\, x > 0\}$. Substituting (2.18) into (2.2), we obtain

$$f(e^{s+t}) = f(e^s) + f(e^t).$$

Defining

$$A(s) = f(e^s) \tag{2.20}$$

and using the last equation we have

$$A(s+t) = A(s) + A(t)$$

for all $s, t \in \mathbb{R}$. Hence from (2.20) we have

$$f(x) = A(\ln x) \qquad \forall x \in \mathbb{R}_+. \tag{2.21}$$

Since $f(t) = f(-t)$, we see that the general solution of (2.2) is

$$f(x) = A(\ln |x|) \qquad \forall x \in \mathbb{R} \setminus \{0\}$$

and the proof is now complete. □

The following corollaries are consequences of the last theorem.

Corollary 2.3. *The general solution of the functional equation (2.2), that is, $f(xy) = f(x) + f(y)$, holding for all $x, y \in \mathbb{R}_+$ is given by*

$$f(x) = A(\ln x), \tag{2.22}$$

where $A : \mathbb{R} \to \mathbb{R}$ is an additive function.

The following result is also trivial.

Corollary 2.4. *The general solution of the functional equation (2.2) holding for all $x, y \in \mathbb{R}$ is given by*

$$f(x) = 0 \qquad \forall x \in \mathbb{R}. \tag{2.23}$$

Proof. Substitute $y = 0$ in (2.2) to get $f(0) = f(x) + f(0)$ and hence we have the asserted solution. □

Corollary 2.5. *The general continuous solution of the functional equation (2.2), that is, $f(xy) = f(x) + f(y)$, holding for all $x, y \in \mathbb{R} \setminus \{0\}$ is given by*

$$f(x) = c \ln |x| \qquad \forall x \in \mathbb{R} \setminus \{0\}, \tag{2.24}$$

where c is an arbitrary real constant.

Definition 2.2. *A function $f : \mathbb{R}_+ \to \mathbb{R}$ is called a logarithmic function if it satisfies $f(xy) = f(x) + f(y)$ for all $x, y \in \mathbb{R}_+$.*

2.4 Solution of the Multiplicative Cauchy Equation

Now we treat the last Cauchy equation (2.3). This equation is the most complicated of the three equations considered in this chapter. In the following theorem we need the notion of the signum function. The signum function is denoted by sgn (x) and defined as

$$\operatorname{sgn}(x) = \begin{cases} 1 & \text{if } x > 0 \\ 0 & \text{if } x = 0 \\ -1 & \text{if } x < 0. \end{cases} \tag{2.25}$$

Theorem 2.4. *The general solution of the multiplicative functional equation (2.3), that is,*

$$f(xy) = f(x)f(y),$$

holding for all $x, y \in \mathbb{R}$ is given by

$$f(x) = 0, \tag{2.26}$$
$$f(x) = 1, \tag{2.27}$$

$$f(x) = e^{A(\ln |x|)} |sgn(x)|, \tag{2.28}$$

and

$$f(x) = e^{A(\ln |x|)} sgn(x), \tag{2.29}$$

where $A : \mathbb{R} \to \mathbb{R}$ is an additive function and e is the Napierian base of logarithm.

Proof. Letting $x = 0 = y$ in (2.3), we obtain $f(0)[1 - f(0)] = 0$ and hence either

$$f(0) = 0 \quad \text{or} \quad f(0) = 1. \tag{2.30}$$

Similarly, substituting $x = 1 = y$ in (2.3), we have $f(1)[1 - f(1)] = 0$ and hence either

$$f(1) = 0 \quad \text{or} \quad f(1) = 1. \tag{2.31}$$

Let x be a positive real number, that is $x > 0$. Then (2.3) implies

$$f(x) = f(\sqrt{x})^2 \geq 0. \tag{2.32}$$

Suppose there exists an $x_0 \in \mathbb{R}$, $x_0 \neq 0$ such that $f(x_0) = 0$. Let $x \in \mathbb{R}$ be an arbitrary real number. Then from (2.3) we have

$$f(x) = f\left(x_0 \frac{x}{x_0}\right) = f(x_0) f\left(\frac{x}{x_0}\right) = 0$$

for all $x \in \mathbb{R}$ and we obtain the solution (2.26).

From now on we suppose that $f(x) \neq 0$ for all $x \in \mathbb{R} \setminus \{0\}$.

From (2.30) we have either $f(0) = 0$ or $f(0) = 1$. If $f(0) = 1$, then letting $y = 0$ in (2.3), we obtain

$$f(0) = f(x) f(0)$$

and hence

$$f(x) = 1.$$

for all $x \in \mathbb{R}$. Thus we have the asserted solution (2.27).

Next we consider the case $f(0) = 0$. In this case we claim that f is nowhere zero in $\mathbb{R} \setminus \{0\}$. Suppose not. Then there exists a y_o in $\mathbb{R} \setminus \{0\}$ such that $f(y_o) = 0$. Letting $y = y_o$ in (2.3), we have

$$f(xy_o) = f(x)f(y_o) = 0.$$

Hence

$$f(x) = 0 \quad \forall x \in \mathbb{R} \setminus \{0\}$$

which is a contradiction to our assumption that f is not identically zero. Thus f is nowhere zero in $\mathbb{R} \setminus \{0\}$.

From the fact that f is nowhere zero in $\mathbb{R} \setminus \{0\}$ and (2.32), we have

$$f(x) > 0 \qquad \text{for} \qquad x > 0. \tag{2.33}$$

Let

$$x = e^s \qquad \text{and} \qquad y = e^t \tag{2.34}$$

so that

$$s = \ln x \qquad \text{and} \qquad t = \ln y. \tag{2.35}$$

Note that $s, t \in \mathbb{R}$ since $x, y \in \mathbb{R}_+$. Substituting (2.34) into (2.3), we obtain

$$f(e^{s+t}) = f(e^s)f(e^t).$$

Since $f(t) > 0$ for all $t > 0$, taking the natural logarithm of both sides of the last equation, we have

$$A(s + t) = A(s) + A(t),$$

where

$$A(s) = \ln f(e^s) \quad \forall s \in \mathbb{R}. \tag{2.36}$$

Thus A is an additive function. From (2.36) and (2.35), we obtain

$$f(x) = e^{A(\ln |x|)} \quad \forall x \in \mathbb{R}_+. \tag{2.37}$$

From (2.31) we see that either $f(1) = 0$ or $f(1) = 1$. If $f(1) = 0$, then letting $y = 1$ in (2.3), we obtain

$$f(x) = 0 \qquad \forall\, x \in \mathbb{R} \setminus \{0\}$$

contrary to our assumption that f is not identically zero on $\mathbb{R} \setminus \{0\}$. Hence $f(1) = 1$. Now letting $x = -1 = y$ in (2.3), we get $f(1) = f(-1)^2$ and hence

$$f(-1) = 1 \qquad \text{or} \qquad f(-1) = -1. \tag{2.38}$$

If $f(-1) = 1$, then letting $y = -1$ in (2.3), we have

$$f(-x) = f(x)f(-1) = f(x)$$

for all $x \in \mathbb{R} \setminus \{0\}$. Thus (2.37) yields

$$f(x) = e^{A(\ln |x|)}$$

for all $x \in \mathbb{R} \setminus \{0\}$. Since $f(0) = 0$, we have

$$f(x) = \begin{cases} e^{A(\ln |x|)} & \text{if } x \in \mathbb{R} \setminus \{0\} \\ 0 & \text{if } x = 0 \end{cases}$$

which is the asserted solution (2.28).

If $f(-1) = -1$, then letting $y = -1$ in (2.3), we have

$$f(-x) = f(x)f(-1) = -f(x)$$

for all $x \in \mathbb{R} \setminus \{0\}$. Hence (2.37) yields

$$f(x) = \begin{cases} e^{A(\ln |x|)} & \text{if } x > 0 \\ -e^{A(\ln |x|)} & \text{if } x < 0 \end{cases}$$

for all $x \in \mathbb{R} \setminus \{0\}$. Together with the fact that $f(0) = 0$, we have

$$f(x) = \begin{cases} e^{A(\ln |x|)} & \text{if } x > 0 \\ 0 & \text{if } x = 0 \\ -e^{A(\ln |x|)} & \text{if } x < 0 \end{cases}$$

which is the asserted solution (2.29). Now the proof of the theorem is complete. $\qquad\square$

By virtue of the above theorem we have the following corollary.

Corollary 2.6. *The general continuous solution of the functional equation (2.3), that is, $f(xy) = f(x)f(y)$, holding for all $x, y \in \mathbb{R}$ is given by*

$$f(x) = 0, \tag{2.39}$$

$$f(x) = 1, \tag{2.40}$$

$$f(x) = |x|^{\alpha}, \tag{2.41}$$

and

$$f(x) = |x|^{\alpha} \, sgn(x), \tag{2.42}$$

where α is an arbitrary positive real constant.

Proof. By Theorem 2.4 either $f = 0$, or $f = 1$, or f has the form (2.28) or (2.29), where $A : \mathbb{R} \to \mathbb{R}$ is an additive function. Since f is continuous and

$$A(t) = \ln f(e^t),$$

A is also continuous on $\mathbb{R}$. Therefore

$$A(t) = \alpha \, t,$$

where $\alpha \in \mathbb{R}$ is an arbitrary constant. Hence from (2.28) and (2.29), we get

$$f(x) = |x|^{\alpha}$$

and

$$f(x) = |x|^{\alpha} \, \text{sgn}\,(\text{x}),$$

respectively. The only thing remaining to be shown is $\alpha > 0$. If we had $\alpha = 0$, then (2.41) will yield $f(x) = 1$ for $x \neq 0$, and by continuity of f we must have $f(0) = 1$. Hence we will have $f = 1$, already listed in (2.40). Formula (2.42) with $\alpha = 0$ yields

$$f(x) = 1 \qquad \text{for} \qquad x > 0$$

and

$$f(x) = -1 \qquad \text{for} \qquad x < 0$$

and thus f cannot be continuous. Similarly if $\alpha < 0$, then f given by (2.41) and (2.42) satisfies

$$\lim_{x \to 0^+} f(x) = \infty$$

and hence cannot be continuous at 0. Now the proof of the corollary is complete. $\qquad\square$

Definition 2.3. *A function $f : \mathbb{R} \to \mathbb{R}$ is called a multiplicative function if it satisfies $f(xy) = f(x) \, f(y)$ for all $x, y \in \mathbb{R}$.*

2.5   Concluding Remarks

In Chapter 1, we proved that if $f : \mathbb{C} \to \mathbb{C}$ is a continuous additive function, then there exist complex constants c_1 and c_2 such that

$$f(z) = c_1\, z + c_2\, \overline{z}$$

where $\overline{z}$ denotes the complex conjugate of z. Note that unlike the real-valued continuous additive functions on the reals, the complex-valued continuous additive functions on the complex plane are not linear. The linearity can be restored if one assumes a stronger regularity condition such as analyticity or differentiability instead of continuity. Abel (1826) investigated the exponential functional equation $f(x+y) = f(x)f(y)$ for complex function f of complex variable to prove the Newton's binomial series, namely

$$\sum_{k=0}^{\infty} \binom{m}{k} z^k = (1+z)^m$$

(see Aczél (1989)).

The real-valued continuous nontrivial exponential function on $\mathbb{R}$ is always of the form $f(x) = e^{kx}$, where k is an arbitrary constant. The continuous nontrivial complex solution of the exponential functional equation is of the form $f(z) = e^{c_1 z + c_2 \overline{z}}$, where c_1, c_2 are complex constants. A real-valued rational exponential function f is always of the form $f(x) = e^{kx}$, where $k = \ln f(1)$. However, even on the set of rational numbers, $\mathbb{Q}$, the complex-valued exponential function is not of the form $f(x) = e^{kx}$. Dharmadhikari (1965) proved that every nontrivial solution $f : \mathbb{Q} \to \mathbb{C}$ of the exponential Cauchy functional equation is of the form

$$f(x) = e^{kx + 2\pi i x \alpha(n)} \tag{2.43}$$

for every rational $x = \frac{m}{n}$, where $\alpha(n) \in \mathbb{Z}$ and satisfies $0 \le \alpha(n) < n$ for $n \ge 2$ and n divides $\alpha(kn) - \alpha(n)$ for all k and n in $\mathbb{Z}_+$.

Let $(G, \oplus)$ and $(H, \otimes)$ be two arbitrary groups and let $f : G \to H$ be a mapping from the group G into group H. Then the four Cauchy functional equations can be expressed as $f(x \oplus y) = f(x) \otimes f(y)$ for all $x, y \in G$. The function $f : G \to H$ satisfying the above equation is called a homomorphism from group G into group H.

Let $G_n = \left\{ x \in \mathbb{R} \,|\, x > -\frac{1}{n}, \ n > 0 \right\}$ and $\oplus$ be a binary operation in G_n defined as $x \oplus y = x + y + nxy$. Then $(G_n, \oplus)$ is a subgroup of group $(G, \oplus)$, where $G = \left\{ x \in \mathbb{R} \,|\, x \ne -\frac{1}{n} \right\}$. Nath and Madaan (1976) have studied functional equations of the type $f(x \oplus y) = g(x)\, h(y)$ for all $x, y \in (G_n, \oplus)$.

We close this section with a brief discussion on the following recent developments about the logarithmic Cauchy functional equation

$$f(xy) = f(x) + f(y) \quad \forall x, y \in (0, \infty).$$

In 1999, Heuvers studied the functional equation

$$f(x + y) - f(x) - f(y) = f\left(\frac{1}{x} + \frac{1}{y}\right) \tag{2.44}$$

for all $x, y \in (0, \infty)$ and showed that this functional equation (2.44) is equivalent to the logarithmic Cauchy functional equation. Heuvers and Kannappan (2005) found another functional equation

$$f(x + y) - f(xy) = f\left(\frac{1}{x} + \frac{1}{y}\right) \tag{2.45}$$

for all $x, y \in (0, \infty)$ and proved that it is also equivalent to the logarithmic Cauchy functional equation. The Pexider generalization of the functional equation (2.45) is the following:

$$f(x + y) - g(xy) = h\left(\frac{1}{x} + \frac{1}{y}\right), \quad \forall x, y \in (0, \infty), \tag{2.46}$$

where $f, g, h : (0, \infty) \to \mathbb{R}$ are real-valued functions on the set of positive reals. They proved the following theorem.

Theorem 2.5. *The general solution of* (2.46) *is given by*

$$f(x) = L(x) + a, \quad g(x) = L(x) + b, \quad h(x) = L(x) + a - b,$$

where $L : (0, \infty) \to \mathbb{R}$ *is a logarithmic function and* a, b *are arbitrary real constants.*

The Pexider generalization of the functional equation (2.44) is the functional equation

$$f(x + y) - g(x) - h(y) = k\left(\frac{1}{x} + \frac{1}{y}\right), \quad \forall x, y \in (0, \infty), \tag{2.47}$$

where $f, g, h, k : (0, \infty) \to \mathbb{R}$ are real-valued functions on the set of positive reals. Heuvers and Kannappan (2005) determined the twice differentiable solution of the functional equation (2.47). A particular solution of (2.47) is the following:

$$f(x) = L(x) + A(x) + c_1,$$
$$g(x) = L(x) + A(x) - A_1(1/x) + c_1 + c_3,$$
$$h(x) = L(x) + A(x) - A_1(1/x) - c_2 - c_3,$$
$$k(x) = L(x) + A_1(x) + c_2,$$

where $L : (0, \infty) \to \mathbb{R}$ is a logarithmic function, $A, A_1 : \mathbb{R} \to \mathbb{R}$ are additive functions, and c_1, c_2, c_3 are arbitrary real constants. The most general solution of (2.47) without any regularity assumption on the unknown functions f, g, h, k is an open problem.

2.6   Exercises

1. Find all continuous functions $f : \mathbb{R} \to \mathbb{R}$ that satisfy the functional equation

$$f(xy) = y\,f(x) + x\,f(y)$$

for all $a, y \in \mathbb{R} \smallsetminus \{0\}$.

2. If $f : \mathbb{R} \to \mathbb{R}$ is a solution of the exponential Cauchy functional equation

$$f(x + y) = f(x)\,f(y) \qquad \forall\, x, y \in \mathbb{R},$$

then show that f is either everywhere or nowhere zero.

3. Determine all functions $f : \mathbb{R} \to \mathbb{R}$ that satisfy the functional equation

$$f(x + y) = f(x) + f(y) + \lambda f(x)f(y) \qquad \forall\, x, y \in \mathbb{R},$$

where λ is a real constant.

4. Determine all functions $f : \mathbb{R} \to \mathbb{R}$ that satisfy the functional equation

$$f(x + y) = a^{xy}\,f(x)\,f(y) \qquad \forall\, x, y \in \mathbb{R},$$

where a is a positive real constant.

5. Determine all functions $f : \mathbb{R} \to \mathbb{R}$ that satisfy the functional equation

$$f(\sqrt{x^2 + y^2}) = f(x)\,f(y) \qquad \forall\, x, y \in \mathbb{R}.$$

6. Determine all functions $f : \mathbb{R} \to \mathbb{R}$ that satisfy the functional equation

$$f(\sqrt{x^2 + y^2}) = f(x) + f(y) \qquad \forall\, x, y \in \mathbb{R}.$$

7. Find all continuous solutions $f : \mathbb{C} \to \mathbb{C}$ of the complex exponential functional equation

$$f(z + w) = f(z) + f(w) \qquad \forall\, z, w \in \mathbb{C}.$$

8. Find all functions $f : \mathbb{R} \to \mathbb{R}$ that satisfy the functional equation

$$[f(x) + f(y)] [f(u) + f(v)] = f(xu - yv) + f(xv + yu)$$

for all $x, y, u, v \in \mathbb{R}$.

9. Determine all functions $f : \mathbb{R} \to \mathbb{R}$ that satisfy the two functional equations

$$f(x + y) = f(x) + f(y) \quad \text{and} \quad f(xy) = f(x) f(y)$$

for all $x, y \in \mathbb{R}$.

10. Find all functions $f : \mathbb{R} \to \mathbb{R}$ that satisfy the functional equation

$$f(x + y) + f(x) f(y) = f(xy) + f(x) + f(y)$$

for all $x, y \in \mathbb{R}$.

11. Find all functions $f, g : \mathbb{R} \to \mathbb{R}$ that satisfy the functional equation

$$f(x + y) = f(x) g(y) + f(y)$$

for all $x, y \in \mathbb{R}$.

12. Determine all functions $f : \mathbb{R} \to \mathbb{R}$ that satisfy the functional equation

$$f(x + y + \lambda xy) = f(x) f(y) \qquad \forall\, x, y \in \mathbb{R},$$

where λ is a real constant.

13. Determine all entire functions $f : \mathbb{C} \to \mathbb{C}$ that satisfy the functional equation

$$|f(s + it)| = |f(s) f(it)|$$

for all $s, t \in \mathbb{R}$.

14. Determine all entire functions $f : \mathbb{C} \to \mathbb{C}$ that satisfy the functional equation

$$|f(s + it) f(s - it)| = |f(s)^2|$$

for all $s, t \in \mathbb{R}$.

15. Let α, β be nonzero a priori chosen real numbers. Find all functions $f, g, h : (0, 1) \to \mathbb{R}$ that satisfy the functional equation

$$f(xy) = x^\alpha g(y) + y^\beta h(x)$$

for all $x, y \in (0, 1)$.

16. Find all functions $f : (0, \infty) \to \mathbb{R}$ that satisfy the Heuvers functional equation

$$f(x + y) = f(x) + f(y) + f(x^{-1} + y^{-1})$$

for all $x, y \in (0, \infty)$.

17. Find all functions $f : (0, \infty) \to \mathbb{R}$ that satisfy the functional equation

$$f(x + y) = f(x) f(y) f(x^{-1} + y^{-1})$$

for all $x, y \in (0, \infty)$.

18. Find all functions $f : \mathbb{R} \to \mathbb{R}$ that satisfy the functional equation

$$f(x + y) + f(x) f(y) = f(xy) + f(x) + f(y)$$

for all $x, y \in \mathbb{R}$.

19. Find all functions $f : [0, \infty) \to \mathbb{R}$ that satisfy the functional equation

$$f\left(x^2 + y^2\right) = f\left(x^2 - y^2\right) + f(2xy)$$

for all $x, y \in [0, \infty)$.

20. Find all functions $f : \mathbb{R} \to \mathbb{R}$ that satisfy the functional equation

$$f\left(\sqrt{x^2 + y^2 + 1}\right) = f(x) + f(y)$$

for all $x, y \in \mathbb{R}$.

21. Find all functions $f : \mathbb{R} \to \mathbb{R}$ that satisfy the functional equation

$$f(x + xy) = f(x) + f(x) f(y)$$

for all $x, y \in \mathbb{R}$.

Chapter 3

Cauchy Equations in Several Variables

3.1   Introduction

In this chapter, we will show that a real-valued additive function on $\mathbb{R}^n$ can be expressed as a sum of n additive functions of one variable. A similar result holds for real-valued logarithmic function on $\mathbb{R}^n$ with some appropriate restrictions on the domain. Further, it will be shown that a real-valued multiplicative function on $\mathbb{R}^n$ can be expressed as a product of n multiplicative functions of one variable. A similar result also holds for exponential function on $\mathbb{R}^n$.

3.2   Additive Cauchy Equations in Several Variables

The Cauchy functional equation

$$f(x+y) = f(x) + f(y), \quad \text{for } x, y \in \mathbb{R}, \tag{CE}$$

can be generalized to

$$f(x_1 + y_1, x_2 + y_2, \ldots, x_n + y_n) = f(x_1, x_2, \ldots, x_n) + f(y_1, y_2, \ldots, y_n)$$

for $(x_1, x_2, \ldots, x_n) \in \mathbb{R}^n$ and $(y_1, y_2, \ldots, y_n) \in \mathbb{R}^n$. Here $f : \mathbb{R}^n \to \mathbb{R}$. We want to find the general solution of this functional equation. To make the problem easy to understand we take $n = 2$. Thus our equation reduces to

$$f(x_1 + y_1, x_2 + y_2) = f(x_1, x_2) + f(y_1, y_2) \tag{FE}$$

for all $x_1, x_2, y_1, y_2 \in \mathbb{R}$.

Theorem 3.1. *The general solution $f : \mathbb{R}^2 \to \mathbb{R}$ of the functional equation (FE) is given by*

$$f(x_1, x_2) = A_1(x_1) + A_2(x_2), \tag{3.1}$$

where $A_1, A_2 : \mathbb{R} \to \mathbb{R}$ are additive.

Proof. Letting $x_2 = y_2 = 0$ in (FE), we have

$$f(x_1 + y_1, 0) = f(x_1, 0) + f(y_1, 0). \tag{3.2}$$

Define $A_1 : \mathbb{R} \to \mathbb{R}$ by

$$A_1(x) = f(x, 0). \tag{3.3}$$

Then by (3.3), (3.2) reduces to

$$A_1(x_1 + y_1) = A_1(x_1) + A_1(y_1).$$

Hence $A_1 : \mathbb{R} \to \mathbb{R}$ is an additive function.

Similarly letting $x_1 = y_1 = 0$ in (FE), we get

$$f(0, x_2 + y_2) = f(0, x_2) + f(0, y_2).$$

Defining $A_2 : \mathbb{R} \to \mathbb{R}$ by

$$A_2(x) = f(0, x) \tag{3.4}$$

we get

$$A_2(x_2 + y_2) = A_2(x_2) + A_2(y_2).$$

Therefore $A_2 : \mathbb{R} \to \mathbb{R}$ is an additive function. Next we substitute in (FE)

$$y_1 = 0 = x_2$$

to obtain

$$\begin{aligned} f(x_1, y_2) &= f(x_1, 0) + f(0, y_2) \\ &= A_1(x_1) + A_2(y_2). \end{aligned}$$

Therefore

$$f(x, y) = A_1(x) + A_2(y), \quad \text{for } x, y \in \mathbb{R},$$

where $A_1, A_2 : \mathbb{R} \to \mathbb{R}$ are additive functions on $\mathbb{R}$. The proof is now complete. $\square$

This theorem says that any additive function of two variables on $\mathbb{R}^2$ can be decomposed as the sum of two additive functions in one variable. That is

$$f(x, y) = A_1(x) + A_2(y),$$

where $f : \mathbb{R}^2 \to \mathbb{R}$, and $A_1, A_2 : \mathbb{R} \to \mathbb{R}$.

Theorem 3.1 can also be rephrased as the following theorem.

Theorem 3.2. *If $f : \mathbb{R}^2 \to \mathbb{R}$ is additive on the plane $\mathbb{R}^2$, then there exist additive functions $A_1, A_2 : \mathbb{R} \to \mathbb{R}$ such that*

$$f(x_1, x_2) = A_1(x_1) + A_2(x_2) \tag{3.5}$$

for all $x_1, x_2 \in \mathbb{R}$.

The following theorem follows from Theorem 3.2 and Theorem 1.1.

Theorem 3.3. *If $f : \mathbb{R}^2 \to \mathbb{R}$ is a continuous additive function on the plane $\mathbb{R}^2$, then there exist constants c_1, c_2 such that*

$$f(x_1, x_2) = c_1 x_1 + c_2 x_2 \tag{3.6}$$

for all $x_1, x_2 \in \mathbb{R}$.

This result can be furthered strengthen by weakening the continuity of $f : \mathbb{R}^2 \to \mathbb{R}$.

Lemma 3.1. *If an additive function $f : \mathbb{R}^2 \to \mathbb{R}$ is continuous with respect to each variable, then it is continuous.*

Proof. Since the function $f : \mathbb{R}^2 \to \mathbb{R}$ is additive, by Theorem 3.2, we have

$$f(x, y) = A_1(x) + A_2(y)$$

for all $x, y \in \mathbb{R}$. Since f is continuous with respect to each variable, we see that A_1 and A_2 are continuous. Hence

$$\lim_{x \to x_o} A_1(x) = A_1(x_o) \qquad \text{and} \qquad \lim_{y \to y_o} A_2(y) = A_2(y_o).$$

In order to show f is continuous, we compute

$$\lim_{(x,y) \to (x_o, y_o)} f(x, y) = \lim_{(x,y) \to (x_o, y_o)} [A_1(x) + A_2(y)]$$
$$= \lim_{x \to x_o} A_1(x) + \lim_{y \to y_o} A_2(y)$$
$$= A_1(x_o) + A_2(y_o) = f(x_o, y_o).$$

This shows that f is continuous. Now this proof is complete. $\square$

Theorem 3.4. *If $f : \mathbb{R}^2 \to \mathbb{R}$ is an additive function on the plane $\mathbb{R}^2$ and continuous in each variable, then there exist constants c_1, c_2 such that*

$$f(x_1, x_2) = c_1 x_1 + c_2 x_2 \tag{3.7}$$

for all $x_1, x_2 \in \mathbb{R}$.

Proof. The proof follows from Theorem 3.3 and Lemma 3.1. $\square$

Theorem 3.1 is also true for general n. That is if $f : \mathbb{R}^n \to \mathbb{R}$ satisfies

$$f(x_1 + y_1, x_2 + y_2, \ldots, x_n + y_n) = f(x_1, x_2, \ldots, x_n) + f(y_1, y_2, \ldots, y_n)$$

for all $(x_1, x_2, \ldots, x_n), (y_1, y_2, \ldots, y_n)$ in $\mathbb{R}^n$, then

$$f(x_1, x_2, \ldots, x_n) = \sum_{k=1}^{n} A_k(x_k),$$

where $A_k : \mathbb{R} \to \mathbb{R}$ $(k = 1, 2, \ldots, n)$ are additive functions.

The proof of Theorem 3.1 depends primarily on the number 0. Hence, following Kuczma (1973), we give an alternative proof without the use of the number 0.

Let $a \in \mathbb{R}$ be a fixed element and $f : \mathbb{R}^2 \to \mathbb{R}$ be additive. Then

$$\begin{aligned}
f(x, y) &= f(x, y) + 2f(a, a) - 2f(a, a) \\
&= f(x + a + a, y + a + a) - 2f(a, a) \\
&= f((x + a) + a, a + (y + a)) - 2f(a, a) \\
&= f(x + a, a) + f(a, y + a) - 2f(a, a) \\
&= f(x + a, a) - f(a, a) + f(a, y + a) - f(a, a) \\
&= A_1(x) + A_2(y),
\end{aligned}$$

where

$$A_1(x) := f(x + a, a) - f(a, a)$$

and

$$A_2(y) := f(a, y + a) - f(a, a).$$

Now we show that A_1, A_2 are additive functions on $\mathbb{R}$. Consider

$$\begin{aligned}
A_1(x + y) &= f(x + y + a, a) - f(a, a) \\
&= f(x + y + a, a) + f(a, a) - 2f(a, a) \\
&= f(x + y + a + a, a + a) - 2f(a, a) \\
&= f(x + a, a) + f(y + a, a) - 2f(a, a) \\
&= f(x + a, a) - f(a, a) + f(y + a, a) - f(a, a) \\
&= A_1(x) + A_1(y).
\end{aligned}$$

Hence A_1 is additive. Similarly, one can show that

$$A_2(x + y) = A_2(x) + A_2(y),$$

and hence A_2 is additive. Thus

$$f(x, y) = A_1(x) + A_2(y),$$

where A_1 and A_2 are additive functions.

3.3 Multiplicative Cauchy Equations in Several Variables

The multiplicative Cauchy functional equation

$$f(xy) = f(x)f(y), \quad \text{for } x, y \in \mathbb{R}, \qquad \text{(ME)}$$

can be generalized to

$$f(x_1y_1, x_2y_2, \ldots, x_ny_n) = f(x_1, x_2, \ldots, x_n) \, f(y_1, y_2, \ldots, y_n)$$

for $(x_1, x_2, \ldots, x_n) \in \mathbb{R}^n$ and $(y_1, y_2, \ldots, y_n) \in \mathbb{R}^n$. Here $f : \mathbb{R}^n \to \mathbb{R}$. We want to find the general solution of this functional equation. To make the problem easy to understand we take $n = 2$. Thus the above equation reduces to

$$f(x_1y_1, x_2y_2) = f(x_1, x_2) \, f(y_1, y_2) \qquad \text{(FE1)}$$

for all $x_1, x_2, y_1, y_2 \in \mathbb{R}$.

Theorem 3.5. *The general solution $f : \mathbb{R}^2 \to \mathbb{R}$ of the functional equation*

$$f(x_1y_1, x_2y_2) = f(x_1, x_2) \, f(y_1, y_2) \qquad \text{(FE1)}$$

is given by

$$f(x_1, x_2) = M_1(x_1) \, M_2(x_2), \qquad (3.8)$$

where $M_1, M_2 : \mathbb{R} \to \mathbb{R}$ are multiplicative functions.

Proof. Let $a \in \mathbb{R}$ be a fixed element and $f : \mathbb{R}^2 \to \mathbb{R}$ be multiplicative with $f(a, a) \neq 0$. Then

$$
\begin{aligned}
f(x, y) &= f(x, y) \, f(a, a) \, f(a, a) \, f(a, a)^{-2} \\
&= f(xaa, yaa) \, f(a, a)^{-2} \\
&= f((xa)a, a(ya)) \, f(a, a)^{-2} \\
&= f(xa, a) \, f(a, ya) \, f(a, a)^{-2} \\
&= f(xa, a) \, f(a, a)^{-1} \, f(a, ya) \, f(a, a)^{-1} \\
&= M_1(x) \, M_2(y),
\end{aligned}
$$

where

$$M_1(x) := f(xa, a) \, f(a, a)^{-1}$$

and

$$M_2(y) := f(a, ya) \, f(a, a)^{-1}.$$

Now we show that M_1, M_2 are multiplicative functions of $\mathbb{R}$. Consider

$$
\begin{aligned}
M_1(xy) &= f(xya, a)\, f(a, a)^{-1} \\
&= f(xya, a)\, f(a, a)\, f(a, a)^{-2} \\
&= f(xyaa, aa)\, f(a, a)^{-2} \\
&= f(xa, a)\, f(ya, a)\, f(a, a)^{-2} \\
&= f(xa, a)\, f(a, a)^{-1}\, f(ya, a)\, f(a, a)^{-1} \\
&= M_1(x)\, M_1(y).
\end{aligned}
$$

Hence M_1 is multiplicative. Similarly, one can show that

$$
M_2(xy) = M_2(x)\, M_2(y)
$$

and hence M_2 is multiplicative. Thus

$$
f(x, y) = M_1(x)\, M_2(y),
$$

where M_1 and M_2 are multiplicative functions.   □

This theorem says that any multiplicative function of two variables on $\mathbb{R}^2$ can be written as a product of two multiplicative functions in one variable.

This theorem is also true for general n. That is, if $f : \mathbb{R}^n \to \mathbb{R}$ satisfies

$$
f(x_1 y_1, x_2 y_2, \ldots, x_n y_n) = f(x_1, x_2, \ldots, x_n)\, f(y_1, y_2, \ldots, y_n)
$$

for all $(x_1, x_2, \ldots, x_n), (y_1, y_2, \ldots, y_n)$ in $\mathbb{R}^n$, then

$$
f(x_1, x_2, \ldots, x_n) = \prod_{k=1}^{n} M_k(x_k),
$$

where $M_k : \mathbb{R} \to \mathbb{R}$ $(k = 1, 2, \ldots, n)$ are multiplicative functions.

3.4   Other Two Cauchy Equations in Several Variables

In this section, we consider the remaining two Cauchy equations in several variables. Analogous to Theorem 3.1 and Theorem 3.5, one can establish the following results.

Theorem 3.6. *The general solution* $f : \mathbb{R}^2 \to \mathbb{R}$ *of the functional equation*

$$f(x_1 + y_1, x_2 + y_2) = f(x_1, x_2) \, f(y_1, y_2) \tag{FE2}$$

is given by

$$f(x_1, x_2) = E_1(x_1) \, E_2(x_2), \tag{3.9}$$

where $E_1, E_2 : \mathbb{R} \to \mathbb{R}$ *are exponential functions.*

In general every exponential function $f : \mathbb{R}^n \to \mathbb{R}$ can be written as

$$f(x_1, x_2, \ldots, x_n) = \prod_{k=1}^{n} E_k(x_k),$$

where $E_k : \mathbb{R} \to \mathbb{R}$ $(k = 1, 2, ..., n)$ are expenential functions.

Let $\mathbb{R}_0 = \{x \in \mathbb{R} \mid x \neq 0\}$. The following theorem says that any logarithmic function of two variables on $\mathbb{R}_0^2$ can be written as a sum of logarithmic functions in one variable.

Theorem 3.7. *The general solution* $f : \mathbb{R}_0^2 \to \mathbb{R}$ *of the functional equation*

$$f(x_1 y_1, x_2 y_2) = f(x_1, x_2) + f(y_1, y_2) \tag{FE3}$$

is given by

$$f(x_1, x_2) = L_1(x_1) + L_2(x_2) \tag{3.10}$$

where $L_1, L_2 : \mathbb{R}_0 \to \mathbb{R}$ *are logarithmic functions.*

In general every logarithmic function $f : \mathbb{R}_0^n \to \mathbb{R}$ can be written as

$$f(x_1, x_2, \ldots, x_n) = \sum_{k=1}^{n} L_k(x_k)$$

where $L_k : \mathbb{R}_0 \to \mathbb{R}$ $(k = 1, 2, ..., n)$ are logarithmic functions.

3.5 Concluding Remarks

From the alternative proof of Theorem 3.1, it is clear that Theorem 3.1 also holds on an abelian semigroup. Similarly Theorem 3.5 holds on an abelian semigroup. Recall that a semigroup S is a set together with associative binary operation $\circ : S \times S \to S$. Thus a semigroup need not have an identity element and its elements need not have inverses within the semigroup.

Kuczma (1973) proved the following result concerning Cauchy functional equation in two variables.

Theorem 3.8. *Let $(S, +)$ be a commutative semigroup and $(G, +)$ be an abelian group. If $f : S^2 \to G$ satisfies the functional equation*

$$f(x + u, y + v) = f(x, y) + f(u, v)$$

for all $x, y, u, v \in S$, then there exist homomorphisms $A_1, A_2 : S \to G$ such that

$$f(x, y) = A_1(x) + A_2(y)$$

for all $x, y \in S$.

3.6 Exercises

1. Find all continuous solutions $f : \mathbb{C} \to \mathbb{C}$ of the additive Cauchy functional equation

$$f(x + y) = f(x) + f(y) \qquad \forall\, x, y \in \mathbb{C}$$

by reducing it to a pair of functional equations obtained by substituting

$$x = x_1 + i\, x_2, \ y = y_1 + i\, y_2, \ f(x) = g(x_1, x_2) + i\, h(x_1, x_2).$$

2. Find all continuous solutions $f : \mathbb{C} \to \mathbb{C}$ of the functional equation

$$f(x + y) + f(x - y) = 2\, f(x) \qquad \forall\, x, y \in \mathbb{C}$$

by reducing it to a pair of functional equations obtained by substituting

$$x = x_1 + i\, x_2, \ y = y_1 + i\, y_2, \ f(x) = g(x_1, x_2) + i\, h(x_1, x_2).$$

3. Find general solutions $f : \mathbb{R}^2 \to \mathbb{R}$ of the Sincov functional equation

$$f(x, y) + f(y, z) = f(x, z)$$

for all $x, y, z \in \mathbb{R}$.

4. Find all differentiable functions $f : \mathbb{R}^2 \to \mathbb{R}$ that satisfy the cocycle functional equation

$$f(x, y) + f(x + y, z) = f(x, y + z) + f(y, z)$$

for all $x, y, z \in \mathbb{R}$.

5. Find all symmetric functions $f : \mathbb{R}^n \to \mathbb{R}$ that satisfy the functional equations

$$f(x_1 + y_1, x_2 + y_2, \cdots, x_n + y_n) = f(x_1, x_2, \cdots, x_n) + f(y_1, y_2, \cdots, y_n)$$

and

$$f(x, x, \cdots, x) = x$$

for all $x, x_1, y_1, x_2, y_2, ..., x_n, y_n \in \mathbb{R}$.

6. Find all functions $f : \mathbb{R}^2 \to \mathbb{R}$ that satisfy the Euler functional equation

$$f(xz, yz) = z^k f(x, y) \qquad \forall x, y \in \mathbb{R}, \ z \in \mathbb{R} \smallsetminus \{0\}$$

and satisfying the additional conditions

$$f(0, y) = f(0 \cdot y, 1 \cdot y) = y^k f(0, 1) \quad \text{for } y \neq 0$$

and

$$f(0, 0) = z^k f(0, 0)$$

where k is a real constant.

7. Find all functions $f : \mathbb{R} \to \mathbb{R}$ that satisfy the functional equation

$$f(x + y, z) + f(y + z, x) + f(z + x, y) = 0$$

for all $x, y, z \in \mathbb{R}$. [Hint: Substitute $x = v + \frac{t}{3}$, $y = -u - v + \frac{t}{3}$ and $z = u + \frac{t}{3}$ and define a new function $g(u, t) := f\left(\frac{2t}{3} - u, \frac{t}{3} + u\right)$.]

8. Find all functions $f : \mathbb{R} \to \mathbb{R}$ that satisfy the functional equation

$$f(pr, qs) + f(ps, qr) = f(p, q) f(r, s)$$

for all $p, q, r, s \in \mathbb{R}$.

9. Let $\mathbb{R}^*$ be the set of nonzero real numbers. Find the general solution $f : \mathbb{R}^2 \to \mathbb{R}^*$ of the functional equation

$$f(ux - vy, uy + v(x + y)) = f(x, y) f(u, v)$$

for all $u, v, x, y \in \mathbb{R}$.

10. Find all functions $f : \mathbb{R}^2 \to \mathbb{R}$ that satisfy the functional equation

$$f(ux + vy, uy + vx) = f(x, y) f(u, v)$$

for all $x, y, u, v \in \mathbb{R}$.

Chapter 4

Extension of the Cauchy Functional Equations

4.1   Introduction

The set of all values of the variables, on which the functional equation is supposed to hold, is called the domain of the functional equation. For instance, the domain of the functional equation

$$f(x+y) = f(x) + f(y) \qquad \forall x, y \in (0, \infty) \tag{4.1}$$

is $\mathbb{R}_+^2$. A function satisfying a functional equation on a given domain is called a solution on that domain. In this chapter, we only consider the problem of extending the additive Cauchy functional equation from a smaller domain to a larger domain that contains the smaller domain. The other three Cauchy functional equations can be extended similarly.

4.2   Extension of Additive Functions

Let $[a, b]$ be an interval in $\mathbb{R}$, and let $f : [a, b] \to \mathbb{R}$ be an additive function on the interval $[a, b]$ in the sense that (1.1) holds for all $x, y, x+y$ in $[a, b]$. Does there exist an additive function $A : \mathbb{R} \to \mathbb{R}$ such that $A(x) = f(x)$ for all $x \in [a, b]$ (that is $A_{\big|[a,b]} = f$)?

The following theorem was proved by Aczél and Erdös (1965).

Theorem 4.1. *Let $\alpha \geq 0$, and let $f : [\alpha, \infty) \to \mathbb{R}$ be an additive function on $[\alpha, \infty)$. Then there exists an additive function $A : \mathbb{R} \to \mathbb{R}$ such that*

$$A(x) = f(x)$$

for all $x \in [\alpha, \infty)$.

Proof. Define a function $A : \mathbb{R} \to \mathbb{R}$ by

$$A(x - y) = f(x) - f(y) \tag{4.2}$$

for all $x, y \in [\alpha, \infty)$. Now we show that

(a) A is well-defined;

(b) A is an extension of the additive map f;

(c) A is an additive function on $\mathbb{R}$.

First we show that $A : \mathbb{R} \to \mathbb{R}$ is well-defined (that is, the definition of A given by (4.2) is a valid definition of a function).

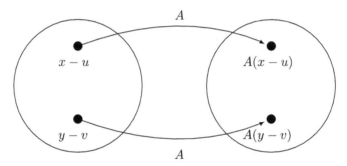

Let $x, u, y, v \in [\alpha, \infty)$ and suppose

$$x - u = y - v. \tag{4.3}$$

Hence

$$x + v = y + u$$
$$\Rightarrow \quad f(x + v) = f(y + u)$$
$$\Rightarrow \quad f(x) + f(v) = f(y) + f(u)$$
$$\Rightarrow \quad f(x) - f(u) = f(y) - f(v)$$
$$\Rightarrow \quad A(x - u) = A(y - v).$$

Therefore A is well-defined.

Next we show that A is an extension of f. For any $t \in [\alpha, \infty)$, there exist $x, y \in [\alpha, \infty)$ such that

$$t = x - y.$$

Hence

$$
\begin{aligned}
A(t) &= A(x - y) \\
&= f(x) - f(y) \\
&= f(y + t) - f(y) \\
&= f(y) + f(t) - f(y) \\
&= f(t)
\end{aligned}
$$

for alll $t \in [\alpha, \infty)$. Thus

$$
A\big|_{[\alpha,\infty)} = f.
$$

Finally, we show that A is additive on the set of reals, $\mathbb{R}$. For any $s, t \in \mathbb{R}$, there exist $x, y, u, v \in [\alpha, \infty)$ such that

$$
s = x - y
$$
$$
t = u - v.
$$

Note that $u + x$ and $y + v$ are in $[\alpha, \infty)$. Also

$$
s + t = (x + u) - (y + v).
$$

Computing $A(s + t)$, we see that

$$
\begin{aligned}
A(s + t) &= A((x + u) - (y + v)) \\
&= f(x + u) - f(y + v) \\
&= f(x) + f(u) - f(y) - f(v) \\
&= f(x) - f(y) + f(u) - f(v) \\
&= A(x - y) + A(u - v) \\
&= A(s) + A(t)
\end{aligned}
$$

for all $s, t \in \mathbb{R}$. This completes the proof of the theorem. $\qquad\square$

This theorem says that the general solution of the Cauchy functional equation

$$
f(x + y) = f(x) + f(y)
$$

for all $x, y \in [\alpha, \infty)$ is the same as the Cauchy functional equation for all $x, y \in \mathbb{R}$.

Remark 4.1. *Note that the domain $[\alpha, \infty)$ is unbounded and therefore if $x, y \in [\alpha, \infty)$ then*

$$
x + y \in [\alpha, \infty).
$$

However, if $x, y \in [a, b]$ where $[a, b]$ is a bounded interval, then $x + y$ is not necessarily in $[a, b]$. Hence the proof of the above theorem does not work for a bounded interval.

The following Theorem is due to Daróczy and Losonczi (1967).

Theorem 4.2. *Let $f : [0,1] \to \mathbb{R}$ satisfy (1.1) for all $x, y, x+y$ in $[0,1]$. Then there exists an additive function $A : \mathbb{R} \to \mathbb{R}$ such that*

$$A(x) = f(x)$$

for all $x \in [0,1]$.

Proof. Let $x \in \mathbb{R}$ be any real number. Then x can be expressed as

$$x = \frac{n}{2} + x', \tag{4.4}$$

where n is an integer and $x' \in \left[0, \frac{1}{2}\right)$. Define a function $A : \mathbb{R} \to \mathbb{R}$ as

$$A(x) := nf\left(\frac{1}{2}\right) + f(x'). \tag{4.5}$$

Clearly A is well-defined. Hence it is enough to show that

(i) $A(x) = f(x)$, $\forall\, x \in [0,1]$, and

(ii) A is additive on $\mathbb{R}$.

To prove (i) we need to consider three cases: (1) $x \in \left[0, \frac{1}{2}\right)$, (2) $x \in \left[\frac{1}{2}, 1\right)$, and (3) $x = 1$ (see figure below).

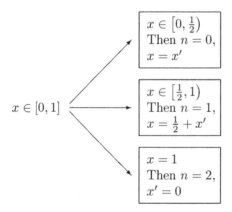

Case 1. Suppose $x \in \left[0, \frac{1}{2}\right)$. Then $n = 0$ and $x = x'$. Hence

$$A(x) = A(x')$$
$$= f(x')$$
$$= f(x)$$

for all $x \in \left[0, \frac{1}{2}\right)$.

Case 2. Suppose $x \in \left[\frac{1}{2}, 1\right)$. Then $n = 1$ and $x = \frac{1}{2} + x'$ where $x' \in \left[0, \frac{1}{2}\right)$. Hence

$$A(x) = A\left(\frac{1}{2} + x'\right)$$
$$= f\left(\frac{1}{2}\right) + f(x')$$
$$= f\left(\frac{1}{2} + x'\right)$$
$$= f(x)$$

for all $x \in \left[\frac{1}{2}, 1\right)$.

Case 3. Suppose $x = 1$. Then $n = 2$ and $x' = 0$.

$$A(x) = A(1)$$
$$= 2f\left(\frac{1}{2}\right)$$
$$= f\left(2 \cdot \frac{1}{2}\right)$$
$$= f(1).$$

Hence we have proved that

$$A(x) = f(x)$$

for all $x \in [0, 1]$. That is, A is an extension of f.

Next we show that $A : \mathbb{R} \to \mathbb{R}$ is an additive function on $\mathbb{R}$. Let s and t be two arbitrary real numbers. Then they can be expressed as

$$s = \frac{m}{2} + s',$$
$$t = \frac{n}{2} + t',$$

where m, n are integers and $s', t' \in \left[0, \frac{1}{2}\right)$. Now we have to consider two cases depending on where the sum $s' + t'$ is:

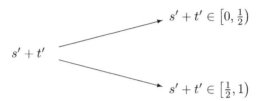

$$s' + t' \in \left[0, \tfrac{1}{2}\right)$$

$$s' + t'$$

$$s' + t' \in \left[\tfrac{1}{2}, 1\right)$$

Case 1. Suppose $s' + t'$ belongs to the interval $\left[0, \frac{1}{2}\right)$. Consider

$$
\begin{aligned}
A(s+t) &= A\left(\frac{m}{2} + s' + \frac{n}{2} + t'\right) \\
&= A\left((m+n)\frac{1}{2} + s' + t'\right) \\
&= (m+n)f\left(\frac{1}{2}\right) + f(s' + t') \\
&= mf\left(\frac{1}{2}\right) + nf\left(\frac{1}{2}\right) + f(s') + f(t') \\
&= mf\left(\frac{1}{2}\right) + f(s') + nf\left(\frac{1}{2}\right) + f(t') \\
&= A(s) + A(t).
\end{aligned}
$$

Case 2. Suppose $s' + t'$ belongs to the interval $\left[\frac{1}{2}, 1\right)$. Then

$$s' + t' = \frac{1}{2} + z',$$

where $z' \in \left[0, \frac{1}{2}\right)$. Hence

$$
\begin{aligned}
A(s+t) &= A\left(\frac{m}{2} + s' + \frac{n}{2} + t'\right) \\
&= A\left((m+n)\frac{1}{2} + s' + t'\right) \\
&= A\left((m+n)\frac{1}{2} + \frac{1}{2} + z'\right) \\
&= A\left((m+n+1)\frac{1}{2} + z'\right) \\
&= (m+n+1)f\left(\frac{1}{2}\right) + f(z')
\end{aligned}
$$

$$= (m+n)f\left(\frac{1}{2}\right) + f\left(\frac{1}{2}\right) + f(z')$$

$$= (m+n)f\left(\frac{1}{2}\right) + f\left(\frac{1}{2} + z'\right)$$

$$= (m+n)f\left(\frac{1}{2}\right) + f(s' + t')$$

$$= (m+n)f\left(\frac{1}{2}\right) + f(s') + f(t')$$

$$= mf\left(\frac{1}{2}\right) + f(s') + nf\left(\frac{1}{2}\right) + f(t')$$

$$= A\left(\frac{m}{2} + s'\right) + A\left(\frac{n}{2} + t'\right)$$

$$= A(s) + A(t).$$

Hence A is additive on $\mathbb{R}$. The proof of the theorem is now complete. $\quad\square$

4.3 Concluding Remarks

The proof of Aczél and Erdös (that is, of Theorem 4.1) can be applied to extensions of more general homomorphisms of subsemigroups. Note that $\mathbb{R}_+ = \{x \in \mathbb{R} \,|\, x > 0\}$ is a subsemigroup of additive groups of reals. Also it is easy to see that $\mathbb{R} = \mathbb{R}_+ - \mathbb{R}_+ = \{x - y \,|\, x, y \in \mathbb{R}_+\}$. Aczél, Baker, Djoković, Kannappan and Rado (1971) proved the following extension theorem for additive functions generalizing the results of Aczél and Erdös (1965). Recall that a subsemigroup S of a group G is a nonempty subset of G which is closed under group operation.

Theorem 4.3. *Let the subsemigroup S generates the abelian group $(G, \cdot)$ and let f be a homomorphism of S into the abelian group $(H, +)$. Then there exists a unique extension A of f to a homomorphism of G into H.*

Proof. It is known that a subsemigroup S generates the abelian group G if and only if

$$G = S \cdot S^{-1} = \{xy^{-1} \,|\, x, y \in S\}.$$

Define a mapping $A : G \to H$ as follows:

$$A(xy^{-1}) = f(x) - f(y) \tag{4.6}$$

for all $x, y \in S$. First we show that A is well-defined. Let $x, u, y, v \in S$ and suppose

$$x\,u^{-1} = y\,v^{-1}. \tag{4.7}$$

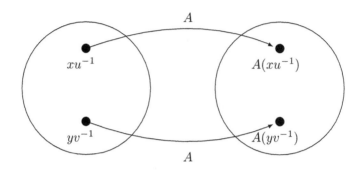

Hence

$$x\,u^{-1} = y\,v^{-1}$$

$\Rightarrow \qquad x\,u^{-1} = v^{-1}\,y \qquad\qquad \text{(since } G \text{ is abelian)}$

$\Rightarrow \qquad x = v^{-1}\,y\,u$

$\Rightarrow \qquad v\,x = y\,u$

$\Rightarrow \qquad x\,v = y\,u \qquad\qquad \text{(since } G \text{ is abelian)}$

$\Rightarrow \qquad f(x\,v) = f(y\,u)$

$\Rightarrow \qquad f(x) + f(v) = f(y) + f(u)$

$\Rightarrow \qquad f(v) + f(x) = f(y) + f(u) \qquad \text{(since } H \text{ is abelian)}$

$\Rightarrow \quad f(v) + f(x) - f(u) = f(y)$

$\Rightarrow \qquad f(x) - f(u) = -f(v) + f(y)$

$\Rightarrow \qquad f(x) - f(u) = f(y) - f(v) \qquad \text{(since } H \text{ is abelian)}$

$\Rightarrow \qquad A(x\,u^{-1}) = A(y\,v^{-1}).$

Therefore A is well-defined.

Next we show that A is a homomorphism of the group G into group H. Let $s, t \in G$ be any two arbitrary elements. Then there exists $x, y, u, v \in S$ such that

$$s = x\,y^{-1} \qquad \text{and} \qquad t = u\,v^{-1}.$$

Thus

$$A(st) = A(xy^{-1}uv^{-1})$$
$$= A(xuy^{-1}v^{-1})$$
$$= A((xu)(vy)^{-1})$$
$$= f(xu) - f(vy)$$
$$= f(x) + f(u) - f(y) - f(v)$$
$$= f(x) - f(y) + f(u) - f(v) \qquad \text{(since } H \text{ is abelian)}$$
$$= A(s) + A(t).$$

Hence $A : G \to H$ is a homomorphism of G into H. Next we show A is an extension of f. Let $x \in S$ be an arbitrary element. Hence

$$A(x) = A(x\,x\,x^{-1})$$
$$= f(x\,x) - f(x)$$
$$= f(x) + f(x) - f(x)$$
$$= f(x)$$

and A is an extension of f.

Finally we show that A is unique. Suppose there is an extension B of f which is also a homomorphism of G into H. Let $z \in G$ be an arbitrary element. Then $z = xy^{-1}$ for some $x, y \in S$. Since

$$A(z) = A(x\,y^{-1})$$
$$= A(x) - A(y) \qquad \text{(since } A \text{ is a homomorphism)}$$
$$= f(x) - f(y) \qquad \text{(since } f = A\big|_S\text{)}$$
$$= B(x) - B(y) \qquad \text{(since } B\big|_S = f \text{)}$$
$$= B(xy^{-1})$$
$$= B(z),$$

the homomorphism A is unique.

This completes the proof of the theorem. $\qquad\qquad\qquad\square$

In the last theorem both the groups $(G, \cdot)$ and $(H, +)$ are abelian. The following theorem due to Aczél, Baker, Djoković, Kannappan and Rado (1971) shows that the groups $(G, \cdot)$ and $(H, +)$ do not have to be abelians.

Theorem 4.4. *Let* $(G, \cdot)$ *and* $(H, +)$ *be groups, and let* $(S, \cdot)$ *be a sub-semigroup of* $(G, \cdot)$ *such that*

$$G = S \cdot S^{-1} = \left\{ xy^{-1} \mid x, y \in S \right\}.$$

Let $f : S \to H$ be a homomorphism. Then there exists a unique homomorphism $A : G \to H$ such that $A\big|_S = f$.

4.4   Exercises

1. Show that the function $f : [0, \infty) \to \mathbb{R}$ defined by

$$f(x) = \begin{cases} 0 & if \ x > 0 \\ 1 & if \ x = 1 \end{cases}$$

is a solution of the exponential Cauchy functional equation

$$f(x + y) = f(x) f(y) \qquad \forall \, x, y \in [0, \infty).$$

Also, show that this function f cannot be extended from $[0, \infty)$ to $\mathbb{R}$.

2. Let $f : (0, \alpha) \to \mathbb{R}$ satisfy

$$f(x + y) = f(x) + f(y)$$

for all $x, y, x + y$ in $(0, \alpha)$, where α is a positive real number. Then show that there exists an additive function $A : \mathbb{R} \to \mathbb{R}$ such that $A(x) = f(x)$ for all $x \in (0, \alpha)$.

3. Let $\mathbb{R}_0$ be the set of nonzero real numbers. Let $f : (\alpha, \infty) \to \mathbb{R}$ satisfy the logarithmic Cauchy functional equation

$$f(xy) = f(x) + f(y)$$

for all x, y in (α, ∞), where α is positive real number. Then show that there exists a logarithmic function $L : \mathbb{R}_0 \to \mathbb{R}$ such that $L(x) = f(x)$ for all $x \in (\alpha, \infty)$.

4. Let $\mathbb{R}_0$ be the set of nonzero real numbers. Let $f : [1, \infty) \to \mathbb{R}$ satisfy the logarithmic Cauchy functional equation

$$f(xy) = f(x) + f(y)$$

for all x, y in $[1, \infty)$. Then show that there exists a logarithmic function $L : \mathbb{R}_0 \to \mathbb{R}$ such that $E(x) = f(x)$ for all $x \in [1, \infty)$.

5. Let $\mathbb{R}_0$ be the set of nonzero real numbers. Let $f : [0, 1] \to \mathbb{R}_0$ satisfy the exponential Cauchy functional equation

$$f(x + y) = f(x) f(y)$$

for all $x, y, x + y$ in $[0, 1]$. Then show that there exists an exponential function $E : \mathbb{R} \to \mathbb{R}_0$ such that $E(x) = f(x)$ for all $x \in [0, 1]$.

6. Let $\mathbb{R}_0$ be the set of nonzero real numbers. Let $f : [1, \infty) \to \mathbb{R}_0$ satisfy the multiplicative Cauchy functional equation

$$f(x + y) = f(x) f(y)$$

for all $x, y, x + y$ in $[0, 1]$. Then show that there exists a multiplicative function $M : (0, \infty) \to \mathbb{R}_0$ such that $M(x) = f(x)$ for all $x \in [1, \infty)$.

7. Let $f : (-\alpha, \alpha) \to \mathbb{R}$ satisfy the Mikusiński functional equation

$$f(x + y) \left[f(x + y) - f(x) - f(y) \right] = 0 \tag{4.8}$$

for all $x, y, x+y$ in $(-\alpha, \alpha)$. Then show that there exists a unique function $H : \mathbb{R} \to \mathbb{R}$ satisfying (4.8) for all $x, y \in \mathbb{R}$ and $H(x) = f(x)$ for all $x \in (-\alpha, \alpha)$, where α is a positive real number.

Chapter 5

Applications of Cauchy Functional Equations

5.1 Introduction

Many functional equations originated from applications. At present, problems in science and engineering are generally modeled by ordinary differential equations (ODE) or partial differential equations (PDE). Before the development of ODE or PDE, physical processes were analyzed using functions. When a physical process is modeled by function, say, f, one uses an input variable x (or several input variables) and the corresponding output variable $f(x)$. The output variable $f(x)$ satisfies some relations corresponding to some properties of the physical process which are often known by observation. This leads to functional equations for the function f. When functional equations are used for modeling, one does not have to assumed the differentiability of the function and consequently, the functional equations lead often to other solutions than those given by ODE or PDE. These other solutions can be of some interest to scientists and engineers.

In this chapter, we present a few applications of Cauchy functional equations. In Section 2, we derive the formula for the area of a rectangle following Legendre (1791). While deriving this formula, we encounter the additive Cauchy functional equation in two variables. In Section 3, using the additive property of definite integral, we show that

$$\int_1^x \frac{1}{t}\, dt = \ln(x)$$

for $x \in (0, \infty)$. While deriving this result, we use the logarithmic Cauchy functional equation. In many calculus books the integral of the left is used for defining the natural logarithm. Section 4 deals with the derivation of formulas for simple and compound interest using functional equations. Since a radioactive substance disintegrates over time, it is useful to have a formula to compute the amount of a radioactive substance present at any time t. Using the exponential Cauchy functional equation, we derive the formula for radioactive disintegration. Sections 6 through 8 present three applications of functional equations in probability theory. In Section 6, the geometric probability distribution is characterized by means of memoryless property. Section 7 deals with the characterization of discrete normal probability distribution. In Section 8, one of the oldest characterization of normal probability distribution is presented. We conclude this chapter with some remarks on further applications of functional equations.

5.2   Area of Rectangles

In 1791, Legendre gave the formula for the area of a rectangle using the additive Cauchy functional equation. To derive the formula for the area of a rectangle, we need the following theorem.

Theorem 5.1. *The function* $f : [0, \infty) \to [0, \infty)$ *satisfies the additive Cauchy functional equation* $f(x + y) = f(x) + f(y)$ *for all* $x, y \in [0, \infty)$ *if and only if* $f(x) = cx$, *where* c *is a nonnegative real constant.*

Proof. Observe that f is bounded from below since $0 \le f(x)$ for $x \ge 0$. Since every bounded below solution of the additive Cauchy functional equation is linear, f is linear and we have $f(x) = cx$ for some arbitrary nonnegative constant $c \in \mathbb{R}$.   □

Consider a rectangle whose length of the base is b and length of the height is a. See figure below.

Figure 5.1. Rectangle with base b and height a.

Clearly the area of this rectangle will depend upon the height a and base b. Thus the area, A, of the rectangle is a function of a and b. That is

$$A = f(a, b).$$

Let us divide this rectangle into two smaller rectangles by drawing a line parallel to its base so that $a = a_1 + a_2$ (see Figure 5.2).

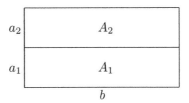

Figure 5.2. Rectangles obtained by dividing along the height.

Then the original area A is the sum of the two areas A_1 and A_2; that is,

$$A = A_1 + A_2$$

which is

$$f(a,b) = f(a_1,b) + f(a_2,b).$$

Since $a = a_1 + a_2$, the equation yields

$$f(a_1 + a_2, b) = f(a_1, b) + f(a_2, b). \tag{5.1}$$

Note that (5.1) holds for all $a_1, a_2, b \in [0, \infty)$. Similarly, dividing the original rectangle into two subrectangles by drawing a line parallel to the height (see Figure 5.3), we obtain

$$f(a, b_1 + b_2) = f(a, b_1) + f(a, b_2) \tag{5.2}$$

for all $a, b_1, b_2 \in [0, \infty)$. Since the area is positive, we have $f(a,b) \geq 0$.

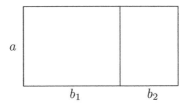

Figure 5.3. Rectangles obtained by dividing along the base.

Using Theorem 5.1, we solve (5.1), that is

$$f(a_1 + a_2, b) = f(a_1, b) + f(a_2, b)$$

for fixed b to get

$$f(a, b) = k\,a, \tag{5.3}$$

where k is a constant that depends on b and $k \geq 0$. Thus

$$f(a, b) = k(b)\,a. \tag{5.4}$$

Using (5.4) in (5.2), we get

$$k(b_1 + b_2)\, a = k(b_1)\, a + k(b_2)\, a;$$

that is,

$$k(b_1 + b_2) = k(b_1) + k(b_2)$$

for all $b_1, b_2 \in \mathbb{R}_+$. Hence by Theorem 5.1, we obtain

$$k(b) = \alpha\, b, \qquad (5.5)$$

where α is an arbitrary real constant. From (5.3) and (5.5) we get

$$f(a, b) = \alpha\, ab. \qquad (5.6)$$

The fact is that $f(a, b) \geq 0$ forces α to be a positive constant and it is associated with the *area unit*.

5.3 Definition of Logarithm

In an elementary calculus course, the logarithm is defined through an integral. Anton (see H. Anton (1992), p. 469) defines natural logarithm as

$$\ln x = \int_1^x \frac{1}{t}\, dt \qquad (5.7)$$

for $x \in (0, \infty)$. We will show that

$$\int_1^x \frac{1}{t}\, dt$$

is really $\ln x$ and we do not have to accept it as a definition. It is rather a consequence of the *properties of the integrals*. We show this by showing that the above integral as a function of x satisfies the logarithmic Cauchy functional equation.

Let us define $\phi : \mathbb{R}_+ \to \mathbb{R}$ by

$$\phi(x) = \int_1^x \frac{1}{t}\, dt, \quad x > 0.$$

Hence, in the case $x, y \in (1, \infty)$, we have

$$
\begin{aligned}
\phi(x) + \phi(y) &= \int_1^x \frac{1}{t} dt + \int_1^y \frac{1}{t} dt \\
&= \int_1^x \frac{1}{t} dt + \int_x^{xy} \frac{1}{z} dz, \quad \text{where } z = tx, \\
&= \int_1^{xy} \frac{1}{w} dw \quad \text{(additive property of the integral)} \\
&= \phi(xy).
\end{aligned}
$$

Other cases can be handled similarly. Hence we obtain

$$
\phi(xy) = \phi(x) + \phi(y) \tag{5.8}
$$

for all $x, y \in \mathbb{R}_+$. By the Fundamental Theorem of Calculus ϕ is differentiable and hence continuous. Therefore the equation (5.8) yields

$$
\phi(x) = c \ln x,
$$

where c is a constant.

Using the Riemann sum one can show that

$$
\phi(e) = \int_1^e \frac{1}{t} dt = 1.
$$

Hence we have $c = 1$ and

$$
\phi(x) = \ln x.
$$

Thus we have shown that

$$
\int_1^x \frac{1}{t} dt = \ln x.
$$

5.4   Simple and Compound Interest

Next we derive the formula for the simple interest using the additive Cauchy functional equation. Let $f(x, t)$ be the future value of capital x having been invested for a period of time of length t. Then, for the simple interest, the function $f(x, t)$ satisfies

$$
f(x + y, t) = f(x, t) + f(y, t)
$$

and

$$f(x, t+s) = f(x, t) + f(x, s)$$

for all $x, y, t, s \in \mathbb{R}_+$. Hence

$$f(x, t) = kxt,$$

where k is an arbitrary positive constant depending on the unit.

Now we derive the formula for the compound interest. Let $f(x, t)$ be the future value of capital x having been invested for a period of time of length t. Then for the compound interest, the function $f(x, t)$ satisfies the equations

$$f(x + y, t) = f(x, t) + f(y, t) \tag{5.9}$$

and

$$f(x, t+s) = f(f(x, t), s) \tag{5.10}$$

for all $x, y, t, s \in \mathbb{R}_+$. The first equation says that the future value of capital $x + y$ after having been invested for a period t is equal to the sum of future values of the capital x after having been invested for a period t and the capital y after having been invested for a period t. The second equation says that the future value of the capital x invested for a period $t + s$ is equal to the future value of the capital $f(x, t)$ invested for a period s. It is natural to assume that $f(x, t)$ is continuous in each variable. Then the solution of equation (5.9) is given by

$$f(x, t) = c(t) x, \tag{5.11}$$

where $c : \mathbb{R}_+ \to \mathbb{R}$. Using this form of f in (5.10), we obtain

$$c(t+s) x = c(t) c(s) x. \tag{5.12}$$

Hence we have

$$c(t+s) = c(t) c(s) \tag{5.13}$$

for all $s, t \in \mathbb{R}_+$. The continous solution of (5.13) is given by $c(t) = e^{\lambda t}$, where λ is an arbitrary constant. Writing $\lambda = \ln(1 + r)$ we obtain

$$f(x, t) = x (1 + r)^t \qquad \text{(where } r \geq 0\text{)}$$

which is the well-known formula for the compound interest.

5.5 Radioactive Disintegration

Let m_0 (in grams) be the initial amount of a radioactive element. Let $m(t)$ be the amount present at time t. Let us assume that the rate of change of $m(t)$ is proportional to $m(t)$. That is,

$$m'(t) \propto m(t).$$

From this assumption we have

$$m'(t) = -\lambda m(t).$$

Therefore

$$m(t) = \alpha e^{-\lambda t}$$

or

$$m(t) = m_0 e^{-\lambda t}. \tag{5.14}$$

Hence (5.14) gives a formula for finding the amount present at time t in terms of the initial amount m_0 and the time period t. Here λ is a decay constant.

Now we derive the formula (5.14) using a functional equation. Let $f(t)$ denote the relationship between the amount present at time t and the initial amount m_0, so that

$$m(t) = m_0 f(t).$$

The amount of radioactive substance at time $t + h$ can be expressed in two different ways, as (see figure below):

$$m(t + h) = m_0 f(t + h)$$

and

$$m(t + h) = m_0 f(t) f(h).$$

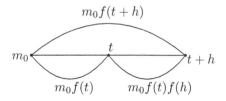

Hence

$$m_0 f(t+h) = m_0 f(t) f(h)$$

for all $t, h \in \mathbb{R}_+$. Thus

$$f(t+h) = f(t)f(h).$$

From an application point of view f can be assumed to be continuous. Then the continuous solution of the above functional equation is given by

$$f(t) = e^{\alpha t},$$

where α is a real constant. Hence we have

$$m(t) = m_0 \, f(t)$$
$$= m_0 \, e^{\alpha t}.$$

Since $m(t)$ decreases over time t, the constant α must be negative and writing

$$\alpha = -\lambda$$

for $\lambda > 0$, we get

$$f(t) = m_0 \, e^{-\lambda t}.$$

The constant λ is called the decay constant. For more on this application see Smital (1988).

5.6 Characterization of Geometric Distribution

In this section using the exponential Cauchy functional equation we give the characterization of geometric distribution in terms of memoryless property.

A random variable X is said to be geometric if its probability density function (pdf) is given by

$$f(x) = (1-p)^{x-1}p, \qquad x = 1, 2, 3, \ldots,$$

where $p \in [0, 1]$ is a parameter. Here p usually denotes the probability of success. If X is a geometric random variable, then it represents the trial number on which the first success occurs.

A random variable X is said to be memoryless if it satisfies

$$P(X > m + n \mid X > n) = P(X > m)$$

for all $m, n \in \mathbb{N}$.

Now we prove that a random variable X is geometric if and only if it satisfies the memoryless property.

The memoryless property yields

$$P(X > m + n \mid X > n) = P(X > m).$$

Since

$$P((X > m + n) \mid (X > n)) = \frac{P((X > m + n) \cap (X > n))}{P(X > n)},$$

we obtain

$$P((X > m + n) \cap (X > n)) = P(X > m)\, P(X > n)$$

which is

$$P(X > m + n) = P(X > m)\, P(X > n)$$

for all $m, n \in \mathbb{N}$.

If X is geometric, that is,

$$X \sim (1 - p)^{x-1} p,$$

then

$$
\begin{aligned}
P(X > m + n) &= \sum_{x=m+n+1}^{\infty} (1-p)^{x-1} p \\
&= (1-p)^{n+m} \\
&= (1-p)^n\, (1-p)^m \\
&= P(X > n)\, P(X > m).
\end{aligned}
$$

Hence the geometric distribution has the lack of memory property.

Next, let X be any random variable that satisfies the lack of memory property,

$$P(X > m + n) = P(X > m)\, P(X > n)$$

for all $m, n \in \mathbb{N}$. We want to show that X is geometric.

Define $g : \mathbb{N} \to \mathbb{R}$ by

$$g(n) = P(X > n).$$

Hence we obtain

$$g(m + n) = g(m)\, g(n)$$

for all $m, n \in \mathbb{N}$. The general solution (even without continuity) is given by

$$g(n) = a^n,$$

where a is a constant. Hence

$$P(X > n) = a^n$$

or

$$1 - F(n) = a^n,$$

where $F(n)$ is the cumulative distribution function (cdf). Hence

$$F(n) = 1 - a^n.$$

Since $F(n)$ is the cdf, we have

$$1 = \lim_{n \to \infty} F(n)$$

or

$$1 = \lim_{n \to \infty} (1 - a^n).$$

From above, we conclude that $0 < a < 1$. Renaming a to be $(1 - p)$, we obtain

$$F(n) = 1 - (1 - p)^n.$$

The pdf of the random variable X is then given by

$$
\begin{aligned}
f(1) &= F(1) = p \\
f(2) &= F(2) - F(1) \\
     &= 1 - (1 - p)^2 - p \\
     &= (1 - p)p \\
f(3) &= F(3) - F(2) \\
     &= 1 - (1 - p)^3 - 1 + (1 - p)^2 \\
     &= (1 - p)^2 p.
\end{aligned}
$$

Thus by induction, we get

$$f(x) = (1 - p)^{x-1} p$$

for $x = 1, 2, 3, \ldots, \infty$. Therefore

$$X \sim \text{Geo}(p)$$

and the proof is complete.

5.7 Characterization of Discrete Normal Distribution

In this section, we treat a functional equation that arises in connection with the characterization of discrete normal distribution or while studying circularly symmetric functions on the integer lattices. The functional equation we are interested in is the following:

$$f(x_1^2 + x_2^2 + \cdots + x_n^2) = f(x_1^2) + f(x_2^2) + \cdots + f(x_n^2)$$

for all $x_1, x_2, \ldots, x_n \in \mathbb{Z}$ (the set of integers).

If $n = 2$, then the above equation reduces to

$$f(x_1^2 + x_2^2) = f(x_1^2) + f(x_2^2) \tag{5.15}$$

for all $x_1, x_2 \in \mathbb{Z}$. One solution of this equation is

$$f(x) = k\,x, \quad x \in \mathbb{Z}. \tag{5.16}$$

However, (5.16) is not the only solution. For example,

$$f(x) = \begin{cases} 0 & \text{if } x = 0 \ (\text{mod } 4) \\ 1 & \text{if } x = 1 \ (\text{mod } 4) \\ 2 & \text{if } x = 2 \ (\text{mod } 4) \end{cases}$$

is also a solution of (5.15). Similarly, the functional equation

$$f(x_1^2 + x_2^2 + x_3^2) = f(x_1^2) + f(x_2^2) + f(x_3^2) \tag{5.17}$$

also has a nonlinear solution (see Dasgupta (1993))

$$f(x) = \begin{cases} 0 & \text{if } x = 0 \ (\text{mod } 4) \\ 1 & \text{if } x = 1 \ (\text{mod } 4) \\ 2 & \text{if } x = 2 \ (\text{mod } 4) \\ 3 & \text{if } x = 3 \ (\text{mod } 4) \end{cases}$$

besides the linear solution $f(x) = kx$.

If $n \geq 4$, then we will show that every solution of the functional equation

$$f(x_1^2 + x_2^2 + \cdots + x_n^2) = f(x_1^2) + f(x_2^2) + \cdots + f(x_n^2) \tag{5.18}$$

for all $x_1, x_2, \ldots, x_n \in \mathbb{Z}$ is linear.

We need the following theorem due to Lagrange in order to find the general solution.

Theorem 5.2. *Every positive integer n is the sum of at most four squares of positive integers, that $n = a^2 + b^2 + c^2 + d^2$, $a, b, c, d \in \mathbb{Z}$*

Example 3.1   The integers $1, 2, 3, 4, 5$ can be represented as

$$\left.\begin{array}{l} 1 = 1^2 + 0^2 \\ 2 = 1^2 + 1^2 \\ 3 = 1^2 + 1^2 + 1^2 \\ 4 = 1^2 + 1^2 + 1^2 + 1^2 \\ = 2^2 \\ 5 = 2^2 + 1^2 \\ \text{etc.} \end{array}\right\} \tag{5.19}$$

Theorem 5.2 first appeared in *Arithmetica* by Diophantus with the Latin translation done by Bachet in 1621. This theorem was unproved until Joseph Louis Lagrange proved it in 1770 (see Milne (1996)). The above Theorem 5.2 is called the Four Squares Theorem or Bachet's conjecture.

Following Dasgupta (1993), we present the solution of the functional equation (5.18) in the next theorem.

Theorem 5.3. *Let $n \geq 4$ be an integer. The function $f : \mathbb{Z} \to \mathbb{R}$ satisfies the equation*

$$f(x_1^2 + x_2^2 + \cdots + x_n^2) = f(x_1^2) + f(x_2^2) + \cdots + f(x_n^2) \tag{5.20}$$

for all $x_1, x_2, \ldots, x_n \in \mathbb{Z}$ if and only if

$$f(x) = k\,x,$$

where k is an arbitrary constant.

Proof. Observe that

$$f(0) = 0.$$

Now letting $x_5 = \cdots = x_n = 0$ in (5.20), we obtain

$$f(x_1^2 + x_2^2 + x_3^2 + x_4^2) = f(x_1^2) + f(x_2^2) + f(x_3^2) + f(x_4^2) \tag{5.21}$$

for all $x_1, x_2, x_3, x_4 \in \mathbb{Z}$. Using (5.19) and (5.21), we see that

$$
\begin{aligned}
f(1) &= 1f(1) \\
f(2) &= 2f(1) \\
f(3) &= 3f(1) \\
f(4) &= 4f(1) \\
f(5) &= f(2^2) + f(1^2) = f(4) + f(1) = 4f(1) + f(1) \\
&= 5f(1).
\end{aligned}
$$

Hence we obtain

$$
f(m) = k\, m \tag{5.22}
$$

for $m \leq 5$. Here $k = f(1)$.

Now we show by induction that (5.22) holds for all positive integers. Assume

$$
f(m) = km
$$

for all integers $m \leq q - 1$, and we shall show

$$
f(q) = kq.
$$

Since q is a positive integer by Theorem 5.2

$$
q = x_1^2 + x_2^2 + x_3^2 + x_4^2, \tag{5.23}
$$

where at least two x_i's are nonzero (since $q > 5$). Then

$$
x_i^2 \leq q - 1 \tag{5.24}
$$

for $i = 1, 2, 3, 4$. Hence

$$
\begin{aligned}
f(q) &= f(x_1^2 + x_2^2 + x_3^2 + x_4^2) \\
&= f(x_1^2) + f(x_2^2) + f(x_3^2) + f(x_4^2) \\
&= k\,(x_1^2 + x_2^2 + x_3^2 + x_4^2) \qquad \text{(by induction hypothesis)} \\
&= kq.
\end{aligned}
$$

Therefore

$$
f(q) = k\, q.
$$

Hence

$$
f(x) = k\, x
$$

for all $x \in \mathbb{Z}$. This completes the proof. $\qquad\square$

5.8 Characterization of Normal Distribution

It is well known that, if $x_1, x_2, ..., x_n$ is a random sample from a normal distribution with mean μ and variance σ^2, then the maximum likelihood estimate (MLE) of the location parameter μ is given by the sample mean $\bar{x} = \sum_{i=1}^{n} x_i/n$. If the maximum likelihood estimate of a location parameter of a population is given by the sample mean, is it true that the distribution of the population is normal? The answer to this question is affirmative and the proof was given by Gauss (1809). In this section, using the additive Cauchy functional equation, we present the oldest characterization of normal distribution. Teicher (1961) characterized normal distribution via MLE by weakening the conditions required by Gauss (1809). Marshall and Olkin (1993) extended Teicher's result to multidimensional normal distribution. Stadje (1993) studied this characterization problem as well, but among others conditions assumed the sample size $n = 2, 3, 4$ simultaneously. We have adapted a recent proof by Azzalini and Genton (2007) that uses only one value of the sample size n, provided $n \geq 3$.

Theorem 5.4. *Consider a parametric location family for a one-dimensional continuous random variable, such that for any choice of $\mu \in \mathbb{R}$ the corresponding probability density function at the point $x \in \mathbb{R}$ is $f(x - \mu)$. Assume that a random sample of size $n \geq 3$ is drawn from a member of this parametric family, and that the following conditions hold:*

1. *$f(x)$ is differentiable function of x and its derivative $f'(x)$ is continuous at least at one point $x \in \mathbb{R}$;*

2. *for each set of sample value, $x_1, x_2, ..., x_n$, the sample mean $\bar{x} = \sum_{i=1}^{n} x_i/n$ is a solution of the likelihood equation for the location parameter μ.*

Then the probability density function $f(x - \mu)$ is the one-dimensional normal density function given by

$$f(x - \mu) = \frac{1}{\sqrt{2\pi\sigma^2}} e^{-\frac{1}{2}\left(\frac{x-\mu}{\sigma}\right)^2}$$

for some positive σ^2.

Proof. Let the likelihood function associated with the sample $x_1, x_2, ..., x_n$ be

$$L(\mu) = f(x_1 - \mu)f(x_2 - \mu) \cdots f(x_n - \mu).$$

Then the log-likelihood function is then given by

$$\ln L(\mu) = \sum_{i=1}^{n} \ln f(x_i - \mu). \tag{5.25}$$

Differentiating $\ln L(\mu)$ in (5.25) with respect to μ, we see that

$$\frac{d}{d\mu} \ln L(\mu) = \sum_{i=1}^{n} \frac{d}{d\mu} \ln f(x_i - \mu). \tag{5.26}$$

Defining $g : \mathbb{R} \to \mathbb{R}$ by

$$g(x) = \frac{d}{dx} \ln f(x) \tag{5.27}$$

from (5.26), we obtain

$$\frac{d}{d\mu} \ln L(\mu) = \sum_{i=1}^{n} g(x_i - \mu).$$

To determine the MLE of μ, we equate $\frac{d}{d\mu} \ln L(\mu)$ to 0 to get

$$\sum_{i=1}^{n} g(x_i - \mu) = 0. \tag{5.28}$$

Note that we allow $f(x) = 0$ and in that case we adopt the convention that $\ln 0 = -\infty$. However, since $f(x) > 0$ must hold true for a range of x values, (5.27) has to be searched for this set.

Using the condition 2 of the theorem, we see that

$$\sum_{i=1}^{n} g(x_i - \bar{x}) = 0 \tag{5.29}$$

for all possible choices of the sample values. Letting $x_1 = x_2 = \cdots = x_n = u$ in (5.28) for some constant u, we get

$$n\, g(u - \mu) = 0 \tag{5.30}$$

and hence $g(0) = 0$ since $\hat{\mu} = u$ by assumption of the theorem.

Next, consider the sample $x_1 = 2u$, $x_2 = 0$, $x_3 = u$, $x_4 = u, ..., x_n = u$. Then

$$g(2u - \mu) + g(-\mu) + g(u - \mu) = 0. \tag{5.31}$$

Since $\hat{\mu} = u$, we get

$$g(u) + g(-u) = 0. \tag{5.32}$$

for all $u \in \mathbb{R}$. Therefore $g(u)$ is an odd function on $\mathbb{R}$.

For any two points u and v in $\mathbb{R}$, consider the sample $x_1 = u$, $x_2 = v$, $x_3 = -u - v$, $x_4 = 0 = x_5 = \cdots = x_n$ such that $\hat{\mu} = 0$. Therefore, from (5.28), we have

$$g(u) + g(v) + g(-u - v) + (n - 3)g(0) = 0,$$

and using the fact that $g(0) = 0$ and g is odd, we have

$$g(u) + g(v) = g(u + v) \tag{5.33}$$

for all $u, v \in \mathbb{R}$. Since $f'(x)$ is continuous at a point x, therefore $g(x)$ is continuous at x and hence by Theorem 1.4, we have

$$g(x) = -c\,x, \tag{5.34}$$

where $c \in \mathbb{R}$ is a constant. From (5.27) and (5.34), we have

$$\frac{d}{dx} \ln f(x) = -cx,$$

and therefore

$$f(x - \mu) = e^{d - \frac{1}{2}c(x - \mu)^2}, \tag{5.35}$$

where d is some real constant. Since $f(x - \mu)$ is a probability density function, $c > 0$; otherwise f would not be integrable over real line $\mathbb{R}$. Letting $c = \frac{1}{\sigma^2}$, we have

$$f(x - \mu) = e^{d - \frac{1}{2}\left(\frac{x - \mu}{\sigma}\right)^2}. \tag{5.36}$$

Since $f(x - \mu)$ is a probability density function, it must integrate over $\mathbb{R}$ to one. Therefore, we have

$$e^d = \frac{1}{\sqrt{2\pi\sigma^2}}$$

and hence

$$f(x - \mu) = \frac{1}{\sqrt{2\pi\sigma^2}}\, e^{-\frac{1}{2}\left(\frac{x - \mu}{\sigma}\right)^2}. \tag{5.37}$$

This completes the proof of the theorem. $\qquad\square$

5.9 Concluding Remarks

We have seen that the additive Cauchy functional equation appeared in the derivation of the formula for area of rectangles. It also appeared in the characterization of discrete and continuous normal probability distributions as well as discrete geometric distribution. Functional equations are used for characterizing other probability distributions. The interested reader should refer to the book by Azlarov and Volodin (1989) for an account on the characterization of exponential distribution by functional equations.

The four Cauchy functional equations have many applications. The functional equation $f(x + y) = f(x)f(y)$ is used by the founders of the non-Euclidean geometry for derivation of the relationship between the arc lengths of two oricycles with same center. Systems of functional equations were used by G. Stokes (1860) to determine the intensities of reflected and absorbed light. Weierstrass (1886) and many others used a system of functional equations to characterize determinant of matrices. For an account of various applications of functional equations the interested reader should referred to Aczél (1966).

Applications of functional equations to characterizing various probability laws and statistics can be found in Ramachandran and Lau (1991), and Rao and Shanbhag (1994). Recently, Castillo, Cobo, Gutiérrez and Pruneda (1999) introduced the functional networks using functional equations. These functional networks have found many applications like the neural networks. For an account on functional networks the interested reader should refer to the book *Functional Networks with Applications: A Neural-Based Paradim* by Castillo, Cobo, Gutiérrez and Pruneda.

Information measures play an important role in information theory and also in coding theory. Various functional equations are used for characterizing information measures. For an account of these applications, the reader is referred to the book *Characterizations of Information Measures* by Ebanks, Sahoo and Sander (1998) (see also Aczél and Daróczy (1975)). For an application of functional equations in the characterization of C. R. Rao's quadratic entropy, we refer the interested reader to Lau (1985) and Chung, Ebanks, Ng and Sahoo (1994).

Functional equations have found applications in program verification and program checking in computer science. Interested reader should refer to Rubinfeld (1999); Ergün, Kumar and Rubinfeld (2001); Kiwi, Magniez and Santha (2003); and Magniez (2005) .

Chapter 6

More Applications of Functional Equations

6.1   Introduction

In this chapter, we give some more applications of Cauchy functional equations. Using the additive Cauchy functional equation, we determine the sum of the kth power of the first n natural numbers for $k = 1, 2, 3$. We show that the number of possible pairs among n things can be determined using the additive Cauchy functional equation. We further illustrate that the additive Cauchy functional equation can be used to find the sum of certain finite series. The materials of this chapter is based on the works of D. R. Snow (1978).

6.2   Sum of Powers of Integers

Let

$$f_k(n) = 1^k + 2^k + \cdots + n^k, \tag{6.1}$$

where n is a positive integer and k is a nonnegative integer. $f_k(n)$ denotes the sum of the kth power of the first n natural numbers. Finding formulas for $f_k(n)$ has interested mathematicians for more than 300 years since the time of James Bernoulli (1655–1705). Several methods were used to find the sum $f_k(n)$ (see, for example, Vakil (1996)). These lead to several recurrence relations. We provide a few methods to determine the sum of powers. Many of them involve functional equations. By using functional equations, we give formulas for $f_k(n)$ for $k = 1, 2, 3$, and for an arbitrary k, we suggest a functional relation for finding a formula. Note that $f_k : \mathbb{N} \to \mathbb{N}$ is a function from $\mathbb{N}$ to $\mathbb{N}$, where $k = 0, 1, 2, \ldots$.

6.2.1   Sum of the first n natural numbers

The function f_1 satisfies

$$
\begin{aligned}
f_1(m+n) &= 1+2+3+\cdots+m+(m+1)+\cdots+(m+n)\\
&= f_1(m)+(m+1)+(m+2)+\cdots+(m+n)\\
&= f_1(m)+f_1(n)+mn
\end{aligned}
\tag{6.2}
$$

for all $m, n \in \mathbb{N}$. Define $g_1 : \mathbb{N} \to \mathbb{R}$ by

$$
g_1(x) = f_1(x) - \frac{1}{2}x^2, \quad \text{for } x \in \mathbb{N}.
\tag{6.3}
$$

Then (6.2) reduces to

$$
g_1(m+n) = g_1(m) + g_1(n), \quad \text{for } m, n \in \mathbb{N}.
\tag{6.4}
$$

The solution of the additive Cauchy functional equation (6.4) on $\mathbb{N}$ is given by

$$
g_1(n) = cn,
\tag{6.5}
$$

where c is a constant. From (6.5) and (6.3), we have

$$
f_1(n) = cn + \frac{1}{2}n^2.
\tag{6.6}
$$

Since $f_1(1) = 1$, we get

$$
1 = c + \frac{1}{2}
$$

which is

$$
c = 1 - \frac{1}{2} = \frac{1}{2}.
$$

Therefore

$$
\begin{aligned}
f_1(n) &= \frac{n}{2} + \frac{n^2}{2}\\
&= \frac{n(n+1)}{2}.
\end{aligned}
$$

Thus

$$
f_1(n) = 1 + 2 + 3 + \cdots + n = \frac{n(n+1)}{2}.
$$

This formula can be established using Gauss's trick.

6.2.2 Sum of square of the first n natural numbers

The function f_2 satisfies

$$
\begin{aligned}
f_2(m+n) &= 1^2 + 2^2 + \cdots + m^2 + (m+1)^2 + \cdots + (m+n)^2 \\
&= f_2(m) + [1^2 + 2^2 + \cdots + n^2] + 2m[1 + 2 + \cdots + n] + m^2 n \\
&= f_2(m) + f_2(n) + 2m f_1(n) + m^2 n \\
&= f_2(m) + f_2(n) + mn^2 + m^2 n + mn
\end{aligned} \tag{6.7}
$$

for all $m, n \in \mathbb{N}$. Defining $g_2 : \mathbb{N} \to \mathbb{R}$ by

$$
g_2(n) = f_2(n) - \frac{n^2}{2} - \frac{n^3}{3}, \quad \text{for } n \in \mathbb{N}
$$

we see that (6.7) reduces to

$$
g_2(m+n) = g_2(m) + g_2(n)
$$

for all $m, n \in \mathbb{N}$. Thus

$$
g_2(n) = cn
$$

and hence

$$
f_2(n) = cn + \frac{n^2}{2} + \frac{n^3}{3}. \tag{6.8}
$$

Using the condition $f_2(1) = 1$, we get

$$
1 = c + \frac{1}{2} - \frac{1}{3}.
$$

Hence

$$
c = \frac{1}{6}.
$$

Therefore

$$
\begin{aligned}
f_2(n) &= \frac{n}{6} + \frac{n^2}{2} + \frac{n^3}{3} \\
&= \frac{n + 3n^2 + 2n^3}{6} \\
&= \frac{n(n+1)(2n+1)}{6}.
\end{aligned}
$$

6.2.3 Sum of kth power of the first n natural numbers

For arbitrary k, we have, using the Binomial Theorem, the following functional equation (recurrence relation):

$$
f_k(n+m) = 1^k + 2^k + \cdots + n^k + (n+1)^k + \cdots + (n+m)^k
$$

$$= f_k(n) + \sum_{i=0}^{k} \binom{k}{i} n^i 1^{k-i} + \cdots + \sum_{i=0}^{k} \binom{k}{i} n^i m^{k-i}$$

$$= f_k(n) + \sum_{i=0}^{k} \binom{k}{i} n^i [1^{k-i} + \cdots + m^{k-i}]$$

$$= f_k(n) + \sum_{i=0}^{k} \binom{k}{i} n^i f_{k-i}(m)$$

$$= f_k(n) + f_k(m) + \sum_{i=1}^{k} \binom{k}{i} n^i f_{k-i}(m) \quad \text{for } m, n, k \in \mathbb{N}.$$

Hence we have

$$f_k(m+n) - f_k(m) - f_k(n) = \sum_{i=1}^{k} \binom{k}{i} n^i f_{k-i}(m) \quad \text{for } m, n \in \mathbb{N}. \quad (6.9)$$

There are several ways of solving (6.9). We will discuss some. Note that $f_k(1) = 1$ for all $k \in \mathbb{N}$ and $f_0(m) = m$.

(A) <u>Evaluation at $n = 1$</u>. This is probably the most direct and simplest method of all. Letting $n = 1$ in (6.9), we have

$$f_k(m+1) - f_k(m) - f_k(1) = \sum_{i=1}^{k} \binom{k}{i} f_{k-i}(m),$$

that is,

$$(m+1)^k - 1 = \sum_{i=1}^{k} \binom{k}{i} f_{k-i}(m) \quad \text{for } m \in \mathbb{N},$$

a simple recurrence relation.

Set $k = 2$ to get

$$m^2 + 2m = 2f_1(m) + f_0(m) = 2f_1(m) + m$$

or

$$f_1(m) = \frac{m(m+1)}{2}.$$

Set $k = 3$ to have

$$m^3 + 3m^2 + 3m = 3f_2(m) + 3f_1(m) + f_0(m)$$

$$= 3f_2(m) + \frac{3m(m+1)}{2} + m$$

or

$$3f_2(m) = \frac{m(m+1)(2m+1)}{2}.$$

(B) <u>General case</u>. The left side of (6.9) is symmetric with respect to m and n; therefore we obtain

$$\sum_{i=1}^{k} \binom{k}{i} n^i f_{k-i}(m) = \sum_{i=1}^{k} \binom{k}{i} m^i f_{k-i}(n) \quad \text{for } m, n \in \mathbb{N}.$$

Substituting $m = 1$ and using the fact that $f_k(1) = 1$, we have

$$\sum_{i=1}^{k} \binom{k}{i} n^i f_{k-i}(1) = \sum_{i=1}^{k} \binom{k}{i} f_{k-i}(n),$$

that is,

$$\sum_{i=1}^{k} \binom{k}{i} f_{k-i}(n) = (1+n)^k - 1.$$

From this we get

$$k f_{k-1}(n) = (1+n)^k - 1 - \sum_{i=2}^{k} \binom{k}{i} f_{k-i}(n) \quad \text{for } n \in \mathbb{N}$$

or

$$f_{k-1}(n) = \frac{(1+n)^k - 1 - \sum\limits_{i=2}^{k} \binom{k}{i} f_{k-i}(n)}{k} \quad \text{for } k, n \in \mathbb{N}, \qquad (6.10)$$

a recurrence relation. Using $f_0(n) = n$, we can determine the rest.

For example, letting $k = 2$ in (6.10), we have

$$f_1(n) = \frac{n^2 + 2n - f_0(n)}{2} = \frac{n(n+1)}{2}.$$

Similarly, $k = 3$ in (6.10) gives

$$
\begin{aligned}
f_2(n) &= \frac{n^3 + 3n^2 + 3n - 3f_1(n) - f_0(n)}{3} \\
&= \frac{1}{3}\left[n^3 + \frac{3}{2}n^2 + \frac{n}{2} \right] \\
&= \frac{1}{6} n(n+1)(2n+1).
\end{aligned}
$$

	n^1	n^2	n^3	n^4	n^5	n^6	n^7
$f_0(n)$	1						
$f_1(n)$	1/2	1/2					
$f_2(n)$	1/6	1/2	1/3				
$f_3(n)$		1/4	1/2	1/4			
$f_4(n)$	−1/30		1/3	1/2	1/5		
$f_5(n)$	−1/12			5/12	1/2	1/6	
$f_6(n)$	1/42		−1/6		1/2	1/2	1/7

Table 6.1. The coefficients of the sums of the powers of the integers

The table above contains the coefficients of the respective closed-form formulas. To generate this table enter a 1 in the upper left-hand corner, since $f_0(n) = n$, and leave the rest of that row blank to represent 0's for the coefficients of the higher order terms. To get the rest of the coefficients in the table, every cell diagonally down (next row and next column) from any $C(i-1, j-1)$ is computed by $C(i,j) = C(i-1, j-1)(i/j)$. The following template illustrates this.

		n^{j-1}	n^j	
$f_{i-1}(n)$		$(i/j)(\quad)$		
$f_i(n)$			$(\quad)$	

Table 6.2. The diagonal process

So you multiply by the new row subscript and divide by the new column power. Everything except the first column is generated by this diagonal process. To get the elements in the first column, add the cells to the right of column 1 and subtract that sum from 1. That's it! This simple procedure generates all the formulas for the sums of the powers of the integers.

6.3 Sum of Powers of Numbers on Arithmetic Progression

For positive integers $n, k \in \mathbb{N}$ and $h \in \mathbb{R}$ define

$$s_k(n; h) = 1^k + (1+h)^k + \cdots + (1 + (n-1)h)^k, \tag{6.11}$$

the sum of kth powers of numbers on arithmetic progression.

As before we derive a recurrence relation:

$$
\begin{aligned}
s_k(m+n;h) &= 1^k + (1+h)^k + \cdots + (1+(n-1)h)^k \\
&\quad + (1+nh)^k + \cdots + (1+(m+n-1)h)^k \\
&= s_k(n;h) + (1+nh)^k + (1+h+nh)^k + \cdots \\
&\quad + (1+(m-1)h+nh)^k \\
&= s_k(n;h) + s_k(m;h) + \sum_{i=1}^{k} \binom{k}{i} s_{k-i}(m;h)(nh)^i;
\end{aligned}
$$

that is, $s_k(n;h)$ satisfy the functional equation

$$
s_k(m+n;h) = s_k(n;h) + s_k(m;h) + \sum_{i=1}^{k} \binom{k}{i} s_{k-i}(m;h)(nh)^i \quad (6.12)
$$

for $k \in \mathbb{N}$, $h \in \mathbb{R}$, $m,n = 1,2,3,\ldots$.

Note that $s_0(n;h) = n$, $s_k(1;h) = 1$. We will determine $s_1(n;h)$ and $s_2(n;h)$.

First let $n = 1$ in (6.12) to obtain

$$
s_k(m+1;h) - s_k(m;h) = s_k(1;h) + \sum_{i=1}^{k} \binom{k}{i} s_{k-i}(m;h)h^i;
$$

that is,

$$
(1+mh)^k = 1 + \sum_{i=1}^{k} \binom{k}{i} s_{k-i}(m;h)h^i \quad (6.13)
$$

for $m = 1,2,\ldots$, $h \in \mathbb{R}$, $k \in \mathbb{N}$.

Similarly, $k = 2$ in (6.13) yields

$$
m^2 h^2 + 2mh = 2 s_1(m;h)h + s_0(m;h)h^2;
$$

that is,

$$
s_1(m;h) = \left(1 - \frac{h}{2}\right)m + \frac{h}{2}m^2.
$$

Let $k = 3$ in (6.13) to have

$$
m^3 h^3 + 3m^2 h^2 + 3mh = 3 s_2(m;h)h + 3 s_1(m;h)h^2 + s_0(m;h)h^3;
$$

that is,

$$
s_2(m;h) = \left(1 - h + \frac{h^2}{6}\right)m + h\left(1 - \frac{h}{2}\right)m^2 + \frac{h^2}{3}m^3.
$$

6.4   Number of Possible Pairs Among n Things

Let $f_2(n)$ denote the number of possible pairs among n things. Consider two sets with n and m things, respectively.

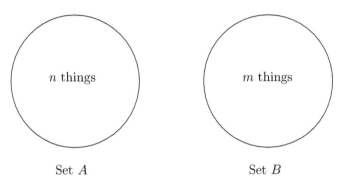

Set A Set B

Then the number of possible pairs among $m+n$ things equals the number of pairs in set A plus the number of pairs in set B plus one item from each set. Hence we have

$$f_2(m + n) = f_2(m) + f_2(n) + mn$$

which reduces to

$$g_2(m + n) = g_2(m) + g_2(n),$$

where

$$g_2(n) = f_2(n) - \frac{n^2}{2}.$$

Hence

$$f_2(n) = cn + \frac{n^2}{2}.$$

Since

$$f_2(2) = 1,$$

we get

$$1 = 2c + 2$$

or

$$c = -\frac{1}{2}.$$

Therefore

$$f_2(n) = \frac{n(n-1)}{2} = \binom{n}{2}.$$

If $f_3(n)$ denotes the number of possible triples among n things, then we shall show that $f_3(n) = \binom{n}{3}$. Now considering two sets with n and m objects, respectively, $f_3(m+n)$ will be the number of triples in set A plus the number of triples in set B plus a combining term of the number of triples with some elements from each set. So

$$f_3(m+n) = f_3(m) + f_3(n) + m f_2(n) + n f_2(m)$$

$$= f_3(m) + f_3(n) + \frac{1}{2}(mn^2 + nm^2) - mn.$$

Defining $g_3 : \mathbb{N} \to \mathbb{R}$ by

$$g_3(n) = f_3(n) - \frac{n^3}{6} + \frac{n}{2} \quad \text{for } n \in \mathbb{N},$$

we have

$$g_3(m+n) = g_3(m) + g_3(n).$$

Therefore

$$f_3(n) = cn - \frac{n^2}{2} + \frac{n^3}{6}.$$

Since

$$f_3(3) = 1,$$

we get

$$c = \frac{1}{3}$$

and

$$f_3(n) = \frac{n(n-1)(n-2)}{6} = \binom{n}{3}.$$

6.5   Cardinality of a Power Set

Let $f(n)$ denote the number of subsets from n elements set (including the empty set). Consider

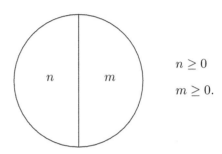

$$n \geq 0$$
$$m \geq 0.$$

Then by the multiplication rule

$$f(m + n) = f(m)f(n).$$

Letting $m = n$, we see that $f(2m) = f(m)^2$. By induction it can be easily shown that $f(mn) = f(m)^n$ for all $n \in \mathbb{N}$. Hence

$$f(n) = f(1)^n.$$

Since $f(1) = 2$, we get
$$f(n) = 2^n.$$

6.6   Sum of Some Finite Series

(i) Let
$$S(n) = 1 \cdot 2 + 2 \cdot 3 + \cdots + n(n+1) \quad \text{for } n \in \mathbb{N}, \qquad (6.14)$$

where $S : \mathbb{N} \to \mathbb{N}$. Hence

$$S(m + n) = S(n) + S(m) + mn^2 + nm^2 + 2mn.$$

Therefore
$$f(m + n) = f(m) + f(n),$$

where

$$f(n) = S(n) - n^2 - \frac{n^3}{3} \quad \text{for } n \in \mathbb{N},$$

and $f : \mathbb{N} \to \mathbb{R}$. Hence f is additive and $f(n) = cn$. Thus

$$S(n) = cn + n^2 + \frac{n^3}{3}.$$

Since $S(1) = 2$, we have

$$S(n) = \frac{n(n+1)(n+2)}{3}. \tag{6.15}$$

(ii) Let

$$t(n) = 1 \cdot 3 + 2 \cdot 5 + \cdots + n(n+2) \quad \text{for } n \in \mathbb{N}, \tag{6.16}$$

where $t : \mathbb{N} \to \mathbb{N}$. Note that $t(1) = 3$. Now

$$t(m+n) = t(n) + t(m) + mn^2 + m^2 n + 3nm \quad \text{for } m, n \in \mathbb{N}.$$

Defining $f : \mathbb{N} \to \mathbb{R}$ by

$$f(n) = t(n) - \frac{1}{3}n^3 - \frac{3}{2}n^2, \quad \text{for } n \in \mathbb{N},$$

the above recurrence relation becomes

$$f(m+n) = f(m) + f(n), \quad \text{for } m, n \in \mathbb{N}.$$

That is, f is additive and $f(n) = cn$ for $n \in \mathbb{N}$. Since $t(1) = 3$, we have

$$t(n) = \frac{n(n+1)(2n+1)}{6} \quad \text{for } n \in \mathbb{N}. \tag{6.17}$$

(iii) Finally we find the sum of a triple product

$$s(n) = 1 \cdot 2 \cdot 3 + 2 \cdot 3 \cdot 4 + \cdots + n(n+1)(n+2), \quad \text{for } n \in \mathbb{N}, \tag{6.18}$$

for $s : \mathbb{N} \to \mathbb{N}$. Note that $s(1) = 6$. Now for $m, n \geq 2$,

$$
\begin{aligned}
s(n + m - 1) \\
= s(n-1) &+ n(n+1)(n+2) + \{(n+1)(n+2)(n+3) \\
&+ \cdots + (n+m-1)(n+m)(n+m+1)\} \\
= s(n-1) &+ (n^3 + 3n^2 + 2n) + (m-1)n^3 + n^2(6 + \cdots + 3m) \\
&+ n\{[1 \cdot 2 + \cdots + (m-1)m] + [1 \cdot 3 + \cdots + (m-1)(m+1)] \\
&+ [2 \cdot 3 + \cdots + m(m+1)]\} + s(m-1)
\end{aligned}
$$

$$= s(n-1) + s(m-1) + mn^3 + 3n^2(1 + 2 + \cdots + m)$$
$$+ n\{[1 \cdot 2 + \cdots + (m-1)m]$$
$$+ [1 \cdot 3 + \cdots + (m-1)(m+1)][1 \cdot 2 + \cdots + m(m+1)]\}$$

$$= s(n-1) + s(m-1) + mn^3 + 3n^2 \frac{m(m+1)}{2}$$
$$+ n\frac{(m-1)m(m+1)}{3} + n\frac{m(m-1)(2m+5)}{6}$$
$$+ n\frac{m(m+1)(m+2)}{3} \qquad \text{(using (6.15), (6.16))}$$

$$= s(n-1) + s(m-1) + mn^3 + \frac{3}{2}m^2n^2 + m^3n$$
$$+ \frac{3}{2}n^2m + \frac{3}{2}nm^2 - \frac{1}{2}mn.$$

As before define $f : \mathbb{N} \to \mathbb{R}$ by

$$f(n) = s(n-1) - \frac{n^4}{4} - \frac{1}{2}n^3 + \frac{1}{4}n^2 \quad \text{for } n \in \mathbb{N}$$

to get $f(m+n) = f(m) + f(n)$ (additivity) and $f(n) = cn$. Using $s(1) = 6$, we have

$$s(n-1) = \frac{1}{4}\left[n^4 + 2n^3 - n^2 - 2n\right];$$

that is,

$$s(n) = \frac{1}{4}n(n+1)(n+2)(n+3) \quad \text{for } n \in \mathbb{N}. \qquad (6.19)$$

6.7   Concluding Remarks

Recently functional equations have found many applications in enumerative combinatorics. For instance, Bousquet-Mélou (2002) using linear functional equations studied planar walks that start from a given point, take their steps in a finite set and are confined in the first quadrant. About the functional equations approaches in combinatorics, Bousquet-Mélou (2001) wrote, "They not only solve (some) problems but they often teach us a lot too. The proofs they provide for a specific problem might be less nice than more combinatorial proofs. But functional equation approaches sometimes give a unified description of apparently distinct problems, and the efforts we make to solve one specific functional equation often teach us what to do in a more generic case."

Functional equations occur in the study of computer graphics. Motivated by the analysis of some geometrical aspects of some functions used in computer graphics for modeling real-world objects, Monreal and Tomás (1998) have studied some functional equations, which under some assumptions characterize these functions.

Functional equations have also found application in the mathematics of signal processing. Sahoo and Székelyhidi (2001, 2002, 2003) studied some functional equations that arise in connection with digital image filtering. The functional equations studied in Sahoo (1990) and Kannappan and Sahoo (1992) also related to digital signal processing.

Functional equations are being used to studied certain models in geometry. For example, using functional equations Powers, Riedel and Sahoo (1993) solved an open problem of Faber, Kuczma and Mycielski (1991). Powers, Riedel and Sahoo (1996) also studied another model in geometry using a functional equation.

Chapter 7

The Jensen Functional Equation

7.1 Introduction

In this chapter, first we present a brief introduction to convex functions. Then we determine the general solution of the Jensen functional equation when it holds for all real numbers in $\mathbb{R}$. We also find the continuous solution of the Jensen functional equation when it holds for all real numbers in a closed and bounded interval $[a, b]$. This chapter concludes with the solution of a Jensen type functional equation that arises from Popoviciu's inequality.

7.2 Convex Function

A function $f : \mathbb{R} \to \mathbb{R}$ is said to be convex if and only if it satisfies the inequality

$$f\left(\frac{x+y}{2}\right) \leq \frac{f(x) + f(y)}{2} \tag{7.1}$$

for all $x, y \in \mathbb{R}$ (see Figure 7.1).

Convex functions were first introduced by J.L.W.V. Jensen in 1905, although functions satisfying the condition (7.1) had been treated by Hadamard (1893) and Hölder (1889). In 1905, Jensen wrote

> It seems to me that the notion of convex functions is just as fundamental as positive or increasing functions. If I am not mistaken in this, the notion ought to find its place in elementary expositions of the theory of real functions.

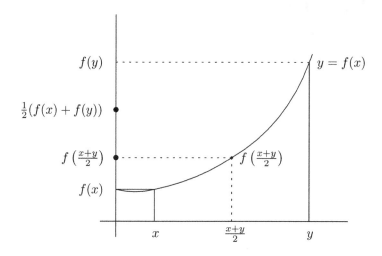

Figure 7.1. Geometrical interpretation on convexity.

Example. The followings are examples of convex functions:

(a) $f(x) = mx + c$ on $\mathbb{R}$ for any $m, c \in \mathbb{R}$

(b) $f(x) = x^2$ on $\mathbb{R}$

(c) $f(x) = e^{\alpha x}$ on $\mathbb{R}$ for any $\alpha \geq 1$ or $\alpha \leq 0$

(d) $f(x) = |x|^\alpha$ on $\mathbb{R}$ for any $\alpha \geq 1$

(e) $f(x) = x \log x$ on $\mathbb{R}_+$

(f) $f(x) = \tan x$ on $\left[0, \frac{\pi}{2}\right]$

A finite sum of convex functions is also a convex function. However, the product of convex functions is not necessarily convex. For example,

$$f(x) = x^2 \quad \text{and} \quad g(x) = e^x$$

are convex functions on $\mathbb{R}$ but their product

$$h(x) = x^2 e^x$$

is not a convex function on $\mathbb{R}$.

If $A : \mathbb{R} \to \mathbb{R}$ is an additive function, then A is also a convex function. Since

$$A\left(\frac{x + y}{2}\right) = \frac{1}{2} A(x + y) = \frac{1}{2}(A(x) + A(y)),$$

A satisfies

$$A\left(\frac{x+y}{2}\right) \leq \frac{A(x) + A(y)}{2}.$$

Therefore A is a convex function.

If $A : \mathbb{R} \to \mathbb{R}$ is an additive function and $f : \mathbb{R} \to \mathbb{R}$ is a convex function, then their composition $f(A(x))$ is a convex function.

7.3 The Jensen Functional Equation

The following functional equation

$$f\left(\frac{x+y}{2}\right) = \frac{1}{2}(f(x) + f(y))$$

for all $x, y \in \mathbb{R}$ is called the *Jensen functional equation.*

Definition 7.1. *A function $f : \mathbb{R} \to \mathbb{R}$ is said to be Jensen if it satisfies*

$$f\left(\frac{x+y}{2}\right) = \frac{f(x) + f(y)}{2} \quad \forall\, x, y \in \mathbb{R}.$$

Definition 7.2. *A function $f : \mathbb{R} \to \mathbb{R}$ is said to be affine if it is of the form*

$$f(x) = cx + a,$$

where c, a are arbitrary constants.

We want to show that every continuous Jensen function on $\mathbb{R}$ is affine.

Theorem 7.1. *The function $f : \mathbb{R} \to \mathbb{R}$ satisfies the Jensen functional equation*

$$f\left(\frac{x+y}{2}\right) = \frac{f(x) + f(y)}{2} \tag{JE}$$

for all $x, y \in \mathbb{R}$ if and only if

$$f(x) = A(x) + a, \tag{7.2}$$

where $A : \mathbb{R} \to \mathbb{R}$ is an additive function and a is an arbitrary constant.

Proof. It is easy to verify that (7.2) satisfies the Jensen equation (JE).

Letting $y = 0$ in (JE), we get

$$f\left(\frac{x}{2}\right) = \frac{f(x)}{2} + \frac{a}{2}, \tag{7.3}$$

where $a = f(0)$. Putting (7.3) in (JE) we see that

$$\frac{f(x+y) + a}{2} = \frac{f(x) + f(y)}{2}$$

which is

$$f(x+y) + a = f(x) + f(y). \tag{7.4}$$

Define $A : \mathbb{R} \to \mathbb{R}$ by

$$A(x) = f(x) - a. \tag{7.5}$$

Then from (7.4), we see that

$$A(x+y) = A(x) + A(y).$$

Hence we have the asserted solution

$$f(x) = A(x) + a,$$

where $A : \mathbb{R} \to \mathbb{R}$ is an additive function. $\qquad \square$

The following theorem is obvious from the last theorem.

Theorem 7.2. *Every continuous Jensen function is affine.*

The proof of this result does not extend to functions defined on a closed and bounded interval. Next we determine the general continuous solution of (JE) on a closed and bounded interval $[a, b]$ for some a, b in $\mathbb{R}$. First we need the following definition.

Definition 7.3. *Let m and n be two positive integers. A rational number of the form*

$$\frac{m}{2^n}$$

is called a dyadic rational number.

Theorem 7.3. *The continuous solution of*

$$f\left(\frac{x+y}{2}\right) = \frac{f(x) + f(y)}{2} \tag{JE}$$

for all $x, y \in [a, b]$ is given by

$$f(x) = \alpha + \beta x, \tag{7.6}$$

where α and β are arbitrary constants.

Proof. Define a new function $F : [0, 1] \to \mathbb{R}$ as

$$F(y) = f((b-a)y + a) \quad \text{for } y \in [0, 1]. \tag{7.7}$$

Note that since $(b-a)y + a \in [a, b]$, therefore $y \in [0, 1]$. Hence the domain of the function F is $[0, 1]$. Next we show that F satisfies (JE). For this we compute $F\left(\frac{x+y}{2}\right)$ as

$$
\begin{aligned}
F\left(\frac{x+y}{2}\right) &= f\left((b-a)\left(\frac{y+x}{2}\right) + a\right) \\
&= f\left(\frac{[(b-a)x + a] + [(b-a)y + a]}{2}\right) \\
&= \frac{f((b-a)x + a) + f((b-a)y + a)}{2} \\
&= \frac{F(x) + F(y)}{2}, \quad \forall\, x, y \in [0, 1].
\end{aligned}
$$

Thus F satisfies the Jensen functional equation on $[0, 1]$. Letting $x = 0$ and $y = 1$ in (JE), we get

$$F\left(\frac{1}{2}\right) = \frac{F(0) + F(1)}{2} = \frac{c+d}{2} = c + \frac{1}{2}(d - c),$$

where $c = F(0)$ and $d = F(1)$. Similarly, letting $x = 0$ and $y = \frac{1}{2}$ in (JE), we have

$$F\left(\frac{1}{4}\right) = \frac{F(0) + F\left(\frac{1}{2}\right)}{2} = \frac{c + c + \frac{1}{2}(d - c)}{2} = c + \frac{1}{4}(d - c).$$

Now letting $x = \frac{1}{2}$ and $y = 1$ in (JE), we obtain

$$F\left(\frac{3}{4}\right) = \frac{F\left(\frac{1}{2}\right) + F(1)}{2} = c + \frac{3}{4}(d - c).$$

Next we will show that if x is any real number of the form $\frac{m}{2^k}$ where m and k positive integers satisfying $0 \le m \le 2^k$, then

$$F(x) = c + x(d - c). \tag{7.8}$$

We proceed by induction on k. We have already shown that the assertion is true for $k = 1, 2$. Assume that (7.8) holds for $k = n$ and consider two cases:

case (a) $x = \frac{2m}{2^{n+1}}$

cases

case (b) $x = \frac{2m+1}{2^{n+1}}$

In case (a) we have

$$F\left(\frac{2m}{2^{n+1}}\right) = F\left(\frac{m}{2^n}\right) = c + \frac{m}{2^n}(d-c) = c + \frac{2m}{2^{n+1}}(d-c),$$

and in the case (b)

$$F\left(\frac{2m+1}{2^{n+1}}\right) = F\left(\frac{1}{2}\left[\frac{m}{2^n} + \frac{m+1}{2^n}\right]\right)$$

$$= \frac{F\left(\frac{m}{2^n}\right) + F\left(\frac{m+1}{2^n}\right)}{2}$$

$$= \frac{1}{2}\left[c + \frac{m}{2^n}(d-c) + c + \frac{m+1}{2^n}(d-c)\right]$$

$$= c + \frac{2m+1}{2^{n+1}}(d-c).$$

Hence (7.8) is satisfied for all dyadic rationals x in $[0,1]$. Since F is continuous and the subset of all dyadic rationals in $[0,1]$ is dense in $[0,1]$, we have

$$F(x) = c + x(d-c)$$

for all $x \in [0,1]$. This yields

$$f(x) = \alpha + \beta x,$$

where α, β are arbitrary constants. The proof of the theorem is now complete.   □

Remark 7.1. *We have seen in the proof of the above theorem that the function F defined by $F(x) = f((b-a)x + a)$ satisfies the Jensen functional equation on the interval $[0,1]$. Following the proof of Theorem 7.1, one can easily show that $F(x) = A(x) + \alpha$, where $A : [0,1] \to \mathbb{R}$ is an additive function and α is an arbitrary constant. By Theorem 4.2, the additive function can be extended from $[0,1]$ to $\mathbb{R}$. Thus the general solution $f : [a,b] \to \mathbb{R}$ of the Jensen equation can be given by*

$$f(x) = A\left(\frac{x-a}{b-a}\right) + \alpha,$$

where $A : \mathbb{R} \to \mathbb{R}$ is an additive function.

Hence we have the following theorem.

Theorem 7.4. *The general solution of*

$$f\left(\frac{x+y}{2}\right) = \frac{f(x) + f(y)}{2} \tag{JE}$$

for all $x, y \in [a, b]$ *is given by*

$$f(x) = A\left(\frac{x-a}{b-a}\right) + \alpha, \tag{7.9}$$

where α *is an arbitrary constant and* $A : \mathbb{R} \to \mathbb{R}$ *is an additive function.*

7.4   A Related Functional Equation

Popoviciu (1965) demonstrated that if I is a nonempty interval and $f : I \to \mathbb{R}$ is a convex function, then f satisfies the inequality

$$3f\left(\frac{x+y+z}{3}\right) + f(x) + f(y) + f(z)$$

$$\geq 2\left[f\left(\frac{x+y}{2}\right) + f\left(\frac{y+z}{2}\right) + f\left(\frac{z+x}{2}\right)\right]$$

for all $x, y, z \in I$. If we change the inequality sign to an equality sign in the above inequality, then we have a functional equation of Jensen type. In this section, our goal is to determine the general solution of this Jensen type functional equation, namely,

$$3f\left(\frac{x+y+z}{3}\right) + f(x) + f(y) + f(z)$$

$$= 2\left[f\left(\frac{x+y}{2}\right) + f\left(\frac{y+z}{2}\right) + f\left(\frac{z+x}{2}\right)\right] \tag{7.10}$$

for all $x, y, z \in \mathbb{R}$.

In Theorem 7.5 the general solution of the functional equation (7.10) is provided following Trif (2000).

Theorem 7.5. *The function* $f : \mathbb{R} \to \mathbb{R}$ *satisfies the functional equation* (7.10) *for all* $x, y, z \in \mathbb{R}$ *if and only if*

$$f(x) = A(x) + b \tag{7.11}$$

for all $x \in \mathbb{R}$, *where* $A : \mathbb{R} \to \mathbb{R}$ *is an additive function and* b *is an arbitrary real constant.*

Proof. It is easy to see that if f is of the form (7.11), then f is a solution of the functional equation (7.10).

Now we prove the converse. That is, every solution of (7.10) is of the form (7.11). First, we define a function $A : \mathbb{R} \to \mathbb{R}$ by

$$A(x) = f(x) - b \tag{7.12}$$

for all $x \in \mathbb{R}$, where $b = f(0)$. Then $A(0) = 0$ and the function A satisfies

$$3A\left(\frac{x+y+z}{3}\right) + A(x) + A(y) + A(z)$$
$$= 2\left[A\left(\frac{x+y}{2}\right) + A\left(\frac{y+z}{2}\right) + A\left(\frac{z+x}{2}\right)\right] \tag{7.13}$$

for all $x, y, z \in \mathbb{R}$. Substitute $y = x$ and $z = -2x$ in (7.10) to obtain

$$A(-2x) = 4A\left(-\frac{x}{2}\right) \tag{7.14}$$

for all $x \in \mathbb{R}$. Replacing x by $-x$ in (7.14), we have

$$A(2x) = 4A\left(\frac{x}{2}\right) \tag{7.15}$$

for all $x \in \mathbb{R}$. Again replacing x by $2x$ in (7.15), we have

$$A(4x) = 4A(x) \tag{7.16}$$

for all $x \in \mathbb{R}$. Putting $y = z = 0$ in (7.13) and taking account of (7.15), we obtain

$$3A\left(\frac{x}{3}\right) = A(2x) - A(x) \tag{7.17}$$

for all $x \in \mathbb{R}$. Substituting $y = x$ and $z = 0$ in (7.13) and taking account of (7.17), we obtain

$$A(4x) = A(2x) - 4A\left(\frac{x}{2}\right) \tag{7.18}$$

for all $x \in \mathbb{R}$. From (7.15), (7.16) and (7.18) it follows that

$$A(2x) = 2A(x) \tag{7.19}$$

for all $x \in \mathbb{R}$. Putting $y = x$ and $z = -x$ in (7.13) and taking account of (7.17) and (7.18), we obtain

$$A(-x) = -A(x) \tag{7.20}$$

for all $x \in \mathbb{R}$. Finally substituting $z = -x - y$ in (7.13) and taking account of (7.18) and (7.19), we obtain

$$A(x + y) = A(x) + A(y)$$

for all $x, y \in \mathbb{R}$. Therefore $A : \mathbb{R} \to \mathbb{R}$ is an additive function and hence from (7.12) we obtain the asserted solution (7.11). This completes the proof of the theorem. $\quad\square$

7.5 Concluding Remarks

If we let $x = u + v$ and $y = u - v$ in (JE), then we have

$$f(u) = \frac{1}{2}[f(u + v) + f(u - v)]$$

for all $u, v \in \mathbb{R}$. Hence the Jensen functional equation can also be written as

$$f(x + y) + f(x - y) = 2f(x).$$

This representation has some advantages over (JE) while studying the Jensen equation on algebraic structures. For an arbitrary group G, we denote $\cdot$ as its group operation and e as the identity element. To simplify our writing, we write xy, instead of $x \cdot y$. If G is abelian, the group operation and the identity element are denoted by $+$ and 0, respectively. In this case we write xy as $x + y$. Similar notations will be adapted for semigroups.

Sinopoulos (2000) proved the following result concerning the Jensen type functional equation on semigroup.

Theorem 7.6. *Let* $(S, +)$ *be a commutative semigroup,* G *a 2-cancellative abelian group, and* σ *an endomorphism of* S *such that* $\sigma(\sigma x) = x$ *for* $x \in S$. *Then the general solution* $f : S \to G$ *of the Jensen functional equation*

$$f(x + y) + f(x + \sigma y) = 2f(x) \qquad \forall\, x, y \in S \qquad (7.21)$$

is given by

$$f(x) = A(x) + a \qquad \forall\, x \in S, \qquad (7.22)$$

where $a \in G$ *is an arbitrary constant and* $A : S \to G$ *is an arbitrary additive function with* $A(\sigma x) = -A(x)$ *for all* $x \in S$.

Proof. In (7.21) we first replace y by $y + \sigma y$ to obtain

$$f(x + y + \sigma y) = f(x). \qquad (7.23)$$

Similarly, replacing x by $x + z$ in (7.21), we have

$$f(x + z + y) + f(x + z + \sigma y) = 2f(x + z). \qquad (7.24)$$

Interchanging y with z in (7.24), we obtain

$$f(x + y + z) + f(x + y + \sigma z) = 2f(x + y). \qquad (7.25)$$

Adding the equations (7.24) and (7.25) and using (7.21), we have

$$f(x + z + y) + f(x + z + \sigma y) + f(x + y + z) + f(x + y + \sigma z)$$
$$= 2f(x + z) + 2f(x + y)$$

which simplifies to

$$2f(x + y + z) + f(x + (z + \sigma y)) + f(x + \sigma(z + \sigma y))$$
$$= 2f(x + z) + 2f(x + y).$$

Using (7.21), we obtain

$$2f(x + y + z) + 2f(x) = 2f(x + z) + 2f(x + y). \qquad (7.26)$$

Setting $z = \sigma x$ in (7.26) and using (7.23), we get

$$f(y) + f(x) = f(x + \sigma x) + f(x + y). \qquad (7.27)$$

Interchanging x with y, we see that $f(x + \sigma x) = f(y + \sigma y)$; that is, $f(x + \sigma x)$ is a constant, say, a. So (7.27) yields

$$[f(x + y) - a] = [f(x) - a] + [f(y) - a] \qquad (7.28)$$

which leads to (7.22) with $A(x) = f(x) - a$. Substituting (7.22) back into (7.21), we see that $A(\sigma x) = -A(x)$ and this completes the proof of the theorem. □

If $f : G \to H$, where G is a group and H is an abelian group, then Jensen equation can be written as

$$f(xy) + f\left(xy^{-1}\right) = 2\,f(x) \qquad (7.29)$$

and

$$f(xy) + f\left(y^{-1}x\right) = 2\,f(x) \qquad (7.30)$$

for all $x, y \in G$. If G is abelian, then both these equations are equivalent to each other. Corovei (1995) has shown that the first equation (7.29) defines a semi-homomorphism. Ng (1990) has solved the functional equation (7.29) along with the normalization condition $f(e) = 0$ when G is a free group with up to two generators and when $G = GL_2(\mathbb{Z})$. The solution of (7.29) was given in Ng (1999) for all free groups and for $GL_n(\mathbb{Z})$, $n \geq 3$. The solution of (7.30) was given in Ng (2001) together with $f(e) = 0$ for all free groups and for more specific groups including $GL_n(\mathbb{Z})$.

A generalization of the functional equation (7.10) is the following:

$$n\binom{n-2}{k-2}f\left(\frac{x_1+x_2+\cdots+x_n}{n}\right)+\binom{n-2}{k-1}\sum_{i=1}^{n}f(x_i)$$

$$=\sum_{1\le i_1<\cdots<i_k\le n}kf\left(\frac{x_{i_1}+\cdots+x_{i_k}}{k}\right),$$

where n and k are positive integers such that $a \le k \le n-1$. This functional equation was studied by Trif (2002).

Replacing x and y by s and it, respectively, in the functional equation (JE), where s and t are real variables, and taking the absolute values of the resulting equation, one obtains the functional equation

$$\left|f\left(\frac{s+it}{2}\right)\right|=\left|\frac{f(s)+f(it)}{2}\right| \tag{7.31}$$

for all $s, t \in \mathbb{R}$. Haruki and Rassias (1995) investigated the solution of this generalized Jensen functional equation. They proved that the only entire solution of the functional equation (7.31) is of the form $f(z) = Az + B$, where A, B are arbitrary complex constants.

7.6 Exercises

1. Find all functions $f : \mathbb{C} \to \mathbb{C}$ that satisfy the functional equation

$$f\left(\frac{x+y}{2}\right)=\frac{f(x)+f(y)}{2}$$

for all $x, y \in \mathbb{C}$.

2. Find all functions $f : \mathbb{R} \to \mathbb{R}$ that satisfy the functional equation

$$f\left(\frac{x+y+z}{3}\right)=\frac{f(x)+f(y)+f(z)}{3}$$

for all $x, y, z \in \mathbb{R}$.

3. Find all functions $f : [0, 1] \to \mathbb{R}$ that satisfy the functional equation

$$f\left(\frac{x+y+z}{3}\right)=\frac{f(x)+f(y)+f(z)}{3}$$

for all $x, y, z \in [0, 1]$.

4. Find all functions $f : \mathbb{C} \to \mathbb{C}$ that satisfy the functional equation

$$\left| f\left(\frac{x+y}{2}\right) \right| = \frac{|f(x)| + |f(y)|}{2}$$

for all $x, y \in \mathbb{C}$.

5. Find all functions $f : \mathbb{C} \to \mathbb{C}$ that satisfy the functional equation

$$f(x+y) + f(x-y) = 2 f(x)$$

for all $x, y \in \mathbb{C}$.

6. Let p, q, r be three a priori chosen positive integers. Find all functions $f : \mathbb{R} \to \mathbb{R}$ that satisfy the functional equation

$$f\left(\frac{px+qy}{r}\right) = \frac{pf(x) + qf(y)}{r}$$

for all $x, y \in \mathbb{C}$.

7. Find all functions $f : \mathbb{R}^2 \to \mathbb{R}$ that satisfy the functional equation

$$3f\left(\frac{x_1+y_1}{2}, \frac{x_2+y_2}{2}\right) = \frac{f(x_1,x_2) + f(y_1,y_2)}{2}$$

for all $x_1, x_2, y_1, y_2 \in \mathbb{R}$.

8. Find all functions $f : \mathbb{R} \to \mathbb{R}$ that satisfy the functional equation

$$3f\left(\frac{x+y+z}{3}\right) + f(x) + f(y) + f(z)$$

$$= 2\left[f\left(\frac{x+y}{2}\right) + f\left(\frac{y+z}{2}\right) + f\left(\frac{z+x}{2}\right) \right]$$

for all $x, y, z \in \mathbb{R}$.

9. Prove that any function $f : \mathbb{R} \to \mathbb{R}$ that satisfies functional equation $f(x+y) + f(x-y) = 2f(x)$ for all $x, y \in \mathbb{R}$ also satisfies the equation

$$f(x+y+z) + f(x) + f(y) + f(z) = f(x+y) + f(y+z) + f(x+z)$$

for all $x, y, z \in \mathbb{R}$.

10. Let $n \geq 3$ be any positive integer. Find all functions $f : \mathbb{R} \to \mathbb{R}$ that satisfy the functional equation

$$f\left(\frac{x_1+x_2+\cdots+x_n}{n}\right) = \frac{f(x_1) + f(x_2) + \cdots + f(x_n)}{n}$$

for all $x_1, x_2, ..., x_n \in \mathbb{R}$.

11. Find all functions $f : \mathbb{R} \to \mathbb{R}$ that satisfy the functional equation

$$f(x + 2y) + f(x - 2y) = 2f(x)$$

for all $x, y \in \mathbb{R}$.

12. If $A : \mathbb{R} \to \mathbb{R}$ is an additive function and $f : \mathbb{R} \to \mathbb{R}$ is a convex function, then show that their composition $f(A(x))$ is a convex function.

Chapter 8

Pexider's Functional Equations

8.1 Introduction

In 1903, J.V. Pexider considered the following functional equations:

$$f(x + y) = g(x) + h(y), \tag{8.1}$$
$$f(x + y) = g(x)h(y), \tag{8.2}$$
$$f(xy) = g(x) + h(y), \tag{8.3}$$
$$f(xy) = g(x)h(y) \tag{8.4}$$

for all $x, y \in \mathbb{R}$ with $f, g, h : \mathbb{R} \to \mathbb{R}$. These functional equations are generalizations of Cauchy functional equations, namely

$$f(x + y) = f(x) + f(y),$$
$$f(x + y) = f(x)f(y),$$
$$f(xy) = f(x) + f(y),$$
$$f(xy) = f(x)f(y),$$

for all $x, y \in \mathbb{R}$. He has solved each of these functional equations (8.1)-(8.4) for continuous real functions.

In this chapter, we determine the general solution of the functional equations (8.1) and (8.2). We also consider the pexiderization of the Jensen functional equation and provide its general solution. Finally, we end this chapter by providing the general solution of a functional equation treated by Hardy, Littlewood and Polya (1934).

8.2 Pexider's Equations

In this section, we provide the solution of only equations (8.1) and (8.2) and leave the equations (8.3) and (8.4) for the reader.

Theorem 8.1. *The general solution $f, g, h : \mathbb{R} \to \mathbb{R}$ of the functional equation*

$$f(x + y) = g(x) + h(y), \quad \text{for } x, y \in \mathbb{R} \tag{8.1}$$

is given by

$$\left. \begin{array}{l} f(x) = A(x) + \alpha + \beta \\ g(x) = A(x) + \beta \\ h(y) = A(y) + \alpha, \end{array} \right\} \tag{8.5}$$

where $A : \mathbb{R} \to \mathbb{R}$ is an additive function and α, β are arbitrary constants.

Proof. Letting $y = 0$ in (8.1) we obtain

$$g(x) = f(x) - \alpha, \tag{8.6}$$

where $\alpha = h(0)$. Similarly letting $x = 0$ in (8.1), we get

$$h(y) = f(y) - \beta, \tag{8.7}$$

where $\beta = g(0)$. Letting (8.6) and (8.7) into (8.1), we have

$$f(x + y) = f(x) + f(y) - \alpha - \beta. \tag{8.8}$$

Defining $A : \mathbb{R} \to \mathbb{R}$ by

$$A(x) = f(x) - \alpha - \beta \quad \text{for } x \in \mathbb{R}, \tag{8.9}$$

we see that

$$A(x + y) = A(x) + A(y).$$

Hence A is an additive function on $\mathbb{R}$ and from (8.9), we have

$$f(x) = A(x) + \alpha + \beta. \tag{8.10}$$

Using (8.10) and (8.6), we get

$$g(x) = A(x) + \beta$$

and from (8.10) and (8.7), we get

$$h(y) = A(y) + \alpha.$$

The proof is now complete. $\qquad\qquad\qquad\qquad\qquad\qquad\qquad\square$

Corollary 8.1. *The general continuous solution of the Pexider's equation*

$$f(x+y) = g(x) + h(y) \quad for \ x, y \in \mathbb{R}$$

is given by

$$\left. \begin{aligned} f(x) &= mx + \alpha + \beta \\ g(x) &= mx + \beta \\ h(y) &= my + \alpha, \end{aligned} \right\}$$

where m, α, β are arbitrary constants.

Theorem 8.2. *The general solution $f, g, h : \mathbb{R} \to \mathbb{R}$ of the functional equation*

$$f(x+y) = g(x)h(y) \quad for \ x, y \in \mathbb{R} \tag{8.2}$$

is given by

$$\left. \begin{aligned} f(x) &= ab \, E(x) \\ g(x) &= a \, E(x) \\ h(y) &= b \, E(y) \end{aligned} \right\} \tag{8.11}$$

together with the trivial solutions

$$\left. \begin{aligned} f(x) &= 0 \\ g(x) &= 0 \\ h(y) &= arbitrary \end{aligned} \right\} \tag{8.12}$$

and

$$\left. \begin{aligned} f(x) &= 0 \\ g(x) &= arbitrary \\ h(y) &= 0, \end{aligned} \right\} \tag{8.13}$$

where E is exponential and a and b are nonzero constants.

Proof. It is easy to check that f, g, h given in (8.11), (8.12) and (8.13) are the solutions of (8.2).

Letting $y = 0$ in (8.2), we get

$$f(x) = g(x)h(0). \tag{8.14}$$

We have two cases to consider. Case (1): $h(0) = 0$ and Case (2): $h(0) \neq 0$.

Case 1. Suppose $h(0) = 0$. Then from (8.14), we get $f(x) = 0$ for all $x \in \mathbb{R}$. Using this in (8.2), we get $g(x) h(y) = 0$, and thus we have the solutions (8.13) and (8.12) with $h(0) = 0$.

Case 2. Suppose $h(0) \neq 0$. Letting $b = h(0)$, we get from (8.14)

$$g(x) = \frac{f(x)}{b}. \tag{8.15}$$

Similarly, letting $x = 0$ in (8.2), we obtain

$$f(y) = g(0) h(y) \tag{8.16}$$

for all $y \in \mathbb{R}$. If $g(0) = 0$, then $f(y) = 0$ for all $y \in \mathbb{R}$ and thus $g(x) h(y) = 0$. Hence we have the solutions (8.12) and (8.13) with $g(0) = 0$.

Suppose $g(0) \neq 0$. Then

$$h(y) = \frac{f(y)}{a}, \tag{8.17}$$

where $a = g(0)$. Letting (8.15) and (8.17) into (8.2), we get

$$f(x + y) = \frac{f(x)f(y)}{ab}. \tag{8.18}$$

Defining $E : \mathbb{R} \to \mathbb{R}$ by

$$E(x) = \frac{f(x)}{ab}, \tag{8.19}$$

we get from (8.18)

$$E(x + y) = E(x) E(y). \tag{8.20}$$

Hence E is an exponential function on $\mathbb{R}$. Therefore

$$f(x) = ab E(x). \tag{8.21}$$

From (8.21) and (8.15), we have

$$g(x) = a E(x).$$

Similarly, from (8.21) and (8.17), we have

$$h(y) = b E(y).$$

Since no more cases are left, the proof of the theorem is now complete.
□

Corollary 8.2. *The nontrivial continuous solution of* (8.2) *is given by*

$$f(x) = ab\, e^{kx}$$
$$g(x) = a\, e^{kx}$$
$$h(y) = b\, e^{ky},$$

where a, b, k *are arbitrary real constants.*

Let $(G, \oplus)$ and $(H, \otimes)$ be two groups and $f, g, h : G \to H$. Then the Pexider's equations can be written as

$$f(x \oplus y) = g(x) \otimes h(y)$$

for all $x, y \in G$. The Pexider's equations have been studied on certain kind of groups.

8.3 Pexiderization of the Jensen Functional Equation

The Jensen functional equation can be generalized to

$$f\left(\frac{x+y}{2}\right) = \frac{g(x) + h(y)}{2} \quad \text{for } x, y \in \mathbb{R}, \tag{8.22}$$

where $f, g, h : \mathbb{R} \to \mathbb{R}$ are unknown functions to be determined.

The following theorem presents the general solution of (8.22).

Theorem 8.3. *The general solution of* (8.22) *is given by*

$$\left.\begin{aligned} f(x) &= 2A(x) + \alpha + \beta \\ g(x) &= 2A(x) + 2\alpha \\ h(y) &= 2A(y) + 2\beta, \end{aligned}\right\} \tag{8.23}$$

where $A : \mathbb{R} \to \mathbb{R}$ *is additive and* α, β *are arbitrary constants.*

Proof. Define $F, G, H : \mathbb{R} \to \mathbb{R}$ by

$$\left.\begin{aligned} F(t) &= f\left(\frac{t}{2}\right) \\ G(t) &= \frac{g(t)}{2} \\ H(t) &= \frac{h(t)}{2}. \end{aligned}\right\} \tag{8.24}$$

Then by (8.24), we get from (8.22)

$$F(x+y) = G(x) + H(y) \tag{8.25}$$

for all $x, y \in \mathbb{R}$. From Theorem 8.1, we obtain

$$F(x) = A(x) + \alpha + \beta$$

$$G(x) = A(x) + \beta$$
$$H(y) = A(y) + \alpha,$$

where α, β are constants and $A : \mathbb{R} \to \mathbb{R}$ additive. From (8.24), we obtain

$$f\left(\frac{x}{2}\right) = A(x) + \alpha + \beta$$
$$\frac{1}{2}g(x) = A(x) + \beta$$
$$\frac{1}{2}h(y) = A(y) + \alpha.$$

Therefore

$$f(x) = 2A(x) + \alpha + \beta$$
$$g(x) = 2A(x) + 2\beta$$
$$h(y) = 2A(y) + 2\alpha$$

and the proof is now complete. $\square$

8.4   A Related Equation

A direct generalization of the Cauchy equations

$$f(x + y) = f(x) + f(y) \qquad \forall\, x, y \in \mathbb{R}$$

and

$$f(x + y) = f(x) f(y) \qquad \forall\, x, y \in \mathbb{R}$$

is the following:

$$f(x + y) = f(x) h(y) + k(y) \tag{8.26}$$

for all $x, y \in \mathbb{R}$.

This functional equation arises in the characterization of homogeneous means and has been studied by Nagumo (1930), de Finetti (1931) and Jessen (1931). The solution of this functional equation can be found in the classical book *Inequalities* by Hardy, Littlewood and Polya (1934). The solution of this equation can also be found in Aczél (1966).

Now we proceed to determine the general solution of the equation (8.26).

To solve (8.26), substitute $x = 0$ to get

$$f(y) = f(0) h(y) + k(y). \tag{8.27}$$

Subtracting (8.27) from (8.26), we obtain

$$f(x + y) - f(y) = [f(x) - f(0)] h(y). \tag{8.28}$$

Define $\phi : \mathbb{R} \to \mathbb{R}$ by

$$\phi(y) = f(y) - f(0), \quad \text{for } y \in \mathbb{R}. \tag{8.29}$$

Hence (8.28) and (8.29) yield

$$\phi(x + y) = \phi(x) h(y) + \phi(y), \tag{8.30}$$

where

$$\phi(0) = 0. \tag{8.31}$$

Interchanging x with y in (8.30), we get

$$\phi(y + x) = \phi(y) h(x) + \phi(x). \tag{8.32}$$

Comparing (8.30) and (8.32), we have

$$\phi(x)h(y) + \phi(y) = \phi(y) h(x) + \phi(x)$$

which is

$$\phi(x) [h(y) - 1] = \phi(y) [h(x) - 1] \tag{8.33}$$

for all $x, y \in \mathbb{R}$.

If $h(y) = 1$ for all $y \in \mathbb{R}$, then (8.30) reduces to

$$\phi(x + y) = \phi(x) + \phi(y).$$

Hence

$$\phi(x) = A(x),$$

where $A : \mathbb{R} \to \mathbb{R}$ is an additive function. From (8.29) and (8.27), we obtain

$$f(x) = A(x) + f(0)$$

and

$$k(y) = A(y).$$

Hence

$$\left. \begin{array}{l} f(x) = A(x) + a \\ h(x) = 1 \\ k(y) = A(y), \end{array} \right\} \tag{8.34}$$

where a is an arbitrary constant, is a solution of (8.26).

Next we assume $h(y) \not\equiv 1$. Then there exists a y_0 such that $h(y_0) \neq 1$. From (8.33) we obtain (with $y = y_0$)

$$\phi(x)\,[h(y_0) - 1] = \phi(y_0)\,[h(x) - 1]$$

or

$$\phi(x) = \frac{\phi(y_0)}{h(y_0) - 1}\,[h(x) - 1]$$

which is

$$\phi(x) = \alpha[h(x) - 1], \tag{8.35}$$

where α is a constant. Now we consider two cases based on whether $\alpha = 0$ or $\alpha \neq 0$. If $\alpha = 0$, then $\phi(x) = 0$ and by (8.29), we have

$$f(y) = f(0)$$

for all $y \in \mathbb{R}$. That is,

$$f(y) = a,$$

where a is a constant. Hence by (8.27)

$$k(y) = a[1 - h(y)].$$

Thus for this case the solution of (8.26) is

$$\left. \begin{array}{l} f(x) = a \\ h(x) \text{ arbitrary but not identically one} \\ k(y) = a[1 - h(y)]. \end{array} \right\} \tag{8.36}$$

If $\alpha \neq 0$, then (8.35) in (8.30) gives

$$\alpha\,[h(x + y) - 1] = \alpha\,[h(x) - 1]h(y) + \alpha\,[h(y) - 1]$$

which is

$$h(x + y) = h(x)h(y).$$

Hence

$$h(x) = E(x), \tag{8.37}$$

where $E : \mathbb{R} \to \mathbb{R}$ is an exponential function. Hence using (8.37) in (8.35), we have

$$\phi(x) = \alpha[E(x) - 1]. \tag{8.38}$$

From (8.38) and (8.29) we get

$$f(x) = \alpha[E(x) - 1] + f(0)$$

which is
$$f(x) = \alpha[E(x) - 1] + a. \tag{8.39}$$

Using (8.37) and (8.39) in (8.27), we obtain
$$k(y) = (\alpha - a)[E(y) - 1].$$

Therefore
$$\left. \begin{array}{l} f(x) = \alpha[E(x) - 1] + a \\ h(x) = E(x) \\ k(y) = (\alpha - a)[E(x) - 1] \end{array} \right\} \tag{8.40}$$

is a solution of (8.26).

Thus we have proved the following theorem.

Theorem 8.4. *The functions* $f, h, k : \mathbb{R} \to \mathbb{R}$ *satisfy the functional equation* (8.26)
$$f(x + y) = f(x)\,h(y) + k(y)$$
for all $x, y \in \mathbb{R}$ *if and only if* (8.34) *or* (8.36) *or* (8.40) *holds.*

8.5   Concluding Remarks

One of the striking features of functional equations is the fact that, contrary to differential equations, a single equation can determine more than one functions. Pexider functional equations are generalization of Cauchy functional equations having three unknown functions. In this chapter we saw how to determine these unknown functions from one single equation.

Hosszú (1962) has treated the Pexider's equation $f(xy) = g(x) + h(y)$, where x, y are elements of a groupoid Q with binary operation xy and f, g, h are mappings of Q onto another groupoid G whose operation is written additively. Vincze (1962b) has also treated the more general equation $f(x \cdot y) = g(x)h(y)$ where the functions f, g and h are defined in a semigroup S and the values of these functions are elements of a group G possibly enlarged by adding a zero-element. He found the general solution of this equation under the supposition that there exist $a, b \in S$ such that $a \cdot S = S \cdot b = S$. He asked whether in case of abelian semigroups S, the conditions of solvability of the above equation can be replaced by the weaker condition $S \cdot S = S$; that is, the product $x \cdot y$ assumes every value in S (at least once). Aczél (1964) showed that the answer is positive; moreover if the values of the functions f, g, h lie in a group G (that is, no zero-element is allowed), then the condition $S \cdot S = S$ can be left aside. We present Aczél's result below.

Theorem 8.5. *The most general solutions of equation* $f(x \cdot y) = g(x)h(y)$ *among the functions* f, g, h *mapping an abelian semigroup* S *into a group* G *are given by*

$$g(x) = a\, k(x), \qquad h(y) = k(y)\, c, \qquad f(x \cdot y) = a\, k(x \cdot y)\, c,$$

where a *and* c *are constants in* G, *and* k *is a homomorphism; that is,* $k(x \cdot y) = k(x)\, k(y)$.

Deeba and Koh (1990) give the distributional solutions of Pexider's functional equation $f(x + y) = g(x) + h(y)$.

The Pexider functional equations can be generalized to the following equatons:

$$f(x_1 + x_2 + \cdots + x_n) = f_1(x_1) + f_2(x_2) + \cdots + f_n(x_n)$$
$$f(x_1 + x_2 + \cdots + x_n) = f_1(x_1)\, f_2(x_2) \cdots f_n(x_n)$$
$$f(x_1 x_2 \cdots x_n) = f_1(x_1)\, f_2(x_2) \cdots f_n(x_n)$$
$$f(x_1 x_2 \cdots x_n) = f_1(x_1) + f_2(x_2) + \cdots + f_n(x_n)$$

for all $x_1, x_2, ..., x_n \in \mathbb{R}$. These were treated by Stamate (1971).

The functional equation $f(x + y) = f(x)\, h(y) + k(y)$ can be further generalized to

$$f(x + y) = g(x)\, h(y) + k(x)\, \ell(y). \tag{8.41}$$

This functional equation contains the Pexider functional equations as particular cases and was treated by Vincze (1960) when the unknown functions are complex-valued and the variables range over an additive group of complex numbers.

Another generalization of $f(x + y) = f(x)\, h(y) + k(y)$ and two of the Pexider equations is the functional equation

$$f(x + y) = g(x) + h(y) + k(x)\, \ell(y). \tag{8.42}$$

The general solution of this functional equation was found by Daróczy (1961).

The still more general equation

$$f(x + y) = \sum_{i=1}^{n} g_i(x)\, h_i(y), \tag{8.43}$$

containing both (8.41) and (8.42), was solved under differentiability conditions by Levi-Civita (1913) and via distribution theory by Fenyö (1956).

8.6 Exercises

1. Find all functions $f, g, h : \mathbb{R} \to \mathbb{R}$ that satisfy the functional equation

$$f(xy) = g(x) + h(y)$$

for all $x, y \in \mathbb{R}$.

2. Find all functions $f, g, h : \mathbb{R} \to \mathbb{R}$ that satisfy the functional equation

$$f(xy) = g(x) h(y)$$

for all $x, y \in \mathbb{R}$.

3. Find all functions $f, g, h, k : \mathbb{R} \to \mathbb{R}$ that satisfy the functional equation

$$f(xy) = g(x) h(y) + k(y)$$

for all $x, y \in \mathbb{R}$.

4. Find all functions $f, g, h : \mathbb{R} \to \mathbb{R}$ that satisfy the functional equation

$$f(x + y) + g(x - y) = h(y)$$

for all $x, y \in \mathbb{R}$.

5. Find all functions $f, g, h : \mathbb{R} \to \mathbb{R}$ that satisfy the functional equation

$$f(x + y) = e^y g(x) + e^x h(y)$$

for all $x, y \in \mathbb{R}$.

6. Find all functions $f, g, h : [0, \infty) \to \mathbb{R}$ that satisfy the functional equation

$$f(\sqrt{x^2 + y^2}) = g(x) + h(y)$$

for all $x, y \in [0, \infty)$.

7. Let α be a positive real number and $a, b, c \in \mathbb{R}$. Find all functions $f, g, h : \mathbb{R} \to \mathbb{R}$ that satisfy the functional equation

$$f(ax + by + c) = \alpha^{xy} g(x) g(y)$$

for all $x, y \in \mathbb{R}$.

8. Find all functions $f, g : \mathbb{R} \to \mathbb{R}$ that satisfy the functional equation

$$f(x + y) g(x - y) = f(x) g(x)$$

for all $x, y \in \mathbb{R}$.

9. Find all functions $f, g : \mathbb{R} \to \mathbb{R}$ and $h : \mathbb{R}^2 \to \mathbb{R}$ that satisfy the functional equation

$$f(x + y + z) = g(x) h(y, z)$$

for all $x, y, z \in \mathbb{R}$.

10. Find all functions $f, g : \mathbb{R} \to \mathbb{R}$ that satisfy the functional equation

$$f(x + y) = f(x) f(y) g(xy)$$

for all $x, y \in \mathbb{R}$.

11. Find all functions $f, g, h, k : \mathbb{R} \to \mathbb{R}$ that satisfy the functional equation

$$f(x + y) = g(x) h(y) k(xy)$$

for all $x, y \in \mathbb{R}$.

12. Find all functions $f, f_i : \mathbb{R} \to \mathbb{R}$ $(i = 1, 2, ..., n)$ that satisfy the functional equation

$$f(x_1 + x_2 + \cdots + x_n) = f_1(x_1) + f_2(x_2) + \cdots + f_n(x_n)$$

for all $x_1, x_2, ..., x_n \in \mathbb{R}$.

13. Find all functions $f, f_i : \mathbb{R} \to \mathbb{R}$ $(i = 1, 2, ..., n)$ that satisfy the functional equation

$$f(x_1 + x_2 + \cdots + x_n) = f_1(x_1) f_2(x_2) \cdots f_n(x_n)$$

for all $x_1, x_2, ..., x_n \in \mathbb{R}$.

14. Find all functions $f, g, h : \mathbb{R}^2 \to \mathbb{R}$ that satisfy the functional equation

$$f(x + u, y + v) = g(x, y) h(u, v)$$

for all $x, y, u, v \in \mathbb{R}$.

Chapter 9

Quadratic Functional Equation

9.1   Introduction

In this chapter we study biadditive functions, quadratic functions, and the quadratic functional equation. First we show that every continuous biadditive function $f : \mathbb{R}^2 \to \mathbb{R}$ is of the form $f(x, y) = c\,xy$, where c is an arbitrary real constant. Then we give a general representation for the biadditive function in terms of a Hamel basis. The continuous solutions of the quadratic functional equation is determined in section three of this chapter. Section 4 deals with the representation of quadratic functions in terms of the diagonal of symmetric biadditive functions. Finally, the pexiderized version of the quadratic functional equation is treated in Section 5 of this chapter.

9.2   Biadditive Functions

In Chapter 3, we studied some interesting properties of additive functions $f : \mathbb{R}^2 \to \mathbb{R}$ in two variables. An additive function $f : \mathbb{R}^2 \to \mathbb{R}$ in two variables is defined as

$$f(x + y, u + v) = f(x, u) + f(y, v) \tag{9.1}$$

for all $x, y, u, v \in \mathbb{R}$. The continuous general solution of (9.1) is given by

$$f(x, u) = k_1 x + k_2 u, \tag{9.2}$$

where k_1, k_2 are arbitrary constants. The general representation of the additive functions $f : \mathbb{R}^2 \to \mathbb{R}$ is given by

$$f(x, u) = A_1(x) + A_2(u), \tag{9.3}$$

where $A_1, A_2 : \mathbb{R} \to \mathbb{R}$ are additive functions on $\mathbb{R}$.

Definition 9.1. *A function $f : \mathbb{R}^2 \to \mathbb{R}$ is said to be biadditive if and only if f is additive in each variable, that is,*

$$f(x + y, z) = f(x, z) + f(y, z),$$
$$f(x, y + z) = f(x, y) + f(x, z)$$

for all $x, y, z \in \mathbb{R}$.

For example,

$$f(x, y) = cxy \quad \text{for } x, y \in \mathbb{R} \tag{9.4}$$

is biadditive, where c is a constant. To see this consider

$$\begin{aligned}
f(x + y, z) &= c(x + y)z \\
&= cxz + cyz \\
&= f(x, z) + f(y, z).
\end{aligned}$$

Similarly

$$\begin{aligned}
f(x, y + z) &= cx(y + z) \\
&= cxy + cxz \\
&= f(x, y) + f(x, z).
\end{aligned}$$

Hence f given by (9.4) is biadditive.

The following theorem says that there are no other continuous biadditive functions besides $f(x, y) = cxy$.

Theorem 9.1. *Every continuous biadditive map $f : \mathbb{R}^2 \to \mathbb{R}$ is of the form*

$$f(x, y) = cxy$$

for all $x, y \in \mathbb{R}$ for some constant c in $\mathbb{R}$.

Proof. Let $f : \mathbb{R}^2 \to \mathbb{R}$ be a continuous biadditive map. Hence f satisfies

$$f(x + y, z) = f(x, z) + f(y, z) \tag{9.5}$$

for all $x, y, z \in \mathbb{R}$. Letting $x = 0 = y$ in (9.5), we obtain

$$f(0, z) = 0$$

for all $z \in \mathbb{R}$. For fixed z, define

$$\phi(x) = f(x, z). \tag{9.6}$$

Then (9.5) reduces to

$$\phi(x + y) = \phi(x) + \phi(y) \tag{9.7}$$

for all $x, y \in \mathbb{R}$. Since f is continuous in each variable, ϕ is also continuous on $\mathbb{R}$. Hence

$$\phi(x) = k\,x. \tag{9.8}$$

Since z is fixed, k depends on z. Hence we have

$$\phi(x) = k(z)\,x.$$

That is,

$$f(x, z) = x\,k(z). \tag{9.9}$$

Letting (9.9) into

$$f(x, y + z) = f(x, y) + f(x, z),$$

we get

$$x\,k(y + z) = x\,k(y) + x\,k(z)$$

for all $x, y, z \in \mathbb{R}$. Then we have

$$k(y + z) = k(y) + k(z).$$

Since k is continuous, we obtain

$$k(y) = c\,y, \tag{9.10}$$

where c is a constant. Thus

$$f(x, y) = c\,xy \tag{9.11}$$

for all $x, y \in \mathbb{R}$. $\qquad\square$

In the next theorem, we present a general representation for the biadditive function in terms of a Hamel basis.

Theorem 9.2. *Every biadditive map $f : \mathbb{R}^2 \to \mathbb{R}$ can be represented as*

$$f(x, y) = \sum_{k=1}^{n} \sum_{j=1}^{m} \alpha_{kj}\, r_k\, s_j, \tag{9.12}$$

where

$$x = \sum_{k=1}^{n} r_k\, b_k, \qquad y = \sum_{j=1}^{m} s_j\, b_j,$$

the r_k, s_j being rational, while the b_j are elements of a Hamel basis B and the α_{kj} arbitrary depending upon b_k and b_j.

Proof. Let B be a Hamel basis for the set of reals $\mathbb{R}$. Then every real number x can be represented as

$$x = \sum_{k=1}^{n} r_k \, b_k \tag{9.13}$$

with $b_k \in B$ and with rational coefficient r_k. Similarly, any real number y can also be represented as

$$y = \sum_{j=1}^{m} s_j \, b_j \tag{9.14}$$

with $b_j \in B$ and with rational coefficient s_j. Since f is biadditive,

$$f(x_1 + x_2, \, y) = f(x_1, \, y) + f(x_2, \, y), \tag{9.15}$$
$$f(x, \, y_1 + y_2) = f(x, \, y_1) + f(x, \, y_2) \tag{9.16}$$

for all $x_1, x_2, y_1, y_2 \in \mathbb{R}$. From (9.15) and (9.16), using induction, we have

$$f\left(\sum_{k=1}^{n} x_k, \, y\right) = \sum_{k=1}^{n} f(x_k, \, y), \tag{9.17}$$

$$f\left(x, \, \sum_{k=1}^{n} y_k\right) = \sum_{k=1}^{n} f(x, \, y_k). \tag{9.18}$$

Letting $x_1 = x_2 = \cdots = x_n = x$ and $y_1 = y_2 = \cdots = y_n = y$ in (9.17) and (9.18) respectively, we get

$$f(nx, \, y) = n \, f(x, \, y) = f(x, \, ny). \tag{9.19}$$

From (9.19) with $t = \frac{m}{n} x$ (that is, $nt = mx$), we get

$$n \, f(t, \, y) = f(nt, \, y) = f(mx, \, y) = m \, f(x, \, y)$$

or

$$f(t, \, y) = \frac{m}{n} \, f(x, \, y).$$

That is,

$$f\left(\frac{m}{n} x, \, y\right) = \frac{m}{n} \, f(x, \, y). \tag{9.20}$$

Since f is biadditive, we see that

$$f(x, \, 0) = 0 = f(0, \, y) \tag{9.21}$$

for all $x, y \in \mathbb{R}$. Next, substituting $x_2 = -x_1 = x$ in (9.15) and using (9.21), we obtain

$$f(-x, y) = -f(x, y). \tag{9.22}$$

From (9.22) and (9.20) we conclude that (9.19) is valid for all rational numbers. The same argument applies to the second variable, and so we have for all rational numbers r and all real x and y:

$$f(rx, y) = r f(x, y) = f(x, ry). \tag{9.23}$$

Hence by (9.13), (9.14), (9.17), (9.18) and (9.23), we obtain

$$f(x, y) = f\left(\sum_{k=1}^{n} r_k b_k, \sum_{j=1}^{m} s_j b_j\right)$$

$$= \sum_{k=1}^{n} r_k f\left(b_k, \sum_{j=1}^{m} s_j b_j\right)$$

$$= \sum_{k=1}^{n}\sum_{j=1}^{m} r_k s_j f(b_k, b_j)$$

$$= \sum_{k=1}^{n}\sum_{j=1}^{m} r_k s_j \alpha_{kj},$$

where $\alpha_{kj} = f(b_k, b_j)$. This completes the proof of the theorem. $\square$

9.3 Continuous Solution of Quadratic Functional Equation

The following functional equation

$$f(x + y) + f(x - y) = 2f(x) + 2f(y) \quad \text{for all } x, y \in \mathbb{R} \tag{9.24}$$

is known as the *quadratic functional equation*. In this section, we determine its continuous solution.

Theorem 9.3. *Let $f : \mathbb{R} \to \mathbb{R}$ be a function that satisfies*

$$f(x + y) + f(x - y) = 2f(x) + 2f(y)$$

for all $x, y \in \mathbb{R}$. Then f is rationally homogeneous of degree 2. Moreover on the set of rational numbers $\mathbb{Q}$, f has the form

$$f(r) = c r^2$$

for $r \in \mathbb{Q}$, where c is an arbitrary constant.

Proof. Letting $x = 0 = y$ in (9.24), we obtain

$$f(0) = 0. \tag{9.25}$$

Next, replacing y by $-y$ in (9.24), we see that

$$f(x - y) + f(x + y) = 2f(x) + 2f(-y). \tag{9.26}$$

Comparing (9.24) and (9.26), we have

$$f(y) = f(-y)$$

for all $y \in \mathbb{R}$. That is, f is an even function.

Next we show that f is a rationally homogeneous function of degree 2. We put $y = x$ in (9.24) to get

$$f(2x) = 4f(x)$$

or

$$f(2x) = 2^2 f(x) \quad \text{for } x \in \mathbb{R}.$$

Similarly

$$f(2x + x) + f(2x - x) = 2f(2x) + 2f(x)$$

or

$$f(3x) = 2f(2x) + f(x) = 8f(x) + f(x)$$

which is

$$f(3x) = 3^2 f(x) \quad \text{for } x \in \mathbb{R}.$$

Hence by induction, we get

$$f(nx) = n^2 f(x) \tag{9.27}$$

for all positive integers n. Next we show that (9.27) holds for all integers $n \in \mathbb{Z}$.

Suppose n is a negative integer. Then $-n$ is a positive integer. Hence

$$\begin{aligned} f(nx) &= f(-(-n)x) \\ &= f(-nx) \quad \text{since } f \text{ is even} \\ &= (-n)^2 f(x) \\ &= n^2 f(x). \end{aligned}$$

Hence

$$f(nx) = n^2 f(x)$$

holds for all $x \in \mathbb{R}$ and all $n \in \mathbb{Z}$.

Let r be an arbitrary rational number. Hence

$$r = \frac{k}{n}$$

for some integer $k \in \mathbb{Z}$ and some natural number $n \in \mathbb{N}$. Therefore

$$k = r\,n.$$

We consider

$$\begin{aligned}
k^2 f(x) &= f(kx) \\
&= f(rnx) \\
&= n^2 f(rx).
\end{aligned}$$

Therefore

$$f(rx) = \frac{k^2}{n^2} f(x)$$

or

$$f(rx) = r^2 f(x). \tag{9.28}$$

That is, f is rationally homogeneous of degree 2.

Letting $x = 1$ in (9.28), we obtain

$$f(r) = c\,r^2 \quad \text{for } r \in \mathbb{Q}, \tag{9.29}$$

where $c := f(1)$. $\qquad\qquad\square$

Theorem 9.4. *The general continuous solution of*

$$f(x + y) + f(x - y) = 2f(x) + 2f(y) \tag{QE}$$

for all $x, y \in \mathbb{R}$ is given by

$$f(x) = c\,x^2,$$

where c is an arbitrary constant.

Proof. Let f be the solution of (9.24) and suppose f to be continuous. For any real number $x \in \mathbb{R}$ there exists a sequence $\{r_n\}$ of rational numbers such that

$$\lim_{n \to \infty} r_n = x.$$

Since f satisfies (9.24), by previous theorem

$$f(r_n) = c\,r_n^2 \tag{9.30}$$

for all $n \in \mathbb{Z}$. Using the continuity of f, we have

$$
\begin{aligned}
f(x) &= f\left(\lim_{n \to \infty} r_n\right) \\
&= \lim_{n \to \infty} f(r_n) \\
&= \lim_{n \to \infty} (c\, r_n^2) \\
&= c \lim_{n \to \infty} r_n^2 \\
&= c \left(\lim_{n \to \infty} r_n\right)^2 \\
&= c\, x^2.
\end{aligned}
$$

Hence

$$
f(x) = cx^2 \quad \text{for } x \in \mathbb{R}.
$$

$\square$

Definition 9.2. *A mapping $f : \mathbb{R} \to \mathbb{R}$ is called a* quadratic function *if*

$$
f(x + y) + f(x - y) = 2f(x) + 2f(y)
$$

holds for all $x, y \in \mathbb{R}$.

In view of the previous theorem we have proven the following fact: Every continuous quadratic function f is of the form

$$
f(x) = c\, x^2,
$$

where c is an arbitrary constant.

9.4 A Representation of Quadratic Functions

In the following theorem we show that every real-valued quadratic function can be represented as the diagonal of a symmetric biadditive map.

Theorem 9.5. *The function $f : \mathbb{R} \to \mathbb{R}$ is quadratic if and only if there exists a symmetric biadditive map $B : \mathbb{R}^2 \to \mathbb{R}$ such that*

$$
f(x) = B(x, x).
$$

Proof. Suppose $f(x) = B(x, x)$. Then

$$
\begin{aligned}
f(x+y) &+ f(x-y) \\
&= B(x+y, x+y) + B(x-y, x-y) \\
&= B(x, x+y) + B(y, x+y) + B(x, x-y) - B(y, x-y) \\
&= B(x, x) + B(x, y) + B(y, x) + B(y, y) \\
&\quad + B(x, x) - B(x, y) - B(y, x) + B(y, y) \\
&= 2B(x, x) + 2B(y, y) \\
&= 2f(x) + 2f(y).
\end{aligned}
$$

Thus f is a quadratic function.

Now we prove the converse. We suppose $f : \mathbb{R} \to \mathbb{R}$ is a quadratic function, and we define $B : \mathbb{R}^2 \to \mathbb{R}$ as

$$
B(x, y) = \frac{1}{4}[f(x+y) - f(x-y)] \quad \text{for } x, y \in \mathbb{R}. \tag{9.31}
$$

Letting $y = 0$ in

$$
f(x+y) + f(x-y) = 2f(x) + 2f(y), \tag{9.32}
$$

we have $f(0) = 0$ and $x = y$ gives

$$
f(2x) = 4f(x). \tag{9.33}
$$

Therefore

$$
\begin{aligned}
B(x, x) &= \frac{1}{4}[f(2x) - f(0)] \\
&= \frac{1}{4}[4f(x)] \\
&= f(x).
\end{aligned}
$$

Interchanging x with y in (9.32), we get

$$
f(x+y) + f(y-x) = 2f(y) + 2f(x).
$$

Comparing this equation with (9.32), we get

$$
f(x-y) = f(y-x).
$$

Hence f is an even function. Next we obtain

$$
B(x, y) = \frac{1}{4}[f(x+y) - f(x-y)]
$$

$$= \frac{1}{4}[f(y+x) - f(y-x)]$$
$$= B(y,x).$$

Further,

$$B(-x,y) = \frac{1}{4}[f(-x+y) - f(-x-y)]$$
$$= \frac{1}{4}[f(x-y) - f(x+y)]$$
$$= -\frac{1}{4}[f(x+y) - f(x-y)]$$
$$= -B(x,y).$$

Thus B is odd in the first variable. Similarly, one can show that B is odd in the second variable.

Next we show that B is additive in the first variable.

$$4[B(x+y,z) + B(x-y,z)] = f(x+y+z) + f(x-y+z)$$
$$- f(x+y-z) - f(x-y-z)$$
$$= 2f(x+z) + 2f(y) - 2f(x-z) - 2f(y)$$
$$= 2f(x+z) - 2f(x-z)$$
$$= 8B(x,z).$$

Therefore we have shown

$$B(x+y,z) + B(x-y,z) = 2B(x,z) \quad \text{for } x, z \in \mathbb{R}. \tag{9.34}$$

Interchanging x with y, we obtain

$$B(y+x,z) + B(y-x,z) = 2B(y,z). \tag{9.35}$$

Subtracting (9.35) from (9.34), we have

$$B(x-y,z) - B(y-x,z) = 2B(x,z) - 2B(y,z). \tag{9.36}$$

Since B is odd in each variable, (9.36) yields

$$B(x-y,z) = B(x,z) - B(y,z).$$

Replacing $-y$ with y and using the fact that B is an odd function in the first variable, we get

$$B(x+y,z) = B(x,z) + B(y,z).$$

Therefore $B : \mathbb{R}^2 \to \mathbb{R}$ is additive in the first variable. Since B is symmetric, B is also additive in the second variable. Thus B is a biadditive function. The proof of the theorem is now complete. □

9.5   Pexiderization of Quadratic Equation

The quadratic functional equation

$$f(x+y) + f(x-y) = 2f(x) + 2f(y) \qquad \text{(QE)}$$

can be pexiderized to

$$f_1(x+y) + f_2(x-y) = f_3(x) + f_4(y), \qquad (9.37)$$

where $f_1, f_2, f_3, f_4 : \mathbb{R} \to \mathbb{R}$ are unknown functions and x, y are real numbers.

The above pexiderization is useful for characterizing quasi-inner product spaces. To determine the general solution of (9.37) we need the following auxiliary result from Ebanks, Kannappan and Sahoo (1992a).

Lemma 9.1. *The general solution* $f, g, h : \mathbb{R} \to \mathbb{R}$ *of the functional equation*

$$f(x+y) + f(x-y) = g(x) + h(y) + h(-y) \qquad (9.38)$$

is given by

$$\left. \begin{aligned} f(x) &= B(x,x) + \frac{1}{2}A(x) - \frac{b}{2} \\ g(x) &= 2B(x,x) + A(x) - b - a \\ h(x) + h(-x) &= 2B(x,x) + a, \end{aligned} \right\} \qquad (9.39)$$

where $B : \mathbb{R}^2 \to \mathbb{R}$ *is a symmetric biadditive function,* $A : \mathbb{R} \to \mathbb{R}$ *is an additive function and* a, b *are arbitrary elements of* $\mathbb{R}$.

Proof. By letting $y = 0$ in (9.38), we get

$$g(x) = 2f(x) - 2h(0). \qquad (9.40)$$

Using (9.40), we rewrite (9.38) as

$$f(x+y) + f(x-y) - 2f(x) = h(y) + h(-y) - 2h(0) \qquad (9.41)$$

for all $x, y \in \mathbb{R}$. Now, $x = 0$ in (9.41) yields

$$h(y) + h(-y) - 2h(0) = f(y) + f(-y) - 2f(0) \qquad (9.42)$$

so that (9.41) becomes

$$f(x+y) + f(x-y) - 2f(x) = f(y) + f(-y) - 2f(0). \qquad (9.43)$$

We define $B : \mathbb{R}^2 \to \mathbb{R}$ by

$$2B(x, y) = f(x + y) - f(x) - f(y) + f(0) \quad \text{for } x, y \in \mathbb{R}. \qquad (9.44)$$

From (9.44) and (9.43) we obtain

$$
\begin{aligned}
2B(x + u, &y) + 2B(x - u, y) \\
&= f(x + u + y) + f(x - u + y) - \{f(x + u) + f(x - u)\} \\
&\quad - 2f(y) + 2f(0) \\
&= f(x + u + y) + f(x - u + y) - \{2f(x) + f(u) + f(-u) \\
&\quad - 2f(0)\} - 2f(y) + 2f(0) \\
&= 2f(x + y) + f(u) + f(-u) - 2f(0) - \{2f(x) + f(u) \\
&\quad + f(-u) - 2f(0)\} - 2f(y) + 2f(0) \\
&= 2f(x + y) - 2f(x) - 2f(y) + 2f(0) \\
&= 4B(x, y).
\end{aligned}
$$

Hence we have

$$B(x + u, y) + B(x - u, y) = 2B(x, y)$$

for all $x, u, y \in \mathbb{R}$. This says that $B(\cdot, y)$ satisfies the Jensen equation. Hence $B(\cdot, y)$ is additive in the first variable, since $B(0, y) = 0$. Since B defined by (9.44) is symmetric, B is additive in the second variable. Hence $B : \mathbb{R}^2 \to \mathbb{R}$ is a symmetric biadditive function.

Now $y = x$ in (9.44) and (9.43) give

$$
\begin{aligned}
2B(x, x) &= f(2x) - 2f(x) + f(0) \\
&= f(x) + f(-x) - 2f(0). \qquad (9.45)
\end{aligned}
$$

By (9.42), this gives

$$h(x) + h(-x) = 2B(x, x) + a,$$

where $a = 2h(0)$.

Next, we define $\ell : \mathbb{R} \to \mathbb{R}$ by

$$\ell(x) = f(x) - f(-x) \quad \text{for } x \in \mathbb{R}. \qquad (9.46)$$

From (9.46) and (9.43) we conclude that

$$
\begin{aligned}
\ell(x + y) &+ \ell(x - y) \\
&= f(x + y) + f(x - y) - \{f(-y - x) + f(y - x)\}
\end{aligned}
$$

$$= 2f(x) + f(y) + f(-y) - 2f(0) - \{2f(-x) + f(y)$$
$$+ f(-y) - 2f(0)\}$$
$$= 2\ell(x),$$

which is Jensen. Since $\ell(0) = 0$, we see that ℓ is additive, that is,

$$\ell(x) = A(x), \tag{9.47}$$

where $A : \mathbb{R} \to \mathbb{R}$ is additive and by (9.46),

$$f(x) - f(-x) = A(x). \tag{9.48}$$

From (9.48) and (9.45) we have

$$f(x) = B(x, x) + \frac{1}{2}A(x) - \frac{b}{2},$$

where $b := -2f(0)$. Now from (9.40) we get

$$g(x) = 2B(x, x) + a.$$

This completes the proof of the lemma. $\qquad\qquad\qquad\square$

The following corollary is obvious from the above lemma.

Corollary 9.1. *The general solution $f : \mathbb{R} \to \mathbb{R}$ of the functional equation*

$$f(x + y) + f(x - y) = 2f(x) + f(y) + f(-y)$$

is given by

$$f(x) = B(x, x) + A(x),$$

where $B : \mathbb{R} \times \mathbb{R} \to \mathbb{R}$ is a symmetric biadditive function and $A : \mathbb{R} \to \mathbb{R}$ is an additive function.

Remark 9.1. *The above functional equation came from a result of Drygas (1987) who obtained a Jordan and von Neumann type characterization theorem for quasi-inner products. In Drygas' characterization of quasi-inner product, the functional equation*

$$f(x) + f(y) = f(x - y) + 2\left\{f\left(\frac{x + y}{2}\right) - f\left(\frac{x - y}{2}\right)\right\}$$

played an important role. By replacing y with $-y$ in the above equation and adding the resultant to the above equation, one obtains

$$f(x + y) + f(x - y) = 2f(x) + f(y) + f(-y).$$

Drygas (1987) did not discuss the solution of this functional equation.

In an Americal Mathematical Society meeting, E. Y. Deeba of the University of Houston asked to find the solution of the functional equation

$$f(x+y+z)+f(x)+f(y)+f(z) = f(x+y)+f(y+z)+f(x+z). \quad (9.49)$$

Kannappan (1995) determined the general solution of the above functional equation when the unknown function f is defined on a vector space and takes values on field of characteristic different from two (or of characteristic zero). The solution of the functional equation of Deeba can be obtained easily from the above corollary. Historically, Whitehead (1950) was the first mathematician to consider the functional equation (9.49), and we call this equation the Whitehead functional equation.

Corollary 9.2. *The general solution $f : \mathbb{R} \to \mathbb{R}$ of the functional equation (9.49) holding for all $x, y, z \in \mathbb{R}$ is given by*

$$f(x) = B(x, x) + A(x),$$

where $B : \mathbb{R} \times \mathbb{R} \to \mathbb{R}$ is a symmetric biadditive function and $A : \mathbb{R} \to \mathbb{R}$ is an additive function.

Proof. Letting $x = y = z = 0$ in (9.49), we see that $f(0) = 0$. Next replacing z by $-y$ in (9.49), we obtain

$$f(x + y) + f(x - y) = 2 f(x) + f(y) + f(-y)$$

for all $x, y \in \mathbb{R}$. Hence we have the asserted solution by Corollary 9.1. $\square$

Using Lemma 9.1 now we determine the general solution of the functional equation (9.37) following Ebanks, Kannappan and Sahoo (1992a).

Theorem 9.6. *The general solution $f_1, f_2, f_3, f_4 : \mathbb{R} \to \mathbb{R}$ of the functional equation*

$$f_1(x + y) + f_2(x - y) = f_3(x) + f_4(y) \quad \text{for } x, y \in \mathbb{R} \qquad (9.36)$$

is given by

$$\left.\begin{aligned}
f_1(x) &= \frac{1}{2}B(x, x) + \frac{1}{4}(A_1 + A_2)(x) + \left(a - \frac{b}{2}\right) \\
f_2(x) &= \frac{1}{2}B(x, x) + \frac{1}{4}(A_1 - A_2)(x) - \left(a + \frac{b}{2}\right) \\
f_3(x) &= B(x, x) + \frac{1}{2}A_1(x) - (b + c) \\
f_4(x) &= B(x, x) + \frac{1}{2}A_2(x) + c,
\end{aligned}\right\} \qquad (9.36s)$$

where $B : \mathbb{R}^2 \to \mathbb{R}$ *is a symmetric biadditive function,* $A_1, A_2 : \mathbb{R} \to \mathbb{R}$ *are additive functions and* a, b, c *are arbitrary constants.*

Proof. It is easy to verify that the f_i's given in the theorem satisfy (9.37). Now we proceed to demonstrate that the asserted form of f_i's is the only solution of the functional equation (9.37).

Replacing y with $-y$ in (9.37), we obtain

$$f_1(x - y) + f_2(x + y) = f_3(x) + f_4(-y). \tag{9.50}$$

Adding (9.50) to (9.37), we get

$$g(x + y) + g(x - y) = 2f_3(x) + f_4(y) + f_4(-y), \tag{9.51}$$

where

$$g(x) = f_1(x) + f_2(x) \quad \text{for } x \in \mathbb{R}. \tag{9.52}$$

Similarly, subtracting (9.50) from (9.37), we obtain

$$h(x + y) - h(x - y) = f_4(y) - f_4(-y), \tag{9.53}$$

where

$$h(x) = f_1(x) - f_2(x) \quad \text{for } x \in \mathbb{R}. \tag{9.54}$$

Solving (9.37) is equivalent to solving the system of equations (9.51) and (9.53). First, we solve (9.53). Define $k : \mathbb{R} \to \mathbb{R}$ by

$$k(y) = f_4(y) - f_4(-y) \quad \text{for } y \in \mathbb{R}. \tag{9.55}$$

Then (9.53) reduces to

$$h(x + y) - h(x - y) = k(y) \tag{9.56}$$

for all $x, y \in \mathbb{R}$. Replacing y by $-y$ in (9.55) or (9.56), we see that

$$k(y) = -k(-y). \tag{9.57}$$

Hence k is an odd function. Next we substitute $y = x$ in (9.56) to obtain

$$h(2x) = k(x) + h(0). \tag{9.58}$$

From (9.56) we conclude that

$$
\begin{aligned}
k(y + v) &+ k(y - v) \\
&= h(x + y + v) - h(x - v - y) + h(x + y - v) - h(x + v - y) \\
&= h((x + v) + y) - h((x + v) - y) + h((x - v) + y)
\end{aligned}
$$

$$- h((x - v) - y)$$
$$= 2k(y);$$

that is, k satisfies the Jensen equation with $k(0) = 0$. Hence k is additive; that is,

$$k(y) = A_1(y), \quad y \in \mathbb{R}, \tag{9.59}$$

where $A_1 : \mathbb{R} \to \mathbb{R}$ is additive. From (9.59) and (9.58), we obtain

$$h(x) = \frac{1}{2}A_1(x) + a, \tag{9.60}$$

where $a := h(0)$. From (9.54), (9.55) and (9.60), we get

$$f_1(x) - f_2(x) = \frac{1}{2}A_1(x) + a \tag{9.61}$$

and

$$f_4(y) - f_4(-y) = A_1(y). \tag{9.62}$$

Now we return to the functional equation (9.51). By Lemma 9.1 we get

$$g(x) = f_1(x) + f_2(x) = B(x, x) + \frac{1}{2}A_2(x) - \frac{b_1}{2}, \tag{9.63}$$

$$f_3(x) = B(x, x) + \frac{1}{2}A_2(x) - \frac{(b_1 + a_1)}{2}, \tag{9.64}$$

$$f_4(x) + f_4(-x) = 2B(x, x) + a_1. \tag{9.65}$$

From (9.63)–(9.65), we obtain the asserted solution. This completes the proof. □

9.6   Concluding Remarks

The quadratic functional equation $f(x+y) + f(x-y) = 2f(x) + 2f(y)$ is very important as it serves in certain abstract spaces for the definition of norm. It was studied by many authors (and we cite only few earlier references) including Jensen (1878, 1897), Jordan and von Neumann (1935), Kurepa (1959, 1962), Aczél and Vincze (1963), Aczél (1965), and Kannappan (1995, 2000). This quadratic functional equation arises from the definition of norm on a vector space. It is well known that a norm square $||x||^2$ is equal to an inner product $< x, x >$ on a vector space if and only if it satisfies the parallelogram law $||x + y||^2 + ||x - y||^2 = 2||x||^2 + 2||y||^2$ for all x, y in the vector space (see Figure 9.1).

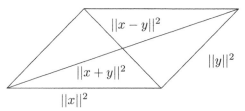

Figure 9.1. Illustration of parallelogram law.

If $f(x) = ||x||^2$, then f satisfies the quadratic functional equation. Unless $f : \mathbb{R} \to \mathbb{R}$ is regular, the general solution of the quadratic functional equation is not of the form $f(x) = c\,x^2$ for some real constant c. Kurepa (1965) asked the following question: Which real functions $f : \mathbb{R} \to \mathbb{R}$ satisfy $f(x + y) + f(x - y) = 2f(x) + 2f(y)$ for all real x, y and $f(x) = x^4 f\left(x^{-1}\right)$ for all nonzero real number x? Grząślewicz (1978) gave a partial solution to Kurepa's problem.

Aczél (1965) proved the result regarding the quadratic functional equation: Let G be an abelian group and let H be an abelian group in which every equation of the form $2x = h \in H$ has one and only one solution $x \in H$. Then any solution $f : G \to H$ of the quadratic functional equation on G is of the form $f(x) = B(x, x)$, where $B : G \times G \to H$ is a symmetric biadditive form.

For an arbitrary group G, we denote $\cdot$ as its group operation and e as the identity element. To simplify our writing, we write xy, instead of $x \cdot y$. If G is abelian, the group operation and the identity element are denoted by $+$ and 0, respectively. In this case we write xy as $x + y$. Similar notations will be adapted for semigroups.

Sinopoulos (2000) proved the following result concerning the quadratic functional equation on semigroup. For the benefit of the interested readers we present the proof.

Theorem 9.7. *Let $(S, +)$ be a commutative semigroup, G a uniquely 2-divisible abelian group, and σ an endomorphism of S such that $\sigma(\sigma x) = x$ for $x \in S$. Then the general solution $f : S \to G$ of the quadratic functional equation*

$$f(x + y) + f(x + \sigma y) = 2f(x) + 2f(y) \qquad \forall\, x, y \in S \qquad (9.66)$$

is given by

$$f(x) = B(x, x) + A(x) \qquad \forall\, x \in S, \qquad (9.67)$$

where $B : S \times S \to G$ is an arbitrary symmetric biadditive function with $B(\sigma x, y) = -B(x, y)$, and $A : S \to G$ is an arbitrary additive function with $A(\sigma x) = A(x)$.

Proof. Replacing y by σy in (9.66), we obtain

$$f(x + \sigma y) + f(x + y) = 2f(x) + 2f(\sigma y) \tag{9.68}$$

for all $x, y \in S$. Comparing (9.66) and (9.68), we see that

$$2f(\sigma y) = f(y) \qquad \forall\, y \in S. \tag{9.69}$$

Next, we replace x by $x + t$ in (9.66) to get

$$f(x + t + y) + f(x + t + \sigma y) = 2f(x + t) + 2f(y) \tag{9.70}$$

for all $x, y \in S$. Again we replace x by $x + \sigma t$ in (9.66) to get

$$f(x + \sigma t + y) + f(x + \sigma t + \sigma y) = 2f(x + \sigma t) + 2f(y) \tag{9.71}$$

for all $x, y \in S$. Subtracting (9.71) from (9.70) and simplifying, we have

$$f(x + y + t) - f(x + y + \sigma t) + f(x + \sigma y + t) - f(x + \sigma t + \sigma y)$$
$$= 2f(x + t) - 2f(x + \sigma t) \tag{9.72}$$

for all $x, y \in S$. Define $h : S \times S \to G$ by

$$h(x, t) = \frac{1}{4}\left[\, f(x + t) - f(x + \sigma t)\,\right] \tag{9.73}$$

for all $x, t \in S$. Then using (9.73) in (9.72), we have

$$h(x + y, t) + h(x + \sigma y, t) = 2\, h(x, t) \tag{9.74}$$

for all $x, y, t \in S$. For a fixed t, the functional equation (9.74) is a Jensen functional equation and its solutions can be found from Theorem 7.6 as

$$h(x, t) = B(x, t) + a(t), \tag{9.75}$$

where $B : S \times S \to G$ is an additive function in the first variable with $B(\sigma x, t) = -B(x, t)$, and $a : S \to G$ is an arbitrary function.

From (9.73) and (9.67), we see that

$$\begin{aligned}
4\, h(\sigma x, t) &= f(\sigma x + t) - f(\sigma x + \sigma t) \\
&= f(\sigma(x + \sigma t)) - f(\sigma(x + t)) \\
&= f(x + \sigma t) - f(x + t) \qquad \text{(by (9.67))} \\
&= -4\, h(x, t).
\end{aligned}$$

Hence

$$h(\sigma x, t) = -h(x, t) \tag{9.76}$$

for all $x, t \in S$. Again from (9.73) and (9.67), we have

$$
\begin{aligned}
4\,h(t, x) &= f(t + x) - f(t + \sigma x) \\
&= f(t + x) - f(\sigma(t + \sigma x)) \qquad \text{(by (9.67))} \\
&= f(t + x) - f(\sigma t + x) \\
&= f(x + t) - f(x + \sigma t) \\
&= 4\,h(x, t).
\end{aligned}
$$

Therefore $h(x, t)$ is a symmetric function, that is,

$$h(x, t) = h(t, x) \tag{9.77}$$

for all $x, t \in S$.

From (9.75), we see that

$$h(\sigma x, t) = B(\sigma x, t) + a(t) = -B(x, t) + a(t).$$

Adding this to (9.75), we have

$$h(x, t) + h(\sigma x, t) = 2\,a(t). \tag{9.78}$$

Using (9.76) in (9.78), we see that $a(t) = 0$ and (9.75) reduces to

$$h(x, t) = B(x, t), \tag{9.79}$$

where B is now a symmetric biadditive function.

Letting $y = x$ in (9.66) and using (9.69), we have

$$f(2x) = 4\,f(x) - f(x + \sigma x). \tag{9.80}$$

Next letting $t = x$ in (9.73), we obtain

$$f(x) = h(x, x) + \frac{1}{2}\,f(x + \sigma x) \tag{9.81}$$

which by (9.79) yields

$$f(x) = B(x, x) + \frac{1}{2}\,f(x + \sigma x). \tag{9.82}$$

Finally we replace x by $x + \sigma x$ and y by $y + \sigma y$ in (9.66) to obtain

$$f(x + \sigma x + y + \sigma y) = f(x + \sigma x) + f(y + \sigma y) \tag{9.83}$$

for all $x, y \in S$. Now defining $A : S \to G$ by

$$A(x) = \frac{1}{2} f(x + \sigma x) \qquad \forall x \in S \qquad (9.84)$$

and using (9.84) in (9.83), we see that

$$A(x + y) = A(x) + A(y);$$

that is, A is an additive function on S. Moreover from (9.68) we have $A(\sigma x) = A(x)$. Using (9.84) in (9.82), we obtain the asserted solution (9.67). This completes the proof of the theorem. $\square$

On an arbitrary group the quadratic functional equation takes the form

$$f(xy) + f(xy^{-1}) = 2f(x) + 2f(y). \qquad (9.85)$$

A generalization of quadratic functional equation is $f(xy) + f(xy^{-1}) = 2f(x) + f(y) + f(y^{-1})$. This equation arises in the characterization of quasi-inner product spaces. This functional equation is treated by Ebanks, Kannappan and Sahoo (1992a). They also studied the Pexider version of quadratic functional equation $f_1(xy) + f_2(xy^{-1}) = f_3(x) + f_4(y)$ and obtained the solution of this functional equation when variables x, y are in a 2-divisible group and values are in a commutative field of characteristic different from 2.

Another generalization of (9.85) is the quadratic-trigonometric functional equation

$$f_1(xy) + f_2(xy^{-1}) = f_3(x) + f_4(y) + f_5(x)f_6(y).$$

The general solution of this functional equation along with

$$f_i(txy) = f_i(tyx), \quad (i = 1, 2)$$

was given by Chung, Ebanks, Ng and Sahoo (1995) when the variables x, y are in a group and values are in the additive field of complex numbers.

Recently, the quadratic functional equation (9.85) has been investigated on nonabelian groups. For example, Yang (2005) solved the quadratic functional equation on free groups and on the general linear group $GL_n(\mathbb{Z})$ over integers. de Place Friis and Stetkaer (2006) solved the quadratic functional equation on various nonabelian groups such as $(ax + b)$-group, the Heisenberg group, and $GL_n(\mathbb{R})$.

9.7   Exercises

1. Find all functions $f : \mathbb{R} \to \mathbb{R}$ that satisfy the functional equation

$$f(x + y + z) + f(x) + f(y) + f(z) = f(x + y) + f(y + z) + f(z + x)$$

for all $x, y, z \in \mathbb{R}$.

2. Find all functions $f : \mathbb{R} \to \mathbb{R}$ that satisfy the functional equation

$$f(x - y - z) + f(x) + f(y) + f(z) = f(x - y) + f(y + z) + f(z - x)$$

for all $x, y, z \in \mathbb{R}$.

3. Find all functions $f : \mathbb{R} \to \mathbb{R}$ that satisfy the functional equation

$$f(x + y)\, f(x - y) = f(x)^2\, f(y)^2$$

for all $x, y, z \in \mathbb{R}$.

4. Find all functions $f : \mathbb{C} \to \mathbb{C}$ that satisfy the functional equation

$$f(x + y)\, f(x - y) = f(x)^2\, f(y)^2$$

for all $x, y, z \in \mathbb{C}$.

5. Let k be a positive integer. Find all functions $f : \mathbb{R} \to \mathbb{R}$ that satisfy the functional equation

$$f(kx + y) + f(kx - y) = 2k^2\, f(x) + 2\, f(y)$$

for all $x, y \in \mathbb{R}$.

6. Find all functions $f : \mathbb{R} \to \mathbb{R}$ that satisfy the functional equation

$$3f(x) + 3f(y) + 3f(z)$$
$$= f(x + y + z) + f(x - y) + f(y - z) + f(z - x)$$

for all $x, y, z \in \mathbb{R}$.

7. Find all functions $f : \mathbb{R}^2 \to \mathbb{R}$ that satisfy the functional equation

$$f(x + u, y + v) + f(x - u, y - v) = 2\, f(x, y) + 2\, f(u, v)$$

for all $x, y, u, v \in \mathbb{R}$.

8. Let a be a nonzero real number. Find all functions $f : \mathbb{R} \to \mathbb{R}$ that satisfy the functional equation

$$f\left(\frac{x+y}{a}\right) + f\left(\frac{x-y}{a}\right) = \frac{2\,f(x) + 2\,f(y)}{a}$$

for all $x, y \in \mathbb{R}$.

9. Find all functions $f : \mathbb{R} \to \mathbb{R}$ that satisfy the functional equation

$$f(2x + y) + f(2x - y) = f(x + y) + f(x - y) + 2f(2x) - 2f(x)$$

for all $x, y \in \mathbb{R}$.

10. Find all functions $f : \mathbb{R}^2 \to \mathbb{R}$ that satisfy the functional equation

$$f(x + y, u + v) + f(x + y, u - v)$$
$$= 2f(x, u) + 2f(x, v) + 2f(y, u) + 2f(y, v)$$

for all $x, y, u, v \in \mathbb{R}$.

11. Let λ be a real number. Find all functions $f : \mathbb{R} \to \mathbb{R}$ that satisfy the functional equation

$$f\left(\frac{x+y}{2}\right) + f\left(\frac{x-y}{2}\right) = 2f\left(\frac{x}{2}\right) + 2f\left(\frac{y}{2}\right) + \lambda f(x)f(y)$$

for all $x, y \in \mathbb{R}$.

12. Let a, b be nonzero real numbers satisfying $a \neq 2b^2$. Find all functions $f : \mathbb{R} \to \mathbb{R}$ that satisfy the functional equation

$$f(ax + by) + f(ax - by)$$
$$= \frac{a}{2}f(x + y) + \frac{a}{2}f(x - y) + (2a^2 - a)f(x) + (2b^2 - a)f(y)$$

for all $x, y \in \mathbb{R}$.

13. Find all functions $f : \mathbb{R} \to \mathbb{R}$ that satisfy the functional equation

$$f(x + y)\,f(x - y) = f(x)^2\,f(y)\,f(-y)$$

for all $x, y \in \mathbb{R}$.

14. Find all functions $f, g : \mathbb{R} \to \mathbb{R}$ that satisfy the functional equation

$$f(x + y)\,f(x - y) = f(x)\,f(y)\,g(x)\,g(-y)$$

for all $x, y \in \mathbb{R}$.

15. Let n be a positive integer. Find all functions $f : \mathbb{R} \to \mathbb{R}$ that satisfy the functional equation

$$f(x + ny) f(x - ny) = 2f(x) + 2n^2 f(y)$$

for all $x, y \in \mathbb{R}$.

16. Find all functions $f : \mathbb{R} \to \mathbb{R}$ that satisfy the functional equation

$$f(x+y+z)+f(x+y-z)+f(y+z-x)+f(z+x-y) = 4[f(x)+f(y)+f(z)]$$

for all $x, y, z \in \mathbb{R}$.

17. Find all functions $f : \mathbb{R} \to \mathbb{R}$ that satisfy the functional equation

$$f(x + y + z) + f(x - y) + f(y - z) + f(z - x) = 3[f(x) + f(y) + f(z)]$$

for all $x, y, z \in \mathbb{R}$.

18. Find all functions $f : \mathbb{R} \to \mathbb{R}$ that satisfy the functional equation

$$f(x + y + z) + f(x - y) + f(x - z) = f(x - y - z) + f(x + y) + f(x + z)$$

for all $x, y, z \in \mathbb{R}$.

19. Find all functions $f : \mathbb{R}^2 \to \mathbb{R}$ that satisfy the functional equation

$$f(u + x, v + y) + f(u - x, v) + f(u, v - y)$$
$$= f(u - x, v - y) + f(u + x, v) + f(u, v + y)$$

for all $x, y, u, v \in \mathbb{R}$.

Chapter 10

d'Alembert Functional Equation

10.1 Introduction

The well-known trigonometric identity

$$\cos(x + y) + \cos(x - y) = 2\cos(x)\,\cos(y)$$

implies the functional equation

$$f(x + y) + f(x - y) = 2f(x)f(y) \qquad \text{(DE)}$$

for all $x, y \in \mathbb{R}$. In this chapter, we study this functional equation together with another functional equation that characterizes the cosine function.

The above functional equation is known as the *d'Alembert functional equation*. It has a long history going back to d'Alembert (1769), Poisson (1804) and Picard (1922, 1928). The equation plays a central role in determining the sum of two vectors in Euclidean and non-Euclidean geometries. Cauchy (1821) determined the continuous solution of the d'Alembert functional equation.

10.2 Continuous Solution of d'Alembert Equation

In the following theorem, we first present the continuous solution of the d'Alembert equation (DE).

Theorem 10.1. *Let $f : \mathbb{R} \to \mathbb{R}$ be continuous and satisfy*

$$f(x + y) + f(x - y) = 2f(x)f(y) \qquad \text{(DE)}$$

for all $x, y \in \mathbb{R}$. Then f is of the form

$$f(x) = 0, \qquad (10.1)$$

$$f(x) = 1, \tag{10.2}$$
$$f(x) = \cosh(\alpha x), \tag{10.3}$$
$$f(x) = \cos(\beta x), \tag{10.4}$$

where α, β are arbitrary real constants.

Proof. Letting $x = 0 = y$ in (DE) we obtain

$$2f(0) = 2[f(0)]^2.$$

Hence

$$f(0) = 0 \quad \text{or} \quad f(0) = 1.$$

If $f(0) = 0$, then letting $y = 0$ in (DE), we have

$$2f(x) = 2f(x)f(0)$$

which is

$$2f(x) = 0$$

and thus we have

$$f(x) = 0, \quad \forall\, x \in \mathbb{R}.$$

This gives the solution (10.1). Hence we assume from now on that f is not identically zero.

Next, we show that any solution of (DE) is an even function. To see this, let $x = 0$ in (DE). Then we obtain

$$f(y) + f(-y) = 2f(0)f(y).$$

Since f is not identically zero, $f(0) \neq 0$ and $f(0) = 1$. Hence the above equation gives

$$f(y) + f(-y) = 2f(y),$$

that is,

$$f(-y) = f(y)$$

for all $y \in \mathbb{R}$. Thus f is an even function. Since f is continuous on $\mathbb{R}$, f is also integrable on any finite interval. Hence, for $t > 0$, we have

$$\int_{-t}^{t} f(x+y)dy + \int_{-t}^{t} f(x-y)dy = 2f(x) \int_{-t}^{t} f(y)dy. \tag{10.5}$$

Now

$$\int_{-t}^{t} f(x+y)dy = \int_{x-t}^{x+t} f(z)dz = \int_{x-t}^{x+t} f(y)dy.$$

Similarly,

$$\int_{-t}^{t} f(x-y)dy = \int_{x+t}^{x-t} f(w)(-dw) = \int_{x-t}^{x+t} f(w)dw = \int_{x-t}^{x+t} f(y)dy.$$

Hence (10.5) becomes

$$\int_{x-t}^{x+t} f(y)dy + \int_{x-t}^{x+t} f(y)dy = 2f(x)\int_{-t}^{t} f(y)dy$$

which is

$$\int_{x-t}^{x+t} f(y)dy = f(x)\int_{-t}^{t} f(y)dy. \tag{10.6}$$

Since f is not identically zero, $f(0) = 1$. Further, since f is continuous, there exists $t > 0$ such that (see Figure 10.1)

$$\int_{-t}^{t} f(y)dy > 0.$$

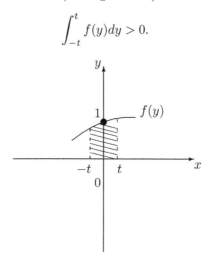

Figure 10.1. Illustration of the existence of $t > 0$.

Note that the left-hand side of (10.6) is differentiable with respect to x by the fundamental theorem of calculus. Hence the right-hand side of (10.6) is also differentiable with respect to the variable x. Then differentiating (10.6) with respect to x, we get

$$\frac{d}{dx}\int_{x-t}^{x+t} f(y)dy = \frac{d}{dx}\left[f(x)\int_{-t}^{t} f(y)dy \right]$$

which is

$$f(x+t) - f(x-t) = f'(x)\int_{-t}^{t} f(y)dy. \tag{10.7}$$

This shows that f is twice differentiable and hence

$$f'(x+t) - f'(x-t) = f''(x) \int_{-t}^{t} f(y)dy.$$

Thus f is 3 times differentiable. Proceeding step by step, we see that any continuous solution of (DE) is infinitely differentiable.

Letting $x = 0$ in (10.7), we obtain

$$f(t) - f(-t) = f'(0) \int_{-t}^{t} f(y)dy. \qquad (10.8)$$

Since f is even, we have $f(t) = f(-t)$ and (10.8) yields

$$f'(0) \int_{-t}^{t} f(y)dy = 0. \qquad (10.9)$$

Since $\int_{-t}^{t} f(y)dy > 0$, (10.9) gives

$$f'(0) = 0. \qquad (10.10)$$

Since $f \in C^{\infty}(\mathbb{R})$, we differentiate (DE) with respect to y twice to get

$$f'(x+y) - f'(x-y) = 2f(x)f'(y)$$
$$f''(x+y) + f''(x-y) = 2f(x)f''(y)$$

for all $x, y \in \mathbb{R}$. Letting $y = 0$, we have

$$2f''(x) = 2f(x)f''(0).$$

Let $k = f''(0)$. Then

$$f''(x) = kf(x)$$

which yields the following initial value problem (IVP)

$$
\boxed{
\begin{array}{l}
\dfrac{d^2y}{dx^2} = ky \\[2mm]
y(0) = 1 \\[1mm]
y'(0) = 0 \\[2mm]
\qquad\qquad \text{(IVP)}
\end{array}
}
$$

To solve this initial value problem we have to consider three cases: $k = 0$, $k > 0$ and $k < 0$.

Case 1. Suppose $k = 0$. Then IVP reduces to

$$\frac{d^2y}{dx^2} = 0.$$

Hence

$$y(x) = c_1 x + c_2.$$

Since $y(0) = 1$, $c_2 = 1$. Again since $y'(0) = 0$, we get $c_1 = 0$. Therefore

$$y(x) = 1$$

is the solution in this case (which is $f(x) = 1 \ \forall \ x \in \mathbb{R}$).

Case 2. Suppose $k > 0$. Letting $y = e^{mx}$ into

$$\frac{d^2y}{dx^2} = ky, \tag{DE$'$}$$

we obtain $m^2 = k$ and hence $m = \pm\sqrt{k}$. Thus

$$y(x) = c_1 e^{\alpha x} + c_2 e^{-\alpha x}, \quad \text{where } \alpha = \sqrt{k}.$$

Now

$$\begin{aligned}
1 &= y(0) \\
&= c_1 e^{\alpha \cdot 0} + c_2 e^{-\alpha \cdot 0} \\
&= c_1 + c_2.
\end{aligned}$$

Hence

$$c_2 = (1 - c_1).$$

Thus

$$y(x) = c_1 e^{\alpha x} + (1 - c_1)e^{-\alpha x}.$$

Now

$$\begin{aligned}
0 &= y'(0) \\
&= c_1 \alpha e^{\alpha x} + (1 - c_1)(-\alpha)e^{-\alpha x}\Big|_{x=0} \\
&= c_1 \alpha + (1 - c_1)(-\alpha) \\
&= c_1 \alpha - \alpha + c_1 \alpha \\
&= 2c_1 \alpha - \alpha.
\end{aligned}$$

Hence

$$2c_1 \alpha = \alpha$$

or

$$c_1 = \frac{1}{2} \quad (\text{since } \alpha \neq 0).$$

Therefore the solution of (DE') is given by

$$y(x) = \frac{e^{\alpha x} + e^{-\alpha x}}{2} = \cosh(\alpha x).$$

Hence in this case we have

$$f(x) = \cosh(\alpha x)$$

which is (10.3).

Case 3. Suppose $k < 0$. Letting $y = e^{mx}$ into

$$\frac{d^2 y}{dx^2} = ky, \quad \text{(DE')}$$

we obtain

$$m^2 y = ky.$$

Hence

$$m = \pm i\beta,$$

where $\beta = \sqrt{-k}$, $i = \sqrt{-1}$. Thus the solution of (DE') is given by

$$y(x) = c_1 e^{i\beta x} + c_2 e^{-i\beta x}.$$

Since

$$1 = y(0) = c_1 + c_2,$$

we have

$$c_2 = 1 - c_1.$$

Hence

$$y(x) = c_1 e^{i\beta x} + (1 - c_1)e^{-i\beta x}.$$

Further, since

$$\begin{aligned}
0 &= y'(0) \\
&= i\beta c_1 - i\beta(1 - c_1) \\
&= 2i\beta c_1 - i\beta
\end{aligned}$$

or

$$i\beta(2c_1 - 1) = 0,$$

we obtain

$$c_1 = \frac{1}{2}.$$

Hence, we have

$$y(x) = \frac{e^{i\beta x} + e^{-i\beta x}}{2} = \cos(\beta x).$$

Therefore the solution of the functional equation is given by

$$f(x) = \cos(\beta x)$$

which is (10.4).

This completes the proof of the theorem. $\quad\square$

Remark 10.1. *Note that the continuity of f together with the functional equation gave infinite differentiability of the solution f. Then by differentiating the functional equation, we obtained a differential equation. By solving this differential equation we found the solution of the functional equation. This is one of the standard methods for solving functional equations when regularity properties like continuity are assumed.*

There is another standard method due to Cauchy. The Cauchy method consists of finding the solution of a functional equation on a dense set (like the set of rationals $\mathbb{Q}$) and then uses continuity to pass to the entire real line $\mathbb{R}$.

10.3 General Solution of d'Alembert Equation

A function $E : \mathbb{R} \to \mathbb{C}$ is said to be *exponential* if E satisfies the equation

$$E(x + y) = E(x)E(y)$$

for all $x, y \in \mathbb{R}$.

If E is a nonzero continuous function, then

$$E(x) = e^{\lambda x},$$

where λ is an arbitrary complex constant.

If $E : \mathbb{R} \to \mathbb{C}$ is a nonzero exponential function, then we denote it by

$$E^*(y) = E(y)^{-1}. \tag{10.11}$$

Now we give some elementary properties of the exponential function.

Proposition 10.1. *If $E : \mathbb{R} \to \mathbb{C}$ is an exponential function and $E(0)$ is zero, then $E(x) \equiv 0$ for all $x \in \mathbb{R}$.*

Proof. Let $E : \mathbb{R} \to \mathbb{C}$ be an exponential function. Hence

$$E(x + y) = E(x)\,E(y) \tag{10.12}$$

for all $x, y \in \mathbb{R}$. Letting $y = 0$ in (10.12), we obtain

$$E(x) = E(x)\,E(0) \quad \text{for } x \in \mathbb{R}. \tag{10.13}$$

Since $E(0) = 0$, (10.13) yields

$$E(x) = 0 \quad \forall\, x \in \mathbb{R}. \tag{10.14}$$

Hence $E(x)$ is identically zero. $\qquad\square$

Proposition 10.2. *Let $E : \mathbb{R} \to \mathbb{C}$ be an exponential function. If $E(x) \not\equiv 0$, then $E(0) = 1$.*

Proof. Let $E : \mathbb{R} \to \mathbb{C}$ be an exponential function. Assume that $E(x)$ is not identically zero. Letting $x = 0 = y$ in (10.12), we get

$$E(0)[1 - E(0)] = 0.$$

Hence either

$$E(0) = 0 \quad \text{or} \quad E(0) = 1.$$

We claim that $E(0) = 1$. Suppose not. Hence $E(0) = 0$. By Proposition 10.1, $E(x) \equiv 0$, is a contradiction. Hence $E(0) = 1$. This completes the proof of the proposition. $\qquad\square$

Proposition 10.3. *Let $E : \mathbb{R} \to \mathbb{C}$ be an exponential function. If $E(x_0) = 0$ for some $x_0 \neq 0$, then $E(x) \equiv 0$ for all $x \in \mathbb{R}$.*

Proof. Let $x\ (\neq x_0) \in \mathbb{R}$. Then, since $E(x_0) = 0$, we have

$$E(x) = E((x - x_0) + x_0) = E(x - x_0)\,E(x_0) = 0.$$

Hence $E(x) \equiv 0$. Thus E is nowhere zero or everywhere zero. $\qquad\square$

Proposition 10.4. *Let $E : \mathbb{R} \to \mathbb{C}$ be an exponential function. If $E(x)$ is not identically zero, then*

$$E^*(-x) = E(x)$$

for all $x \in \mathbb{R}$.

Proof. Let $E : \mathbb{R} \to \mathbb{C}$ be exponential. Next, letting $y = -x$ in (10.12), we get

$$E(0) = E(x) E(-x). \tag{10.15}$$

Since $E(x)$ is not identically zero, by Proposition 10.2, $E(0) = 1$ and (10.15) yields

$$E(-x) = \frac{1}{E(x)}.$$

That is

$$E(-x) = E(x)^{-1}$$

or

$$E(-x) = E^*(x) \tag{10.16}$$

for all $x \in \mathbb{R}$. Hence replacing x by $-x$ in (10.16), we obtain

$$E^*(-x) = E(x) \tag{10.17}$$

and the proof of the proposition is complete. $\qquad\square$

Proposition 10.5. *Let $E : \mathbb{R} \to \mathbb{C}$ be an exponential function. Suppose $E(x)$ is not identically zero. Then*

$$E^*(x + y) = E^*(x)E^*(y) \tag{10.18}$$

for all $x, y \in \mathbb{R}$.

Proof. Since $E(x)$ is not identically zero, $E(x)$ is never zero on $\mathbb{R}$ by Proposition 10.3. Now we consider

$$E^*(x + y) = \frac{1}{E(x + y)}$$

$$= \frac{1}{E(x)\, E(y)} = E(x)^{-1}\, E(y)^{-1} = E^*(x)\, E^*(y).$$

Hence

$$E^*(x + y) = E^*(x)\, E^*(y)$$

for all $x, y \in \mathbb{R}$. $\qquad\square$

Now we prove some elementary properties of the d'Alembert functional equation.

Proposition 10.6. *Every nonzero solution $f : \mathbb{R} \to \mathbb{C}$ of the d'Alembert equation*

$$f(x + y) + f(x - y) = 2f(x)f(y) \tag{DE}$$

is an even function.

Proof. Replacing y by $-y$ in the above equation (DE), we have

$$f(x + y) + f(x - y) = 2f(x)f(-y). \tag{10.19}$$

Subtracting (10.19) from (DE), we obtain

$$f(y) = f(-y)$$

for all $y \in \mathbb{R}$. Hence f is an even function. □

Next we proceed to determine the nontrivial general solution of the functional equation (DE) following Kannappan (1968a).

Theorem 10.2. *Every nontrivial solution $f : \mathbb{R} \to \mathbb{C}$ of the functional equation*

$$f(x + y) + f(x - y) = 2f(x)f(y) \tag{DE}$$

is of the form

$$f(x) = \frac{E(x) + E^*(x)}{2}, \tag{10.20}$$

where $E : \mathbb{R} \to \mathbb{C}^*$ (the set of nonzero complex numbers) is an exponential function.

Proof. Let f be a nontrivial solution of (DE); that is, f is not an identically zero function. Letting $x = 0 = y$ in (DE), we obtain $f(0)[1 - f(0)] = 0$. Hence either $f(0) = 0$ or $f(0) = 1$. Since $f(x)$ is not identically zero,

$$f(0) = 1. \tag{10.21}$$

Letting $y = x$ in (DE), we get

$$f(2x) + f(0) = 2 f(x)^2$$

or

$$f(2x) = 2f(x)^2 - 1 \tag{10.22}$$

by (10.21).

Replacing x by $x + y$ and y by $x - y$ in (DE), we get

$$f(x + y + x - y) + f(x + y - x + y) = 2f(x + y)f(x - y).$$

Hence

$$f(2x) + f(2y) = 2f(x + y)f(x - y) \tag{10.23}$$

for all $x, y \in \mathbb{R}$. Next, we compute

$$[f(x + y) - f(x - y)]^2 = [f(x + y) + f(x - y)]^2 - 4f(x + y)f(x - y)$$

$$= [2f(x)f(y)]^2 - 4f(x+y)f(x-y)$$
$$= 4f(x)^2 f(y)^2 - 2[f(2x) + f(2y)]$$
$$= 4f(x)^2 f(y)^2 - 2[2f(x)^2 - 1 + 2f(y)^2 - 1]$$
$$= 4f(x)^2 f(y)^2 - 4f(x)^2 - 4f(y)^2 + 4$$
$$= 4[f(x)^2 - 1][f(y)^2 - 1].$$

Therefore

$$f(x+y) - f(x-y) = \pm 2\sqrt{[f(x)^2 - 1][f(y)^2 - 1]}.$$

Adding this to (DE), we get

$$f(x+y) = f(x)f(y) \pm \sqrt{[f(x)^2 - 1][f(y)^2 - 1]}.$$

Hence

$$[f(x+y) - f(x)f(y)]^2 = [f(x)^2 - 1][f(y)^2 - 1]. \qquad (10.24)$$

Now we consider two cases based on whether (1) $f(x) \in \{1, -1\}$ for all $x \in \mathbb{R}$ or (2) $f(x) \notin \{1, -1\}$ for some $x \in \mathbb{R}$.

Case 1. Suppose $f(x) \in \{1, -1\}$. Hence by (10.24), we get

$$f(x+y) = f(x)f(y) \qquad (10.25)$$

for all $x, y \in \mathbb{R}$. Since $f(x)$ is either 1 or -1, we have

$$f^*(x) = f(x)$$

and hence

$$f(x) = \frac{f(x) + f^*(x)}{2}$$

is a solution of (DE). Note that

$$f(x) = \frac{E(x) + E^*(x)}{2}$$

with $E(x) \in \{1, -1\}$.

Case 2. Suppose $f(x) \notin \{1, -1\}$ for some x. Hence

$$f(x_0)^2 \neq 1$$

for some $x_0 \in \mathbb{R}$. Let $\alpha = f(x_0)$. Hence $\alpha^2 - 1 \neq 0$. Let us call

$$\beta^2 = \alpha^2 - 1. \qquad (10.26)$$

Now we define

$$E(x) = f(x) + \frac{1}{\beta}[f(x + x_0) - f(x)f(x_0)]$$

$$= \frac{1}{\beta}[f(x + x_0) + (\beta - \alpha)f(x)] \tag{10.27}$$

for all $x \in \mathbb{R}$.

Clearly E is well defined. To see this, let $x_1 = x_2$ and consider

$$E(x_1) = \frac{1}{\beta}[f(x_1 + x_0) + (\beta - \alpha)f(x_1)]$$

$$= \frac{1}{\beta}[f(x_2 + x_0) + (\beta - \alpha)f(x_2)]$$

$$= E(x_2).$$

Hence E is well defined. Next, we compute

$$[E(x) - f(x)]^2 = \frac{1}{\beta^2}[f(x + x_0) - f(x)f(x_0)]^2$$

$$= \frac{1}{\beta^2}[f(x)^2 - 1][f(x_0)^2 - 1] \quad \text{(by (10.24))}$$

$$= \frac{\alpha^2 - 1}{\beta^2}[f(x)^2 - 1]$$

$$= f(x)^2 - 1, \tag{10.28}$$

since $\beta^2 = \alpha^2 - 1$. Hence (10.28) yields

$$E(x)^2 - 2E(x)f(x) + f(x)^2 = f(x)^2 - 1$$

which is

$$E(x)^2 - 2E(x)f(x) + 1 = 0.$$

$E(x) = 0$ leads to the contradiction $1 = 0$.

Therefore, $E(x) \neq 0$, and then

$$f(x) = \frac{E(x)^2 + 1}{2E(x)}$$

$$= \frac{E(x) + E^*(x)}{2}.$$

Next we show that $E(x)$ satisfies

$$E(x + y) = E(x)\,E(y).$$

To show this we need the following:

$$2[f(x_0 + x)f(y) + f(x_0 + y)f(x)]$$
$$= f(x_0 + x + y) + f(x_0 + x - y) + f(x_0 + y + x)$$
$$\quad + f(x_0 + y - x) \quad \text{(by (DE))}$$
$$= 2f(x_0 + x + y) + f(x_0 + x - y) + f(x_0 + y - x)$$
$$= 2f(x_0 + x + y) + f(x_0 + (x - y)) + f(x_0 - (x - y))$$
$$= 2f(x_0 + x + y) + 2f(x_0)f(x - y)$$
$$= 2[f(x_0 + x + y) + f(x_0)\{2f(x)f(y) - f(x + y)\}]$$
$$= 2[f(x_0 + x + y) + \alpha\{2f(x)f(y) - f(x + y)\}] \qquad (10.29)$$

and

$$2f(x_0 + x)f(x_0 + y)$$
$$= f(x_0 + x + x_0 + y) + f(x_0 + x - x_0 - y) \quad \text{(by (DE))}$$
$$= f(x_0 + (x_0 + x + y)) + f(x - y)$$
$$= [2f(x_0)f(x_0 + x + y) - f(x_0 + x + y - x_0)]$$
$$\quad + [2f(x)f(y) - f(x + y)] \quad \text{(by (DE))}$$
$$= [2f(x_0)f(x_0 + x + y) - f(x + y)] + [2f(x)f(y) - f(x + y)]$$
$$= 2[f(x)f(y) + \alpha f(x_0 + x + y) - f(x + y)]. \qquad (10.30)$$

Next, we consider

$$E(x)\,E(y)$$
$$= \frac{1}{\beta^2}[f(x + x_0) + (\beta - \alpha)f(x)][f(y + x_0) + (\beta - \alpha)f(y)]$$
$$= \frac{1}{\beta^2}[f(x + x_0)f(y + x_0) + (\beta - \alpha)\{f(x)f(x_0 + y)$$
$$\quad + f(y)f(x_0 + x)\} + (\beta - \alpha)^2 f(x)f(y)]$$
$$= \frac{1}{\beta^2}[f(x)f(y) + \alpha f(x_0 + x + y) - f(x + y)$$
$$\quad + (\beta - \alpha)\{f(x_0 + x + y) + 2\alpha f(x)f(y) - \alpha f(x + y)\}$$
$$\quad + (\beta - \alpha)^2 f(x)f(y)] \quad \text{(by (10.30) and (10.29))}$$
$$= \frac{1}{\beta^2}[\{(\beta - \alpha)^2 + 2\alpha(\beta - \alpha) + 1\}f(x)f(y) + \beta f(x_0 + x + y)$$
$$\quad - \{1 + (\beta - \alpha)\alpha\}f(x + y)]$$
$$= \frac{1}{\beta^2}[(\beta^2 - \alpha^2 + 1)f(x)f(y) + \beta f(x_0 + x + y) - (\beta\alpha - \beta^2)f(x + y)]$$
$$= \frac{1}{\beta^2}[\beta f(x_0 + x + y) + \beta(\beta - \alpha)f(x + y)]$$

$$= \frac{1}{\beta}[f(x_0 + x + y) + (\beta - \alpha)f(x + y)]$$

$$= E(x + y).$$

Hence $E : \mathbb{R} \to \mathbb{C}^\star$ is an exponential function. This completes the "only if" part.

The "if" part can be shown by direct verification. Consider

$$
\begin{aligned}
f(x + y) + &f(x - y) \\
&= \frac{E(x + y) + E^*(x + y)}{2} + \frac{E(x - y) + E^*(x - y)}{2} \\
&= \frac{E(x)E(y) + E^*(x)E^*(y) + E(x)E(-y) + E^*(x)E^*(-y)}{2} \\
&= \frac{E(x)E(y) + E^*(x)E^*(y) + E(x)E^*(y) + E^*(x)E(y)}{2} \\
&= \frac{E(x)[E(y) + E^*(y)] + E^*(x)[E^*(y) + E(y)]}{2} \\
&= \frac{[E(x) + E^*(x)][E(y) + E^*(y)]}{2} \\
&= 2f(x)f(y).
\end{aligned}
$$

This completes the proof. □

Remark 10.2. *In Theorem 10.2, the function $E : \mathbb{R} \to \mathbb{C}^\star$ is a homomorphism from the additive group of reals, $\mathbb{R}$, to the multiplicative group of nonzero complex numbers, $\mathbb{C}^\star$.*

The last theorem can be generalized to the following theorem which was originally proved by Kannappan (1968a).

Theorem 10.3. *Let G be an arbitrary abelian group and $\mathbb{C}^\star$ be the multiplicative group of nonzero complex numbers. Then every nontrivial solution $f : G \to \mathbb{C}$ of the functional equation* (DE), *that is,*

$$f(x + y) + f(x - y) = 2f(x)f(y),$$

is of the form

$$f(x) = \frac{g(x) + g^*(x)}{2},$$

where $g : G \to \mathbb{C}^\star$ is a homomorphism of the group G into $\mathbb{C}^\star$.

Proof. The proof is almost identical to the proof of the last theorem. □

10.4 A Charcterization of Cosine Functions

In this section, we characterize the cosine function through a functional equation. The following result is adapted from Van Vleck (1910).

Theorem 10.4. *Let $\alpha \in \mathbb{R}$. The continuous function $f : \mathbb{R} \to \mathbb{R}$ satisfies the functional equation*

$$f(x - y + \alpha) - f(x + y + \alpha) = 2f(x)\,f(y) \qquad (10.31)$$

for all $x, y \in \mathbb{R}$, if and only if f is given by

$$f(x) = 0 \qquad \forall\, x \in \mathbb{R} \qquad (10.32)$$

or

$$f(x) = \cos\left(\frac{\pi}{2\alpha}(x - \alpha)\right) \qquad \forall\, x \in \mathbb{R}. \qquad (10.33)$$

Proof. It is easy to check that $f = 0$ is a solution of (10.31). From now on we assume that f is not identically zero.

Replacing y with $-y$ in (10.31), we obtain

$$f(x + y + \alpha) - f(x - y + \alpha) = 2f(x)\,f(-y) \qquad (10.34)$$

for all $x, y \in \mathbb{R}$. Comparing (10.31) and (10.34), we get

$$f(x)\,f(y) = -f(x)\,f(-y)$$

for all $x, y \in \mathbb{R}$. Hence f is an odd function, that is,

$$f(-y) = -f(y) \qquad (10.35)$$

for $y \in \mathbb{R}$. Interchanging x with y in (10.31), we get

$$f(y - x + \alpha) - f(x + y + \alpha) = 2f(y)\,f(x). \qquad (10.36)$$

From (10.31) and (10.36), we see that

$$\begin{aligned}
f(x - y + \alpha) &= f(y - x + \alpha) \\
&= f(-(x - y) + \alpha) \\
&= -f(x - y - \alpha) \qquad \text{(since f is odd)}
\end{aligned} \qquad (10.37)$$

for all $x, y \in \mathbb{R}$. Letting $y = 0$ in (10.37), we get

$$f(x + \alpha) = -f(x - \alpha) \qquad \forall\, x \in \mathbb{R}. \qquad (10.38)$$

Hence

$$f(x + 2\alpha) = -f(x) \qquad \forall\, x \in \mathbb{R} \qquad\qquad (10.39)$$

and

$$f(x + 3\alpha) = -f(x + \alpha) \qquad \forall\, x \in \mathbb{R}. \qquad\qquad (10.40)$$

Thus f is a periodic function of period 4α, that is,

$$f(x + 4\alpha) = f(x) \qquad\qquad (10.41)$$

for all $x \in \mathbb{R}$. Without loss of generality we assume that $\alpha > 0$. Next, we replace x by $x + \alpha$ and y by $y + \alpha$ in (10.31) to obtain

$$f(x - y + \alpha) - f(x + y + 3\alpha) = 2f(x + \alpha)\, f(y + \alpha) \qquad\qquad (10.42)$$

for all $x, y \in \mathbb{R}$. Using (10.40) in (10.42), we get

$$f(x - y + \alpha) + f(x + y + \alpha) = 2f(x + \alpha)\, f(y + \alpha) \qquad\qquad (10.43)$$

for all $x, y \in \mathbb{R}$. Hence, we have

$$g(x + y) + g(x - y) = 2g(x)\, g(y), \qquad\qquad (10.44)$$

where

$$g(x) = f(x + \alpha). \qquad\qquad (10.45)$$

Since f is continuous, so also is g. Hence g is given by $g(x) = 0$, $g(x) = 1$, $g(x) = \cosh(ax)$ or $g(x) = \cos(ax)$, where a is a constant.

If $g(x) = 0$, then $f(x) = 0$ but this is not the case since f is nonzero. If $g(x) = 1$, then $f(x) = 1$. However letting this into (10.31), we see that $f(x) = 1$ is not a solution.

Since f is periodic of period 4α, so also is g. Hence $g(x) = \cosh(ax)$ does not yield a solution of (10.31).

If $g(x) = \cos(ax)$, then $f(x) = \cos(\, a(x - \alpha)\,)$. Hence $f(x + 4\alpha) = f(x)$ implies

$$\cos(a\,(x + 3\alpha)\,) = \cos(a\,(x - \alpha)\,)$$

that is

$$\cos(ax + 4a\alpha)\,) = \cos(ax)$$

for all $x \in \mathbb{R}$. Therefore

$$4a\alpha = 2\pi.$$

Thus

$$f(x) = \cos\left(\frac{\pi}{2\,\alpha}(x - \alpha)\right)$$

and the proof of the theorem is complete. □

Thus every nontrivial continuous function that satisfies the functional equation (10.31) is a cosine function.

10.5   Concluding Remarks

The d'Alembert functional equation $f(x+y) + f(x-y) = 2f(x)f(y)$ has been studied extensively and has a long history going back to d'Alembert (1769), Poisson (1804) and Picard Picard (1922, 1928). As the name suggests this functional equation was introduced by d'Alembert in connection with the composition of forces. First Cauchy (1821) gave the continuous solution of this equation. Kaczmarz (1924) extended Cauchy's result by showing that the result of Cauchy regarding d'Alembert functional equation still holds if the condition of continuity is replaced by measurability, and his argument covers the case in which f takes complex values. Flett (1963) proved that if $f : \mathbb{C} \to \mathbb{C}$ satisfies the d'Alembert functional equation for all $x, y \in \mathbb{C}$ and f is continuous at a point, then f has one of the following form

$$f = 0, \qquad f = 1 \qquad \text{or} \qquad f(x+iy) = \cos h(\alpha x + \beta y)$$

for all $z = x + iy \in \mathbb{C}$, where α, β are complex constants not both zero.

On an arbitrary group, the d'Alembert functional equation can be written as $f(xy) + f(xy^{-1}) = 2f(x)f(y)$ where $f : G \to \mathbb{F}$. The d'Alembert functional equation was studied by Kannappan (1968a) when G is a group and $\mathbb{F}$ is the field of complex numbers $\mathbb{C}$. He proved that any non-zero solution of d'Alembert functional equation which satisfies the condition $f(xyz) = f(xzy)$ for all $x, y, z \in G$ has the form

$$f(x) = \frac{g(x) + g(x)^{-1}}{2}, \tag{10.46}$$

where g is a homomorphism of G into the multiplicative group of $\mathbb{C}$. Kannappan (1968a) asked the question whether all the solutions of d'Alembert equation on arbitrary group have the form (10.46). Studies of d'Alembert functional equation on non-abelian groups have produced solutions that are not of the form (10.46). Interested readers should refer to Penney and Rukhin (1979) and Stetkaer (1994). Corovei (1977) showed that if G is a nilpotent group or a generalized nilpotent group, then every nonzero solution of the d'Alembert equation is of the form (10.46) provided all elements of G have odd orders. Corovei (1999) also studied d'Alembert equation on metabelian groups.

Sinopoulos (2000) determined the general solution $f : S \to \mathbb{F}$ of the d'Alembert equation $f(x+y) + f(x + \sigma y) = 2f(x)f(y)$, where S is a commutative semigroup, $\mathbb{F}$ is a quadratically closed commutative field of characteristic different from 2, and σ is an endomorphism of S with $\sigma(\sigma(x)) = x$. We present his result next.

Theorem 10.5. *Let $(S, +)$ be a commutative semigroup, and let $\mathbb{F}$ be a quadratically closed field of characteristically different from 2. Suppose σ is an endomorphism of S such that $\sigma(\sigma x) = x$ for all $x \in S$. Then, the general solution $g : S \to \mathbb{F}$ of the functional equation*

$$g(x + y) + g(x + \sigma y) = 2g(x)\, g(y) \tag{10.47}$$

for all $x, y \in S$ is given by

$$g(x) = \frac{\chi(x) + \chi(\sigma x)}{2}, \qquad \forall\, x \in S, \tag{10.48}$$

where $\chi : S \to \mathbb{F}$ is an exponential map.

Proof. It is easy to check that the function $g(x)$ given by (10.48) satisfies the functional equation (10.47). Hence we prove that $g(x)$ is the only solution of (10.47).

It is easy to check that $g = 0$ or $g = 1$ is a solution of (10.47). Hence from now on we assume that g is a non-constant function.

Replacing y by σy in (10.47), we obtain

$$g(x + \sigma y) + g(x + y) = 2g(x)\, g(\sigma y) \tag{10.49}$$

for all $x, y \in S$. Comparing (10.47) and (10.49), we have

$$g(y) = g(\sigma y) \quad \text{and} \quad g(x + \sigma y) = g(\sigma x + y). \tag{10.50}$$

Now we consider two cases: (1) $g(x + y) = g(x + \sigma y)$ for all $x, y \in \mathbb{R}$ and (2) $g(x + y) \neq g(x + \sigma y)$ for all $x, y \in S$.

Case 1. Suppose $g(x+y) = g(x+\sigma y)$ for all $x, y \in S$. Then from (10.47) we have

$$g(x + y) = g(x)\, g(y) \tag{10.51}$$

for all $x, y \in S$. Hence $g(x) = \chi(x)$, where $\chi : S \to \mathbb{F}$ is an exponential function. Since $g(x) = g(\sigma x)$, therefore we have

$$g(x) = \frac{\chi(x) + \chi(\sigma x)}{2}, \qquad \forall\, x \in S.$$

Case 2. Next, we suppose $g(x + y) \neq g(x + \sigma y)$ for all $x, y \in S$. Hence there exist $x_o, y_o \in S$ such that

$$g(x_o + y_o) - g(x_o + \sigma y_o) \neq 0. \tag{10.52}$$

Define a function $f : S \to \mathbb{F}$ by

$$f(x) = g(x + y_o) - g(x + \sigma y_o). \qquad (10.53)$$

By (10.52), we see that $f(x_o) \neq 0$. Also by (10.50), we get

$$
\begin{aligned}
f(\sigma x) &= g(x + y_o) - g(x + \sigma y_o) \\
&= g(x + \sigma y_o) - g(x + y_o) \qquad \text{by (10.50)} \\
&= -f(x) \qquad \text{by (10.53).}
\end{aligned}
$$

Hence, we have

$$f(x) = -f(\sigma x) \qquad\qquad f(x + \sigma y) = -f(\sigma x + y). \qquad (10.54)$$

Further using (10.53), we obtain

$$
\begin{aligned}
f(x + y) &= g(x + y + y_o) - g(x + y + \sigma y_o), \\
f(x + \sigma y) &= g(x + \sigma y + y_o) - g(x + \sigma y + \sigma y_o).
\end{aligned}
$$

Adding the last two equations and using (10.47), we obtain

$$f(x + y) + f(x + \sigma y) = 2\, g(x + y_o)\, g(y) - 2\, g(x + \sigma y_o)\, g(y). \qquad (10.55)$$

Hence using (10.53) in (10.55), we see that

$$f(x + y) + f(x + \sigma y) = 2\, f(x)\, g(y). \qquad (10.56)$$

Interchanging x and y in (10.56), we have

$$f(y + x) + f(y + \sigma x) = 2\, f(y)\, g(x). \qquad (10.57)$$

Adding (10.56) and (10.57) and using (10.50), we obtain

$$f(x + y) = f(x)\, g(y) + f(y)\, g(x) \qquad (10.58)$$

for all $x, y \in S$. Now we solve the functional equation (10.58). Using (10.58), we compute $f(x + t + y)$ first as $f((x + t) + y)$ and then as $f(x + (t + y))$:

$$
\begin{aligned}
f(x + t + y) &= f(x + t)\, g(y) + f(y)\, g(x + t) \\
&= [\, f(x)\, g(t) + f(t)\, g(x)\,]\, g(y) + f(y)\, g(x + t), \\
f(x + t + y) &= f(x)\, g(t + y) + f(t + y)\, g(x) \\
&= f(x)\, g(t + y) + [\, f(t)\, g(y) + f(y)\, g(t)\,]\, g(x).
\end{aligned}
$$

Comparing the results, we find

$$[\, g(x + t) - g(x)\, g(t)\,]\, f(y) = [\, g(t + y) - g(t)\, g(y)\,]\, f(x). \qquad (10.59)$$

Setting $y = y_o$ and $h(t) := f(x_o)^{-1}[g(t + x_o) - g(t) g(x_o)]$, we have

$$g(x + t) - g(x) g(t) = h(t) f(x). \tag{10.60}$$

Since the left side of this equation is symmetric in x, t, so is the right side. Hence we have $h(t) f(x) = h(x) f(t)$. Setting $x = x_o$, we have $h(t) = f(x_o)^{-1} h(x_o) f(t)$. Since $\mathbb{F}$ is quadratically closed, there exists an $\alpha \in \mathbb{F}$ such that $\alpha^2 = f(x_o)^{-1} h(x_o)$. Hence the equation (10.60), with y in place of t, becomes

$$g(x + y) = g(x) g(y) + \alpha^2 f(x) h(y). \tag{10.61}$$

Multiplying (10.57) by α and adding it to and subtracting it from (10.61), we obtain respectively

$$g(x + y) + \alpha f(x + y) = [g(x) + \alpha f(x)] [g(y) + \alpha f(y)]$$

and

$$g(x + y) - \alpha f(x + y) = [g(x) - \alpha f(x)] [g(y) - \alpha f(y)].$$

So the functions $\chi_1 := g + \alpha f$ and $\chi_2 := g - \alpha f$ are exponential, and by addition we get

$$g(x) = \frac{\chi_1(x) + \chi_2(x)}{2}. \tag{10.62}$$

Now in view of (10.50), we have

$$\chi_1(\sigma x) = g(\sigma x) + \alpha f(\sigma x) = g(x) - \alpha f(x) = \chi_2(x).$$

So (10.62) leads to the asserted solution (10.48) (after renaming the function χ_1 as χ) and this completes the proof of the theorem. $\qquad\square$

Remark 10.3. *The general solution* $f : \mathbb{R} \to \mathbb{C}$ *of the functional equation* (10.58), *that is,* $f(x + y) = f(x) g(y) + f(y) g(x)$, *will be treated in Theorem 11.3 of Chapter 11. However, here we have included a proof for the convenience of the reader.*

A generalization of the d'Alembert functional equation is the quadratic-trigonometric functional equation

$$f_1(xy) + f_2(xy^{-1}) = f_3(x) + f_4(y) + f_5(x) f_6(y).$$

The general solution of this functional equation along with

$$f_i(txy) = f_i(tyx), \quad (i = 1, 2)$$

was determined by Chung, Ebanks, Ng and Sahoo (1995) when the variables x, y are in a group and values are in the additive field of complex numbers.

10.6 Exercises

1. Find all functions $f, g : \mathbb{R} \to \mathbb{R}$ that satisfy the functional equation

$$f(x + y) + f(x - y) = 2f(x)\, g(y)$$

for all $x, y \in \mathbb{R}$.

2. Find all functions $f, g : \mathbb{R} \to \mathbb{R}$ that satisfy the functional equation

$$g(x + y) + g(x - y) = 2f(x)\, f(y)$$

for all $x, y \in \mathbb{R}$.

3. Let α be a nonzero real constant. Find all continuous functions $f, g : \mathbb{R} \to \mathbb{R}$ that satisfy the functional equation

$$f(x + y + \alpha) - f(x - y + \alpha) = 2f(x)\, g(y)$$

for all $x, y \in \mathbb{R}$.

4. Let λ be a nonzero real constant. Find all functions $f, g : \mathbb{R} \to \mathbb{R}$ that satisfy the functional equation

$$f(x + y) + g(x - y) = \lambda\, f(x)\, g(y)$$

for all $x, y \in \mathbb{R}$.

5. Let λ be a nonzero real constant. Find all functions $f, g : \mathbb{R} \to \mathbb{R}$ that satisfy the functional equation

$$f(x + y) + g(x - y) = \lambda\, g(x)\, f(y)$$

for all $x, y \in \mathbb{R}$.

6. Find all functions $f, g, h : \mathbb{R} \to \mathbb{R}$ that satisfy the functional equation

$$f(x + y) + f(x - y) = 2f(x) + g(x)\, h(y)$$

for all $x, y \in \mathbb{R}$.

7. Find all functions $f, g, h, k : \mathbb{R} \to \mathbb{R}$ that satisfy the functional equation

$$f(x + y) + g(x - y) = h(x)\, k(y)$$

for all $x, y \in \mathbb{R}$.

8. Let α be a nonzero real constant. Determine all continuous functions $f, g : \mathbb{R} \to \mathbb{R}$ that satisfy the functional equation

$$f(x + y + \alpha) - g(x - y + \alpha) = 2f(x)\,f(y)$$

for all $x, y \in \mathbb{R}$.

9. Find all functions $f : \mathbb{R} \to \mathbb{R}$ that satisfy $f(0) \geq 0$ and the functional equation

$$f(x+y+z) + f(x+y-z) + f(y+z-x) + f(z+x-y) = 4f(x)\,f(y)\,f(z)$$

for all $x, y \in \mathbb{R}$.

10. Find all functions $f : \mathbb{R} \to \mathbb{R}$ that satisfy the functional equation

$$f(x+y) + f(x-y) = f(x)f(y) + f(x)f(-y)$$

for all $x, y \in \mathbb{R}$.

11. Find all functions $f, g : \mathbb{R} \to \mathbb{R}$ that satisfy the functional equation

$$f(x+y) + f(x-y) = f(x)g(y) + f(x)g(-y)$$

for all $x, y \in \mathbb{R}$.

12. Find all functions $f : \mathbb{R}^2 \to \mathbb{R}$ that satisfy the functional equation

$$f(x+y, u+v) + f(x+y, u-v) = 2f(x, u)\,f(y, v)$$

for all $x, y, u, v \in \mathbb{R}$.

Chapter 11

Trigonometric Functional Equations

11.1 Introduction

The trigonometric functions $f(x) = \cos x$ and $g(x) = \sin x$ satisfy the functional equation

$$f(x - y) = f(x)f(y) + g(x)g(y) \tag{11.1}$$

for all $x, y \in \mathbb{R}$. Hence the functional equation (11.1) is called a trigonometric functional equation.

The functions $f(x) = \cos x$ and $g(x) = \sin x$ also satisfy other functional equations such as

$$g(x + y) = g(x)f(y) + f(x)g(y), \tag{11.2}$$
$$f(x + y) = f(x)f(y) - g(x)g(y), \tag{11.3}$$
$$g(x - y) = g(x)f(y) - g(y)f(x) \tag{11.4}$$

for all $x, y \in \mathbb{R}$.

The primary goals of this chapter is to present some important results concerning these trigonometric functional equations and some other related functional equations. In Section 2, we first present the solution of the cosine-sine functional equation (11.1). Then in Section 3, we consider the sine-cosine functional equation (11.2) and determine its general solutions without any regularity assumptions on the unknown complex-valued functions f, g defined on $\mathbb{R}$. Section 4 deals with the sine functional equation $f(x + y)f(x - y) = f^2(x) - f^2(y)$ and its generalization $f(x + y)g(x - y) = f(x)g(x) - f(y)g(y)$. In Section 5, we treat a functional inequality that is associated with the sine function and in Section 6, we consider an elementary problem of Steven Butler (2003) related to the functional equation $f(x + y) = f(x)f(y) - d\sin(x)\sin(y)$, where d is an appropriate real constant, and provide a solution due to M.Th. Rassias (2004).

11.2   Solution of a Cosine-Sine Functional Equation

In this section, we find the general solution of the functional equation (11.1). We find the solution of (11.1) following the ideas due to Gerretsen (1939).

Note that if $f(x) = c$, a constant, then (11.1) yields

$$c = c^2 + g(x)g(y) \tag{11.5}$$

for all $x, y \in \mathbb{R}$. If g is a zero function, then $c = 0$ or $c = 1$. If $g(x)$ is a nonzero function, then g is a constant function equal to $k(c - c^2)$ with $k^2(c - c^2) = 1$. This is the trivial system of solutions of (11.1).

Hence, from now on, we assume that f is nonconstant.

Theorem 11.1. *If a function $f : \mathbb{R} \to \mathbb{C}$ satisfies with some $g : \mathbb{R} \to \mathbb{C}$ the functional equation*

$$f(x - y) = f(x)f(y) + g(x)g(y) \tag{11.1}$$

and is not constant, then it satisfies also the d'Alembert equation (DE)

$$f(x + y) + f(x - y) = 2f(x)f(y). \tag{DE}$$

Further f and g are given by

$$f(x) = \frac{E(x) + E^*(x)}{2}, \quad g(x) = k\frac{E(x) - E^*(x)}{2}, \tag{11.6}$$

where $E : \mathbb{R} \to \mathbb{C}^*$ (the set of nonzero complex numbers) is exponential, with $k^2 = -1$.

Proof. Interchanging x and y in (11.1), we get

$$f(y - x) = f(y)f(x) + g(y)g(x). \tag{11.7}$$

Comparing (11.1) and (11.7), we obtain

$$f(x - y) = f(y - x) \tag{11.8}$$

for all $x, y \in \mathbb{R}$. Letting $y = 0$ in (11.8), we see that

$$f(x) = f(-x) \tag{11.9}$$

for all $x \in \mathbb{R}$. Hence f is an even function on $\mathbb{R}$. Now we substitute $-x$ in place of x and $-y$ in place of y in (11.1) to get

$$f(y - x) = f(-x)f(-y) + g(-x)g(-y). \tag{11.10}$$

Since f is even, (11.10) reduces to

$$f(x - y) = f(x)f(y) + g(-x)g(-y). \tag{11.11}$$

Hence, comparing (11.1) and (11.11), we see that

$$g(x)g(y) = g(-x)g(-y) \tag{11.12}$$

for all $x, y \in \mathbb{R}$. Suppose g is constant, say c. Then (11.1) gives

$$f(x - y) = f(x)f(y) + c^2.$$

Replacing y by $-y$ and using the fact that f is an even function, we see that $f(x+y) = f(x-y)$. That is, f is constant, which is not the case. So g is nonconstant. Since g is nonconstant, choose y_0 such that $g(y_0) \neq 0$.

Then from (11.12) with $y = y_0$, we get

$$g(x) = c\,g(-x), \tag{11.13}$$

where c is a nonzero constant. Then $g(x) = c^2 g(x)$. Since g is nonconstant, $c^2 = 1$, that is, $c = \pm 1$. We claim that $c = -1$. Suppose not. Then $c = 1$ and from (11.13), we get

$$g(x) = g(-x). \tag{11.14}$$

That is, g is even. Hence replacing y by $-y$ in (11.1), we get

$$\begin{aligned} f(x + y) &= f(x)f(-y) + g(x)g(-y) \\ &= f(x)f(y) + g(x)g(y) \\ &= f(x - y). \end{aligned}$$

That is,

$$f(x + y) = f(x - y)$$

for all $x, y \in \mathbb{R}$. Then

$$f(u) = f(v) \tag{11.15}$$

for all $u, v \in \mathbb{R}$. Thus (11.15) implies

$$f(u) = c \quad \forall\, u \in \mathbb{R},$$

where c is a constant. This is a contradiction since f is nonconstant. Therefore $c = -1$ and

$$g(x) = -g(-x) \tag{11.16}$$

for $x \in \mathbb{R}$. That is, g is an odd function.

Finally, replacing y by $-y$ in (11.1), we get

$$f(x+y) = f(x)f(-y) + g(x)g(-y)$$
$$= f(x)f(y) - g(x)g(y)$$

by using f even and g odd. Therefore

$$f(x+y) = f(x)f(y) - g(x)g(y). \qquad (11.17)$$

Adding (11.1) and (11.17), we get

$$f(x+y) + f(x-y) = 2f(x)f(y) \qquad \text{(DE)}$$

for all $x, y \in \mathbb{R}$. That is, f satisfies (DE).

Now by Theorem 10.3, the function f has the form

$$f(x) = \frac{E(x) + E^*(x)}{2} \qquad (11.18)$$

for $x \in \mathbb{R}$, where E is nonzero exponential. Substituting this value of f in (11.1), we get

$$-\frac{E(x) - E^*(x)}{2} \cdot \frac{E(y) - E^*(y)}{2} = g(x)g(y).$$

From this it is easy to obtain (using g nonconstant)

$$g(x) = k\frac{E(x) - E^*(x)}{2}$$

with $k^2 = -1$. Thus (11.6) holds.

This completes the proof of this theorem. $\qquad \square$

Remark 11.1. *The statement of the above theorem also implies that* $f(x)^2 + g(x)^2 = 1$ *which can be obtained as follows. Since g is an odd function*

$$g(0) = 0. \qquad (11.19)$$

Letting $y = 0$ in (11.1), we see that

$$f(x) = f(x)f(0) + g(x)g(0)$$

which yields (by (11.19))

$$f(x) = f(x)f(0). \qquad (11.20)$$

Since f is not identically zero, we get

$$f(0) = 1. \qquad (11.21)$$

Letting $y = x$ in (11.1) and then using (11.21), we get

$$f(0) = f(x)^2 + g(x)^2$$

which is

$$f(x)^2 + g(x)^2 = 1. \tag{11.22}$$

In the following theorem we present the continuous solution of the functional equation (11.1).

Theorem 11.2. *The continuous solution of the functional equation*

$$f(x - y) = f(x)f(y) + g(x)g(y) \tag{11.1}$$

is given by

$$\left. \begin{array}{l} f(x) = c \\ g(x) = \sqrt{c(1 - c)} \end{array} \right\} \tag{11.23}$$

$$\left. \begin{array}{l} f(x) = c \\ g(x) = -\sqrt{c(1 - c)} \end{array} \right\} \tag{11.24}$$

$$\left. \begin{array}{l} f(x) = \cos(\alpha x) \\ g(x) = \pm \sin(\alpha x), \end{array} \right\} \tag{11.25}$$

where α and c are arbitrary constants.

Proof. We obtain (11.23) and (11.24) because of our discussion prior to Theorem 11.1. Since f is continuous and satisfies the equation

$$f(x + y) + f(x - y) = 2f(x)f(y),$$

therefore (nonconstant) solutions of f are given by

$$f(x) = \cos(\alpha x) \tag{11.26}$$

and

$$f(x) = \cos h(\beta x). \tag{11.27}$$

If $f(x) = \cos(\alpha x)$, then by (11.22) we get

$$g(x)^2 = 1 - \cos^2(\alpha x)$$

or

$$g(x) = \pm \sin(\alpha x), \tag{11.28}$$

which can also be obtained from (11.6). Hence (11.26) and (11.28) yield the solution (11.25).

If $f(x) = \cos h(\beta x)$, then we observe that

$$|f(x)| = |\cos h(\beta x)| \geq 1.$$

For $x \neq 0$, $|f(x)| > 1$ since $\beta \neq 0$; otherwise f is a constant function. However, by (11.22), we know that

$$|f(x)| \leq 1$$

for all $x \in \mathbb{R}$. Hence $f(x) = \cos h(\beta x)$ is not a solution of (11.1).

This completes the proof of this theorem.   □

Remark 11.2. *Using (11.6) and the continuity of E, we get (11.25).*

11.3   Solution of a Sine-Cosine Functional Equation

In this section, we consider a sine-cosine functional equation, namely, the equation

$$f(x + y) = f(x)\, g(y) + f(y)\, g(x) \tag{11.29}$$

for all $x, y \in \mathbb{R}$. We determine the general solution of this equation without any regularity condition on the unknown functions f and g.

Theorem 11.3. *Let $f, g : \mathbb{R} \to \mathbb{C}$ satisfy the functional equation (11.29) for all $x, y \in \mathbb{R}$. Then f and g are of the form*

$$\left.\begin{array}{l} f(x) = 0 \\ g \text{ arbitrary}; \end{array}\right\} \tag{11.30}$$

$$\left.\begin{array}{l} f(x) = A(x)\, E(x) \\[6pt] g(x) = E(x); \end{array}\right\} \tag{11.31}$$

$$\left.\begin{array}{l} f(x) = \dfrac{E_1(x) - E_2(x)}{2\,\alpha} \\[6pt] g(x) = \dfrac{E_1(x) + E_2(x)}{2}, \end{array}\right\} \tag{11.32}$$

where $E_1, E_2 : \mathbb{R} \to \mathbb{C}^\star$ are exponential functions, $A : \mathbb{R} \to \mathbb{C}$ is an additive function, and α is a nonzero complex constant.

Proof. It is easy to check that $f = 0$ and g arbitrary are solutions of the functional equation (11.29). Hence from now on we assume that $f \neq 0$.

Next we compute

$$f(x + y + z) = f(x) g(y + z) + f(y + z) g(x)$$
$$= f(x) g(y + z) + f(y) g(z) g(x) + f(z) g(y) g(x) \quad (11.33)$$

and also

$$f(x + y + z) = f(x + y) g(z) + f(z) g(x + y)$$
$$= f(x) g(y) g(z) + f(y) g(x) g(z) + f(z) g(x + y). \quad (11.34)$$

From (11.33) and (11.34), we obtain

$$f(x) [g(y + z) - g(y)g(z)] = f(z) [g(x + y) - g(x)g(y)]. \quad (11.35)$$

Since $f \neq 0$, there exists a $z_0 \in \mathbb{R}$ such that $f(z_0) \neq 0$. Letting $z = z_0$ in (11.35), we see that

$$g(x + y) - g(x)g(y) = f(x) \ell(y), \quad (11.36)$$

where

$$\ell(y) = \frac{g(y + z_0) - g(y)g(z_0)}{f(z_0)}. \quad (11.37)$$

Interchanging x with y in (11.36), we obtain

$$g(x + y) - g(x)g(y) = f(y) \ell(x). \quad (11.38)$$

Thus from (11.36) and (11.38), we have

$$f(x) \ell(y) = f(y) \ell(x) \quad (11.39)$$

for all $x, y \in \mathbb{R}$. Since $f \neq 0$, (11.39) yields

$$\ell(x) = \alpha^2 f(x), \quad (11.40)$$

where α is some constant. Inserting (11.40) into (11.36), we get

$$g(x + y) = g(x) g(y) + \alpha^2 f(x) f(y) \quad (11.41)$$

for all $x, y \in \mathbb{R}$.

Now we consider two cases.

Case 1. Suppose $\alpha = 0$. Then (11.41) yields

$$g(x + y) = g(x) g(y) \quad (11.42)$$

for all $x, y \in \mathbb{R}$. Hence

$$g(x) = E(x), \quad (11.43)$$

where $E : \mathbb{R} \to \mathbb{C}$ is an exponential function. Since $f \neq 0$, we have $E \neq 0$. Since every exponential function that is not identically zero is nowhere zero, we have

$$f(x + y) = f(x)\,E(y) + f(y)\,E(x)$$

which is

$$\frac{f(x + y)}{E(x + y)} = \frac{f(x)}{E(x)} + \frac{f(y)}{E(y)}. \tag{11.44}$$

Defining

$$A(x) = \frac{f(x)}{E(x)},$$

we have

$$A(x + y) = A(x) + A(y).$$

Hence $A : \mathbb{R} \to \mathbb{C}$ is an additive function. In this case the solution of (11.29) is given by

$$\left. \begin{aligned} f(x) &= A(x)\,E(x) \\ g(x) &= E(x). \end{aligned} \right\}$$

Case 2. Suppose $\alpha \neq 0$. Multiplying the equation (11.29) by α and then adding and subtracting the resulting equation to and from (11.41), we have

$$g(x + y) + \alpha\, f(x + y) = [g(x) + \alpha\, f(x)]\,[g(y) + \alpha\, f(y)] \tag{11.45}$$

and

$$g(x + y) - \alpha\, f(x + y) = [g(x) - \alpha\, f(x)]\,[g(y) - \alpha\, f(y)], \tag{11.46}$$

respectively. Therefore

$$\left. \begin{aligned} g(x) + \alpha\, f(x) &= E_1(x) \\ g(x) - \alpha\, f(x) &= E_2(x), \end{aligned} \right\} \tag{11.47}$$

where $E_1, E_2 : \mathbb{R} \to \mathbb{C}$ are exponential functions. Hence by adding and subtracting the above expressions, we have the asserted solution

$$\left. \begin{aligned} f(x) &= \frac{E_1(x) - E_2(x)}{2\,\alpha} \\ g(x) &= \frac{E_1(x) + E_2(x)}{2}. \end{aligned} \right\}$$

Now the proof of the theorem is complete. $\square$

Remark 11.3. *If the function* $f : \mathbb{R} \to \mathbb{C}$ *is an odd function and the function* $g : \mathbb{R} \to \mathbb{C}$ *is an even function, then*

$$E_1^*(x) = E_1(-x) = g(-x) + \alpha\, f(-x) = g(x) - \alpha\, f(x) = E_2(x).$$

Hence the solution (11.32) *can be expressed as*

$$f(x) = \frac{E_1(x) - E_1^*(x)}{2\,\alpha} \quad \text{and} \quad g(x) = \frac{E_1(x) + E_1^*(x)}{2}.$$

11.4   Solution of a Sine Functional Equation

In this section, first we consider a sine functional equation

$$f(x + y) f(x - y) = f(x)^2 - f(y)^2 \tag{11.48}$$

for all $x, y \in \mathbb{R}$. We prove the following theorem.

Theorem 11.4. *The function* $f : \mathbb{R} \to \mathbb{C}$ *satisfies the functional equation* (11.48) *for all* $x, y \in \mathbb{R}$ *if and only if* f *is of the form*

$$\left. \begin{array}{l} f(x) = 0 \\ f(x) = A(x) \\ f(x) = \frac{E(x) - E^*(x)}{2\,\alpha}, \end{array} \right\} \tag{11.49}$$

where $E : \mathbb{R} \to \mathbb{C}^*$ *is an exponential function,* $A : \mathbb{R} \to \mathbb{C}$ *is a nonzero additive function, and* α *is a nonzero constant.*

Proof. It is easy to check that $f = 0$ is a solution of (11.48). Hence, from now, on we assume that $f \neq 0$. Therefore there exists a $x_0 \in \mathbb{R}$ such that $f(x_0) \neq 0$. Define

$$\phi(x) = \frac{f(x + x_0) - f(x - x_0)}{2\, f(x_0)} \tag{11.50}$$

for all $x \in \mathbb{R}$. Then from (11.50) and (11.48), we have

$$2\phi(x)\,\phi(y)$$
$$= \frac{1}{2\, f(x_0)^2}\, [f(x + x_0) - f(x - x_0)]\,[f(y + x_0) - f(y - x_0)]$$
$$= \frac{1}{2\, f(x_0)^2}\, [f(x + x_0)f(y + x_0) - f(x - x_0)f(y + x_0)$$
$$\quad - f(x + x_0)f(y - x_0) + f(x - x_0)f(y - x_0)]$$

$$= \frac{1}{2\,f(x_0)^2} \left[f\left(\frac{x+y}{2} + x_0 + \frac{x-y}{2}\right) f\left(\frac{x+y}{2} + x_0 - \frac{x-y}{2}\right) \right.$$

$$- f\left(\frac{x+y}{2} + \frac{x-y}{2} - x_0\right) f\left(\frac{x+y}{2} - \frac{x-y}{2} + x_0\right)$$

$$- f\left(\frac{x+y}{2} + \frac{x-y}{2} + x_0\right) f\left(\frac{x+y}{2} - \frac{x-y}{2} - x_0\right)$$

$$\left. + f\left(\frac{x+y}{2} - x_0 + \frac{x-y}{2}\right) f\left(\frac{x+y}{2} - x_0 - \frac{x-y}{2}\right) \right]$$

$$= \frac{1}{2\,f(x_0)^2} \left[f\left(\frac{x+y}{2} + x_0\right)^2 - f\left(\frac{x-y}{2}\right)^2 - f\left(\frac{x+y}{2}\right)^2 \right.$$

$$+ f\left(\frac{x-y}{2} - x_0\right)^2 - f\left(\frac{x+y}{2}\right)^2 + f\left(\frac{x-y}{2} + x_0\right)^2$$

$$\left. + f\left(\frac{x+y}{2} - x_0\right)^2 - f\left(\frac{x-y}{2}\right)^2 \right]$$

$$= \frac{1}{2\,f(x_0)^2} \left[f\left(\frac{x+y}{2} + x_0\right)^2 - f\left(\frac{x+y}{2}\right)^2 + f\left(\frac{x-y}{2} + x_0\right)^2 \right.$$

$$- f\left(\frac{x-y}{2}\right)^2 - f\left(\frac{x+y}{2}\right)^2 + f\left(\frac{x+y}{2} - x_0\right)^2$$

$$\left. - f\left(\frac{x-y}{2}\right)^2 + f\left(\frac{x-y}{2} - x_0\right)^2 \right]$$

$$= \frac{1}{2\,f(x_0)^2} \left[f(x+y+x_0)\,f(x_0) + f(x-y+x_0)\,f(x_0) \right.$$

$$\left. - f(x+y-x_0)\,f(x_0) - f(x-y-x_0)\,f(x_0) \right]$$

$$= \frac{1}{2\,f(x_0)} \left[f(x+y-x_0) + f(x-y+x_0) \right.$$

$$\left. + f(x-y+x_0) - f(x-y-x_0) \right]$$

$$= \phi(x+y) + \phi(x-y).$$

Thus the function $\phi : \mathbb{R} \to \mathbb{C}$ satisfies the d'Alembert equation

$$\phi(x+y) + \phi(x-y) = 2\phi(x)\,\phi(y)$$

for all $x, y \in \mathbb{R}$. The general solution of this equation can be obtained from Theorem 10.2 as

$$\phi(x) = \frac{E(x) + E^*(x)}{2}, \tag{11.51}$$

where $E : \mathbb{R} \to \mathbb{C}^\star$ is an exponential function. The functional equation

(11.48) can be rewritten as

$$f(u)\, f(v) = f\left(\frac{u+v}{2}\right)^2 - f\left(\frac{u-v}{2}\right)^2 \qquad (11.52)$$

for all $u, v \in \mathbb{R}$. If $f(y) \neq 0$ but y otherwise arbitrary, then

$$
\begin{aligned}
&\frac{f(x+y) - f(x-y)}{2\, f(y)} \\
&= \frac{f(x+y)f(a) - f(x-y)f(a)}{2\, f(y)\, f(a)} \\
&= \frac{f\left(\frac{x+y+a}{2}\right)^2 - f\left(\frac{x+y-a}{2}\right)^2 - f\left(\frac{x-y+a}{2}\right)^2 + f\left(\frac{x-y-a}{2}\right)^2}{2\, f(y)\, f(a)} \\
&= \frac{f\left(\frac{x+y+a}{2}\right)^2 - f\left(\frac{x+a-y}{2}\right)^2 - f\left(\frac{x-a+y}{2}\right)^2 + f\left(\frac{x-a-y}{2}\right)^2}{2\, f(y)\, f(a)} \\
&= \frac{f(x+a)\, f(y) - f(x-a)\, f(y)}{2\, f(y)\, f(a)} \\
&= \frac{f(x+a) - f(x-a)}{2\, f(a)} \\
&= \phi(x).
\end{aligned}
$$

Hence we obtain

$$f(x+y) - f(x-y) = 2\, f(y)\, \phi(x). \qquad (11.53)$$

By (11.51) it is easy to see that

$$\phi(0) = 1. \qquad (11.54)$$

Letting $x = 0$ in (11.53) and using (11.54), we obtain

$$f(-y) = -f(y). \qquad (11.55)$$

Thus (11.53) yields

$$f(y+x) + f(y-x) = 2\, f(y)\, \phi(x). \qquad (11.56)$$

Interchanging x with y in (11.56), we see that

$$f(x+y) + f(x-y) = 2\, f(x)\, \phi(y). \qquad (11.57)$$

Adding (11.56) and (11.57) and then using (11.55), we obtain

$$f(x+y) = f(x)\, \phi(y) + f(y)\, \phi(x) \qquad (11.58)$$

for all $x, y \in \mathbb{R}$. Since f is odd and ϕ is even, from Theorem 11.3 and the remark following Theorem 11.3, we have the asserted solution

$$f(x) = A(x) \tag{11.59}$$

and

$$f(x) = \frac{E(x) - E^\star(x)}{2\,\alpha}, \tag{11.60}$$

where $E : \mathbb{R} \to \mathbb{C}^\star$ is an exponential function, $A : \mathbb{R} \to \mathbb{C}$ is an additive function, and α is a nonzero constant. This completes the proof of the theorem. $\square$

In the next theorem, we present the general solution of a functional equation which is a generalization of the functional equation (11.48).

Theorem 11.5. *The functions $f, g : \mathbb{R} \to \mathbb{C}$ satisfy the functional equation*

$$f(x + y)\, g(x - y) = f(x)\, g(x) - f(y)\, g(y) \tag{11.61}$$

for all $x, y \in \mathbb{R}$ if and only if

$$f(x) = 0 \quad \text{and} \quad g(x) \text{ arbitrary;} \tag{11.62}$$

$$f(x) \text{ arbitrary} \quad \text{and} \quad g(x) = 0; \tag{11.63}$$

$$f(x) = k \quad \text{and} \quad g(x) = A(x); \tag{11.64}$$

$$f(x) = A(x) + \delta \quad \text{and} \quad g(x) = \beta\, A(x); \tag{11.65}$$

$$\left.\begin{array}{l} f(x) = \gamma\, \frac{E(x) - E^\star(x)}{2} + \delta\, \frac{E(x) + E^\star(x)}{2} \\[2mm] g(x) = \beta\, \frac{E(x) - E^\star(x)}{2}, \end{array}\right\} \tag{11.66}$$

where $A : \mathbb{R} \to \mathbb{C}$ is an additive function, $E : \mathbb{R} \to \mathbb{C}^\star$ is an exponential function, β, δ, k are nonzero arbitrary complex constants and γ is an arbitrary complex constant.

Proof. It is easy to check that the function f and g in (11.62)–(11.66) satisfy the functional equation (11.61). Hence we prove that the functions f and g given in (11.62)–(11.66) are the only solutions of (11.61).

Suppose f is a constant function, say, k. Then (11.61) yields

$$k\, [g(x - y) - g(x) + g(y)] = 0 \tag{11.67}$$

for all $x, y \in \mathbb{R}$.

If $k = 0$, the g can be arbitrary and

$$\left.\begin{array}{l} f(x) = 0 \\ g(x) \text{ arbitrary} \end{array}\right\}$$

is a solution of (11.61) which is (11.62).

If $k \neq 0$, then (11.67) yields

$$g(x - y) = g(x) - g(y) \tag{11.68}$$

for all $x, y \in \mathbb{R}$. Letting $y = x$ in the last equation, we obtain $g(0) = 0$. Further, letting $x = 0$ in (11.68), we see that

$$g(-y) = -g(y) \tag{11.69}$$

for all $y \in \mathbb{R}$. Thus for nonzero k, the function g is an odd function. Interchanging y with $-y$ in (11.68), we get

$$g(x + y) = g(x) - g(-y)$$

and using (11.69), we have

$$g(x + y) = g(x) + g(y). \tag{11.70}$$

Hence $g(x) = A(x)$, where $A : \mathbb{R} \to \mathbb{C}$ is an additive function; that is, A satisfies

$$A(x + y) = A(x) + A(y) \tag{11.71}$$

for all $x, y \in \mathbb{R}$. Thus we have the solution

$$\left. \begin{array}{l} f(x) = k \\ g(x) = A(x) \end{array} \right\}$$

which is (11.64).

Next we suppose g is a constant function, say, k. Then (11.61) yields

$$k\,[f(x + y) - f(x) + f(y)] = 0 \tag{11.72}$$

for all $x, y \in \mathbb{R}$.

If $k = 0$, then f can be arbitrary and

$$\left. \begin{array}{l} f(x) \text{ arbitrary} \\ g(x) = 0 \end{array} \right\}$$

is a solution of (11.61) which is (11.63).

If $k \neq 0$, then (11.72) yields

$$f(x + y) = f(x) - f(y) \tag{11.73}$$

for all $x, y \in \mathbb{R}$. Letting $x = 0$ and $y = 0$ in (11.73), we see that $f(0) = 0$. Further, letting $x = 0$ in (11.73), we have $f(y) = 0$ for all $y \in \mathbb{R}$. Hence

$$\left. \begin{array}{l} f(x) = 0 \\ g(x) = k \end{array} \right\}$$

is a solution of (11.61) and it is included in the solution (11.62).

From now on we assume that both f and g are non-constant functions. Interchanging x with y in (11.61), we obtain

$$f(x + y) \, g(y - x) = f(y) \, g(y) - f(x) \, g(x) \tag{11.74}$$

for all $x, y \in \mathbb{R}$. Comparing (11.74) with (11.61), we obtain

$$f(x + y) \, g(x - y) = -f(x + y) \, g(y - x) \tag{11.75}$$

for all $x, y \in \mathbb{R}$. Letting $u = x + y$ and $v = x - y$, we have

$$f(u) \, g(v) = -f(u) \, g(-v) \tag{11.76}$$

for all $u, v \in \mathbb{R}$. Since f is non-constant, there exists a $u_o \in \mathbb{R}$ such that $f(u_o) \neq 0$. Hence letting $u = u_o$ in (11.76), we obtain

$$g(-v) = -g(v) \tag{11.77}$$

for all $v \in \mathbb{R}$. Therefore g is an odd function.

Define

$$\psi(x) = \frac{f(x) + f(-x)}{2} \tag{11.78}$$

and

$$\phi(x) = \frac{f(x) - f(-x)}{2}. \tag{11.79}$$

Then ψ is an even function and ϕ is an odd function. Moreover

$$f(x) = \psi(x) + \phi(x) \tag{11.80}$$

holds for all $x \in \mathbb{R}$.

Using (11.80) in (11.61), we obtain

$$\psi(x + y) \, g(x - y) + \phi(x + y) \, g(x - y)$$
$$= \psi(x) \, g(x) - \psi(y) \, g(y) + \phi(x) \, g(x) - \phi(y) \, g(y) \tag{11.81}$$

for all $x, y \in \mathbb{R}$. Replacing x by $-x$ and y by $-y$ and using the fact that ψ is even, and ϕ and g are odd, we get

$$-\psi(x + y) \, g(x - y) + \phi(x + y) \, g(x - y)$$

$$= -\psi(x)\,g(x) + \psi(y)\,g(y) + \phi(x)\,g(x) - \phi(y)\,g(y). \tag{11.82}$$

Adding (11.81) and (11.82), we have

$$\phi(x+y)\,g(x-y) = \phi(x)\,g(x) - \phi(y)\,g(y) \tag{11.83}$$

for all $x, y \in \mathbb{R}$. From (11.81) and (11.83), we also get

$$\psi(x+y)\,g(x-y) = \psi(x)\,g(x) - \psi(y)\,g(y) \tag{11.84}$$

for all $x, y \in \mathbb{R}$.

Replace y with $-y$ in (11.83), we get

$$\phi(x-y)\,g(x+y) = \phi(x)\,g(x) - \phi(y)\,g(y) \tag{11.85}$$

for all $x, y \in \mathbb{R}$. Comparing (11.83) and (11.85), we see that

$$\phi(x+y)\,g(x-y) = \phi(x-y)\,g(x+y) \tag{11.86}$$

for all $x, y \in \mathbb{R}$. Letting $u = x + y$ and $v = x - y$ in (11.86), we have

$$\phi(u)\,g(v) = \phi(v)\,g(u) \tag{11.87}$$

for all $u, v \in \mathbb{R}$. Since g is a non-constant, we obtain

$$\phi(x) = \alpha\,g(x), \tag{11.88}$$

where α is a real constant.

Case 1. Suppose $\alpha \neq 0$. Then (11.88) in (11.83) yields

$$\phi(x+y)\,\phi(x-y) = \phi(x)^2 - \phi(y)^2 \tag{11.89}$$

for all $x, y \in \mathbb{R}$. Since g is non-constant so also is ϕ. Hence the non-constant solution of (11.89) is given by

$$\phi(x) = A(x) \tag{11.90}$$

or

$$\phi(x) = \frac{E(x) - E^\star(x)}{2b}. \tag{11.91}$$

Therefore from (11.88), we get

$$g(x) = \frac{1}{\alpha}\,A(x), \tag{11.92}$$

$$g(x) = \frac{E(x) - E^\star(x)}{2\,b\,\alpha}. \tag{11.93}$$

Letting $y = -x$ in (11.84) and using the fact that g is odd, and ψ is even, we get

$$\psi(0)\, g(2x) = 2\, \psi(x)\, g(x) \tag{11.94}$$

for all $x \in \mathbb{R}$.

Subcase 1.1. Suppose $\psi(0) = 0$. Then (11.94) yields

$$\psi(x)\, g(x) = 0 \tag{11.95}$$

for all $x \in \mathbb{R}$. Using (11.95) in (11.84), we obtain

$$\psi(x + y)\, g(x - y) = 0 \tag{11.96}$$

for all $x, y \in \mathbb{R}$. Therefore either $\psi = 0$ or $g = 0$.

The situation $g = 0$ is not possible since g is non-constant. Since f is non-constant and g is non-constant with $\phi = \alpha\, g$, from (11.80) we see that ψ is also non-constant. Therefore $\psi = 0$ is not possible. Hence $\psi(0) = 0$ is not the case.

Subcase 1.2. Suppose $\psi(0) \neq 0$. From (11.94), we get

$$g(2x) = \frac{2}{\delta}\, \psi(x)\, g(x), \tag{11.97}$$

where $\delta := \psi(0)$. Using (11.92) in (11.97), we get

$$\psi(x) = \delta. \tag{11.98}$$

Therefore we have the solution

$$f(x) = \phi(x) + \psi(x) = A(x) + \delta$$

and

$$g(x) = \frac{1}{\alpha}\, A(x) = \beta\, A(x),$$

where β is a nonzero constant. Thus we have

$$\left.\begin{array}{l} f(x) = A(x) + \delta \\ g(x) = \beta\, A(x) \end{array}\right\}$$

which is the asserted solution (11.65). Here δ and β are arbitrary nonzero complex constants.

Next using (11.93) and (11.97), we get

$$\frac{1}{\alpha\, b}\, \frac{E(2x) - E^\star(2x)}{2} = \frac{2}{\delta}\, \psi(x)\, \frac{1}{\alpha\, b}\, \frac{E(x) - E^\star(x)}{2}$$

which is

$$E(x) + E^\star(x) = \frac{2}{\delta}\,\psi(x).$$

Hence

$$\psi(x) = \delta\,\frac{E(x) + E^\star(x)}{2}. \tag{11.99}$$

Therefore

$$f(x) = \phi(x) + \psi(x)$$
$$= \frac{E(x) - E^\star(x)}{2b} + \delta\,\frac{E(x) + E^\star(x)}{2}$$

$$= \gamma\,\frac{E(x) - E^\star(x)}{2} + \delta\,\frac{E(x) + E^\star(x)}{2},$$

where $\gamma = \frac{1}{b}$ and

$$g(x) = \frac{1}{\delta\,b}\,\frac{E(x) - E^\star(x)}{2} = \beta\,\frac{E(x) - E^\star(x)}{2},$$

where $\beta = \frac{1}{b\delta}$ is a nonzero complex constant. Therefore we have

$$\left.\begin{array}{l} f(x) = \gamma\,\frac{E(x) - E^\star(x)}{2} + \delta\,\frac{E(x) + E^\star(x)}{2} \\[2mm] g(x) = \beta\,\frac{E(x) - E^\star(x)}{2}, \end{array}\right\}$$

which is the solution (11.66).

Case 2. Finally we suppose $\alpha = 0$. Then from (11.88) we have $\phi = 0$. Therefore $f = \psi$. Hence we arrive at the equation (11.84); that is,

$$\psi(x + y)\,g(x - y) = \psi(x)\,g(x) - \psi(y)\,g(y)$$

for all $x, y \in \mathbb{R}$. Using $y = -x$, we obtain as before

$$\psi(0)\,g(2x) = 2\,\psi(x)\,g(x)$$

for all $x \in \mathbb{R}$. As before $\psi(0) \neq 0$. Therefore

$$g(2x) = \frac{2}{\delta}\,\psi(x)\,g(x),$$

where $\delta := \psi(0)$.

Next we compute

$g(x + y) \, g(x - y)$

$$= \frac{4}{\delta^2} \left[\psi \left(\frac{x+y}{2} \right) g \left(\frac{x+y}{2} \right) \psi \left(\frac{x-y}{2} \right) g \left(\frac{x-y}{2} \right) \right]$$

$$= \frac{4}{\delta^2} \left[\psi \left(\frac{x}{2} \right) g \left(\frac{x}{2} \right) - \psi \left(\frac{y}{2} \right) g \left(\frac{y}{2} \right) \right] \left[\psi \left(\frac{x}{2} \right) g \left(\frac{x}{2} \right) + \psi \left(\frac{y}{2} \right) g \left(\frac{y}{2} \right) \right]$$

$$= \frac{4}{\delta^2} \left[\psi \left(\frac{x}{2} \right)^2 g \left(\frac{x}{2} \right)^2 - \psi \left(\frac{y}{2} \right)^2 g \left(\frac{y}{2} \right)^2 \right]$$

$$= g(x)^2 - g(y)^2$$

for all $x, y \in \mathbb{R}$. Hence g satisfies

$$g(x + y) \, g(x - y) = g(x)^2 - g(y)^2 \tag{11.100}$$

for all $x, y \in \mathbb{R}$. Therefore g is of the form

$$g(x) = \frac{1}{\alpha} A(x) \tag{11.101}$$

or

$$g(x) = \frac{1}{\alpha b} \frac{E(x) - E^\star(x)}{2}, \tag{11.102}$$

where $A : \mathbb{R} \to \mathbb{C}$ is additive, $E : \mathbb{R} \to \mathbb{C}^\star$ is exponential, and α, b are nonzero complex constants.

Using (11.97) and (11.101)–(11.102), we get

$$\psi(x) = \delta \tag{11.103}$$

or

$$\psi(x) = \delta \frac{E(x) + E^\star(x)}{2}. \tag{11.104}$$

Hence

$$\left. \begin{array}{l} f(x) = \delta \, \frac{E(x) + E^\star(x)}{2} \\[2mm] g(x) = \beta \, \frac{E(x) - E^\star(x)}{2} \end{array} \right\}$$

is the solution of (11.61) which is included in (11.66) with $\gamma = 0$.

Since no more cases are left, the proof of the theorem is complete. $\quad\square$

11.5   Solution of a Sine Functional Inequality

In this section, following Segal (1963), we will show that the solution of the functional inequality

$$f(x + y) f(x - y) \leq f(x)^2 - f(y)^2 \tag{11.105}$$

for all $x, y \in \mathbb{R}$ is same as the solution of the sine functional equation (11.48), that is,

$$f(x + y) f(x - y) = f(x)^2 - f(y)^2$$

Theorem 11.6. *If* $f : \mathbb{R} \to \mathbb{R}$ *satisfies the functional inequality* (11.105) *for all* $x, y \in \mathbb{R}$, *then* f *is a solution of the sine functional equation* (11.48).

Proof. Putting $x = y = 0$ in (11.105), we obtain $f(0)^2 \leq 0$. Hence

$$f(0) = 0. \tag{11.106}$$

Letting $x = -y$ in (11.105), we see that

$$0 \leq f(-y)^2 - f(y)^2.$$

Hence

$$f(y)^2 \leq f(-y)^2. \tag{11.107}$$

Changing y to $-y$, we get

$$f(-y)^2 \leq f(y)^2 \leq f(-y)^2. \tag{11.108}$$

Therefore

$$f(y)^2 = f(-y)^2. \tag{11.109}$$

Hence for each $y \in \mathbb{R}$, we have either

$$f(-y) = -f(y)$$

or

$$f(-y) = f(y).$$

Suppose for some $y_o \in \mathbb{R}$,

$$f(y_0) = f(-y_0). \tag{11.110}$$

Then letting $x = 0$ and $y = y_0$ in (11.105), we obtain

$$f(y_0)f(-y_0) \leq -f(y_0)^2. \tag{11.111}$$

Using (11.110) and (11.111), we get

$$2f(y_0)^2 \le 0.$$

Hence

$$f(y_0) = 0 \tag{11.112}$$

and also

$$f(-y_0) = 0.$$

Therefore, we have

$$f(-y) = -f(y) \tag{11.113}$$

for all $y \in \mathbb{R}$. From (11.105) and (11.113), we have

$$\begin{aligned}
f(y)^2 &\le f(x)^2 - f(x+y)\,f(x-y) \\
&= f(x)^2 + f(x+y)\,f(y-x) \\
&\le f(x)^2 + f(y)^2 - f(x)^2 \\
&= f(y)^2.
\end{aligned}$$

Thus, we have

$$f(x+y)f(x-y) = f(x)^2 - f(y)^2$$

and the proof of the theorem is now complete. $\qquad\square$

Remark 11.4. *The continuous solution of the sine functional equation* (11.48) *is given by*

$$\left.\begin{aligned}
f(x) &= k_1\,x \\
f(x) &= k_2 \sin(k_3 x) \\
f(x) &= k_4 \sinh(k_5 x),
\end{aligned}\right\}$$

where k_i $(i = 1, 2, ..., 5)$ are arbitrary real constants.

11.6   An Elementary Functional Equation

In 2003, Butler asked for the solution of the following problem: Prove that for $d < -1$, there are exactly two solutions $f : \mathbb{R} \to \mathbb{R}$ of the functional equation

$$f(x+y) = f(x)\,f(y) - d\,\sin(x)\,\sin(y). \tag{11.114}$$

Michael Th. Rassias (2004), then a high school student from Athens (Greece), solved this problem. The functional equation (11.114) is a special case of the following functional equation:

$$f(x + y) = f(x) f(y) + \lambda g(x) g(y). \tag{11.115}$$

The solution of this functional equation can be obtained similarly to the functional equation (11.1).

By replacing y with $-y$ in (11.114), we get

$$f(x - y) = f(x) f(-y) + d \sin(x) \sin(y) \tag{11.116}$$

for all $x, y \in \mathbb{R}$. When (11.114) is added to (11.116), we have the following interesting functional equation:

$$f(x + y) + f(x - y) = f(x) [f(y) + f(-y)] \tag{11.117}$$

for all $x, y \in \mathbb{R}$. If f is an even function, then (11.117) reduces to the well-known d' Alembert functional equation that we studied in the previous chapter.

The general solution (11.117) can be obtained from the works of Chung, Ebanks, Ng and Sahoo (1995). Note that the trivial solutions of (11.117) are $f = 0$ and $f = 1$. Using Theorem 3.1 in Chung et. al., we have the following theorem.

Theorem 11.7. *The nontrivial solution $f : \mathbb{R} \to \mathbb{C}$ of the functional equation (11.117) for all $x, y \in \mathbb{R}$ is given by*

$$f(x) = \begin{cases} a E(x) + b E^\star(x) & \text{if } E(x) \not\equiv E^\star(x) \\ E(x) [A(x) + 1] & \text{if } E(x) \equiv E^\star(x), \end{cases} \tag{11.118}$$

where $A : \mathbb{R} \to \mathbb{C}$ is an additive function, $E : \mathbb{R} \to \mathbb{C}$ is a nonzero exponential function, and a, b are arbitrary complex constants such that $a + b = 1$.

Proof. The proof is left to the reader as an exercise. □

If $f : \mathbb{R} \to \mathbb{R}$ is a continuous or measurable nontrivial solution of the functional equation (11.117) holding for all $x, y \in \mathbb{R}$, then f is given by either $f(x) = b \sin(\alpha x) + \cos(\alpha x)$, $f(x) = c \sinh(\beta x) + \cosh(\beta x)$ or $f(x) = dx + 1$, where α, β, b, c, d are arbitrary real constants.

Next, we present the solution provided by Rassias (2004) of the functional equation (11.114).

Theorem 11.8. *Let d be a real number in the interval $(-\infty, -1)$. The function $f : \mathbb{R} \rightarrow \mathbb{R}$ satisfies the functional equation (11.114) for all $x, y \in \mathbb{R}$ if and only if f has one of the following forms*

$$f(x) = a\sin(x) + \cos(x) \tag{11.119}$$

or

$$f(x) = -a\sin(x) + \cos(x), \tag{11.120}$$

where a is a constant given by $a = \sqrt{-d-1}$.

Proof. It is easy to check that the solutions f given in (11.119) and (11.120) satisfy the functional equation (11.114). So we proceed to show that (11.119) and (11.120) are the only solutions of the functional equation (11.114).

Using the functional equation (11.114), we compute $f(x+y+z)$ first as $f((x+y)) + z)$ and then as $f(x + (y + z))$ to get

$$\begin{aligned}
f(x+y+z) &= f(x+y)f(z) - d\sin(x+y)\sin(z) \\
&= f(x)f(y)f(z) - d\sin(x)\sin(y)f(z) \\
&\quad - d\sin(x)\cos(y)\sin(z) - d\cos(x)\sin(y)\sin(z).
\end{aligned}$$

and

$$\begin{aligned}
f(x+y+z) &= f(x)f(y+z) - d\sin(x)\sin(y+z) \\
&= f(x)f(y)f(z) - d\,f(x)\sin(y)\sin(z) \\
&\quad - d\sin(x)\sin(y)\cos(z) - d\sin(x)\cos(y)\sin(z).
\end{aligned}$$

Hence, we have

$$\begin{aligned}
\sin(x)&\sin(y)f(z) + \sin(x)\cos(y)\sin(z) + \cos(x)\sin(y)\sin(z) \\
&= f(x)\sin(y)\sin(z) + \sin(x)\sin(y)\cos(z) + \sin(x)\cos(y)\sin(z)
\end{aligned}$$

for all $x, y, z \in \mathbb{R}$. Letting $y = z = \frac{\pi}{2}$, we obtain

$$f(x) = \alpha\,\sin(x) + \cos(x), \tag{11.121}$$

where $\alpha = f\left(\frac{\pi}{2}\right)$.

Substituting (11.121) into the functional equation (11.114), we see that $\alpha^2 = -d - 1$. Hence

$$\alpha = \sqrt{-d-1} \quad \text{or} \quad \alpha = -\sqrt{-d-1}.$$

Setting a equals to $\sqrt{-d-1}$, we obtain the asserted solutions (11.119) and (11.120). This completes the proof of the theorem. □

11.7 Concluding Remarks

The solution of the functional equation (11.1), that is,

$$f(x - y) = f(x)f(y) + g(x)g(y)$$

for all $x, y \in \mathbb{R}$, was treated by Gerretsen (1939) and Vaughan (1955). The functional equation (11.2), that is,

$$g(x + y) = g(x)f(y) + f(x)g(y)$$

for all $x, y \in \mathbb{R}$, was considered by Vietoris (1944) and also by van der Corput (1941). The general solution of the functional equation (11.2) on abelian groups follows from the works of Vincze (1962b) and Vincze (1963). The most frequent characterization of trigonometric functions is that by the system

$$\left. \begin{array}{l} f(x + y) = f(x)\,f(y) - g(x)\,g(y) \\ g(x + y) = g(x)\,f(y) + f(x)\,g(y). \end{array} \right\} \tag{11.122}$$

Angheluta (1943) solved the system (11.122) under the assumption of continuity of the functions f and g. Ghermanescu (1949) proved that the general real measurable solution of the system (11.122) is given by

$$f(x) = e^{ax} \cos(bx) \quad \text{and} \quad g(x) = e^{ax} \sin(bx),$$

where a, b are arbitrary constants.

The functional equation (11.2) takes the form

$$g(xy) = g(x)f(y) + f(x)g(y)$$

when one replaces the domain of the functions by a group G and the range by a commutative field $\mathbb{F}$. A generalization of this equation is the following:

$$g(xy) = g(x)f(y) + f(x)g(y) + h(x)\,h(y). \tag{11.123}$$

Chung, Kannappan and Ng (1985) determined the general solution $f, g, h : G \to \mathbb{C}$ of the last functional equation when G is an arbitrary group and the $\mathbb{F}$ is the complex field. The following functional equation

$$\frac{g(x + y) + g(x + \sigma y)}{2} = g(x)f(y) + f(x)g(y) + h(x)\,h(y) \tag{11.124}$$

is a generalization of the functional equation (11.123). The general solution $f, g, h \in C(G)$ of this functional equation, when G is an abelian topological group, $\sigma : G \to G$, is a continuous involutive automorphism of order 2, and $C(G)$ is the algebra of continuous, complex-valued functions on G was determined by de Place Friis and Stetkaer (2002). There are six sets of solutions and they are certain exponential polynomials.

The functional equation (11.48) was studied by Carmichael (1909), Rosenbaum and Segal (1960), Kurepa (1960), Segal (1963) and Baker (1970) among others researchers. The functional equation (11.48) on arbitrary groups takes the form

$$f(xy)\, f(xy^{-1}) = f(x)^2 - f(y)^2. \tag{11.125}$$

Kannappan (1968b) found the general solution $f : G \to \mathbb{C}$ of this functional equation when G is a cyclic group. Corovei (1983) showed that if G is a group whose elements are of odd order, $\mathbb{F}$ is a field of characteristic different from 2, and $f : G \to \mathbb{F}$ is a nonzero solution of (11.125), then either f is of the form

$$f(x) = \frac{g(x) - g(x^{-1})}{2b},$$

where $b^2 \in \mathbb{F}^*$ and g is a homomorphism of G into the multiplicative group $\mathbb{F}^*$ of $\mathbb{F}$ or f is a homomorphism from G into the additive group of K. Corovei (2005) also proved the following result. Let G be a group divisible by 2, such that the commutator subgroup G' of G is 2-divisible, and $\mathbb{F}$ be a quadratically closed field with characteristic different from 2. If $f : G \to \mathbb{F}$ is a solution of (11.125), then either f is of the form

$$f(x) = \alpha \frac{g(x) - g(x^{-1})}{2},$$

where α is an arbitrary element of $\mathbb{F}$ and g is a homomorphism of G into the multiplicative group $\mathbb{F}^*$ of $\mathbb{F}$ or f is a homomorphism from G into the additive group of $\mathbb{F}$.

Kannappan (1969) studied a variant of the functional equation (11.48), namely,

$$g(x + y + a)\, g(x - y + a) = g(x)^2 - g(y)^2. \tag{11.126}$$

He proved that if $g : \mathbb{R} \to \mathbb{C}$ satisfies the functional equation (11.126) with $a \neq 0$, then $g(x) = f(x - a)$, where $f : \mathbb{R} \to \mathbb{C}$ is an arbitrary solution of (11.48) with period $2a$.

In the remaining of this section we present a result due to Sahoo (2007).

Theorem 11.9. *Let $(S, +)$ be a commutative semigroup, and let $\mathbb{F}$ be a quadratically closed commutative field of characteristic different from 2. Suppose σ is an endomorphism of S such that $\sigma(\sigma x) = x$ for all $x \in S$. The non-constant functions $f, g : S \to \mathbb{F}$ satisfy the functional equation*

$$f(x + \sigma y) = f(x) f(y) + g(x) g(y) \tag{11.127}$$

for all $x, y \in S$ if and only if f and g have one of the following forms:

$$\left. \begin{array}{l} f(x) = \frac{\chi(x) + \chi(\sigma x)}{2} \\[2mm] g(x) = k \frac{\chi(x) - \chi(\sigma x)}{2} \end{array} \right\} \tag{11.128}$$

or

$$\left. \begin{array}{l} f(x) = a E(x) \\ g(x) = b E(x) \end{array} \right\} \tag{11.129}$$

or

$$\left. \begin{array}{l} f(x) = \psi(x) - \psi(x) A(x) \\ g(x) = k \psi(x) A(x) \end{array} \right\} \tag{11.130}$$

or

$$\left. \begin{array}{l} f(x) = (1 - a) \psi(x) + a \psi(x) \phi(x) \\ g(x) = b \psi(x) - b \psi(x) \phi(x), \end{array} \right\} \tag{11.131}$$

where k, a, b are elements in $\mathbb{F}$ satisfying $k^2 = -1$; $a^2 + b^2 = a$; $A : S \to \mathbb{F}$ is an additive function satisfying $A(\sigma x) = A(x)$; and $\chi, \psi, \phi, E : S \to \mathbb{F}$ are exponential functions such that $\psi(\sigma x) = \psi(x)$, $\phi(\sigma x) = \phi(x)$ and $E(\sigma x) = E(x)$ with $\psi \neq 0$.

Proof. Interchanging x with y in (11.127), we obtain

$$f(y + \sigma x) = f(y) f(x) + g(y) g(x). \tag{11.132}$$

Comparing (11.127) and (11.132) we see that

$$f(x + \sigma y) = f(y + \sigma x) \tag{11.133}$$

for all $x, y \in S$. Hence $f(\sigma x) = f(x)$ for all $x \in S$.

Next letting σx for x and σy for y in (11.127) and using (11.133), we get

$$f(x + \sigma y) = f(x) f(y) + g(\sigma x) g(\sigma y). \tag{11.134}$$

Hence from (11.127) and (11.134), we have

$$g(x) g(y) = g(\sigma x) g(\sigma y) \tag{11.135}$$

for all $x, y \in S$. Since g is non-constant, there exists a $y_o \in S$ such that $g(y_o) \neq 0$. Hence letting $y = y_o$ in (11.135), we obtain

$$g(x) = \alpha \, g(\sigma x), \tag{11.136}$$

where $\alpha := g(y_o)^{-1}g(\sigma y_o)$. Interchanging x with σx in (11.136) and then using (11.136), we see that

$$g(\sigma x) = \alpha\, g(\sigma(\sigma x)) = \alpha\, g(x) = \alpha^2\, g(\sigma x). \tag{11.137}$$

Hence $\alpha^2 = 1$. Therefore $\alpha \in \{-1, 1\}$.

Case 1. Suppose $\alpha = -1$. Then (11.136) yields

$$g(\sigma x) = -\,g(x) \tag{11.138}$$

for all $x \in S$. Interchanging y with σy in (11.127) and using the fact that $f(\sigma y) = f(y)$ and $g(\sigma y) = -\,g(y)$, we have

$$f(x+y) = f(x)\,f(y) - g(x)\,g(y) \tag{11.139}$$

for all $x, y \in S$. Adding (11.127) and (11.139), we obtain

$$f(x+y) + f(x+\sigma y) = 2\,f(x)\,f(y) \tag{11.140}$$

for all $x, y \in S$. The general solution of the functional equation can be obtained from Theorem 10.5 as

$$f(x) = \frac{\chi(x) + \chi(\sigma x)}{2}, \tag{11.141}$$

where $\chi : S \to \mathbb{F}$ is an exponential function. Using (11.141) in (11.137) and simplifying, we obtain

$$g(x)\,g(y) = -\left(\frac{\chi(x) - \chi(\sigma x)}{2}\right)\left(\frac{\chi(y) - \chi(\sigma y)}{2}\right) \tag{11.142}$$

for all $x, y \in S$. Hence

$$g(x) = k\,\frac{\chi(x) - \chi(\sigma x)}{2}, \tag{11.143}$$

where k is an element of $\mathbb{F}$ satisfying $k^2 = -1$. Hence for this case we have the asserted solution (11.128).

Case 2. Suppose $\alpha = 1$. Hence (11.136) yields

$$g(\sigma x) = g(x) \tag{11.144}$$

for all $x \in S$. Letting σy for y in (11.127) and using (11.144) and the fact $f(\sigma y) = f(y)$, we get

$$f(x+y) = f(x)\,f(y) + g(x)\,g(y) \tag{11.145}$$

for all $x, y \in S$. The general solution of the functional equation can be found from Lemma 4 in Chung, Kannappan and Ng (1985). However, for the sake of convenience for the reader, the proof is repeated with some simplifications. Computing $f(x + y + z)$ first as $f((x + y) + z)$ and then as $f(x + (y + z))$, using (11.145), we obtain

$$\begin{aligned} f(x + y + z) &= f(x) \, f(y + z) + g(x) \, g(y + z) \\ &= f(x) \, [f(y) \, f(z) + g(y) \, g(z)] + g(x) \, g(y + z) \\ &= f(x) \, f(y) \, f(z) + f(x) \, g(y) \, g(z) + g(x) \, g(y + z) \end{aligned}$$

and

$$\begin{aligned} f(x + y + z) &= f(x + y) \, f(z) + g(x + y) \, g(z) \\ &= [f(x) \, f(y) + g(x) \, g(y)] \, f(z) + g(x + y) \, g(z) \\ &= f(x) \, f(y) \, f(z) + g(x) \, g(y) \, f(z) + g(x + y) \, g(z). \end{aligned}$$

Comparing the last two expressions, we have

$$g(x + y) \, g(z) - f(x) \, g(y) \, g(z) = g(x) \, g(y + z) - g(x) \, g(y) \, f(z). \quad (11.146)$$

Subtracting $g(x) \, f(y) \, g(z)$ from both sides of (11.146), we get

$$\begin{aligned} [g(x + y) &- f(x) \, g(y) - f(y) \, g(x)] \, g(z) \\ &= [g(y + z) - f(y) \, g(z) - f(z) \, g(y)] \, g(x) \end{aligned} \quad (11.147)$$

for all $x, y, z \in S$. We fix $z = z_o$ with $g(z_o) \neq 0$ and obtain

$$g(x + y) - f(x) \, g(y) - f(y) \, g(x) = g(x) \, k(y), \quad (11.148)$$

where

$$k(y) := g(z_o)^{-1} \, [g(y + z_o) - f(y) \, g(z_o) - f(z_o) \, g(y)]. \quad (11.149)$$

Using (11.148) in (11.145), we see that

$$g(x) \, k(y) \, g(z) = g(y) \, k(z) \, g(x). \quad (11.150)$$

Since $g \neq 0$, it follows that

$$k(y) = 2 \, \alpha \, g(y), \quad (11.151)$$

where α is some element in $\mathbb{F}$. Hence (11.151) in (11.148) yields

$$g(x + y) = f(x) \, g(y) + f(y) \, g(x) + 2 \, \alpha \, g(x) \, g(y) \quad (11.152)$$

for all $x, y \in S$. Multiplying (11.152) by λ and adding the resulting expression to (11.145), we have

$$
\begin{aligned}
f(x + y) &+ \lambda\, g(x + y) \\
&= f(x)\, f(y) + g(x)\, g(y) + \lambda\, f(x)\, g(y) \\
&\qquad + \lambda\, g(x)\, f(y) + 2\,\alpha\,\lambda\, g(x)\, g(y).
\end{aligned}
$$

The equation can be written as

$$
f(x + y) + \lambda\, g(x + y) = [f(x) + \lambda\, g(x)]\,[f(y) + \lambda\, g(y)] \qquad (11.153)
$$

if and only if λ satisfies

$$
\lambda^2 - 2\,\lambda\,\alpha - 1 = 0. \qquad (11.154)
$$

By fixing λ to be such a constant, we get

$$
f(x) + \lambda\, g(x) = \psi(x), \qquad (11.155)
$$

where $\psi : S \to \mathbb{F}$ is an exponential map. From (11.154) it is easy to see that $\lambda \neq 0$. From (11.155) we have

$$
f(x) = \psi(x) - \lambda\, g(x) \qquad (11.156)
$$

and letting this into (11.145) and simplifying, we obtain

$$
\lambda\, g(x + y) = \lambda\, g(x)\, \psi(y) + \lambda\, g(y)\, \psi(x) - (\lambda^2 + 1)\, g(x)\, g(y). \qquad (11.157)
$$

There are two possibilities: (1) $\psi = 0$ and (2) $\psi \neq 0$. If $\psi = 0$, then from (11.157), we get

$$
g(x + y) = -\lambda^{-1}\, (\lambda^2 + 1)\, g(x)\, g(y) \qquad (11.158)
$$

for all $x, y \in S$. Defining $E : S \to \mathbb{F}$ by

$$
E(x) = -\lambda^{-1}\, (\lambda^2 + 1)\, g(x). \qquad (11.159)
$$

Then by (11.159), the equation (11.158) reduces to

$$
E(x + y) = F(x)\, E(y) \qquad (11.160)
$$

for all $x, y \in S$. Therefore $E : S \to \mathbb{F}$ is an exponential map.

Hence from (11.159) and (11.156), we get

$$
g(x) = b\, E(x) \qquad \text{and} \qquad f(x) = a\, E(x), \qquad (11.161)
$$

where $a := \lambda^2\, (\lambda^2 + a)^{-1}$ and $b := -\lambda\, (\lambda^2 + a)^{-1}$ are elements of $\mathbb{F}$. The constants a and b satisfy $a^2 + b^2 = a$. Thus we have the asserted solution (11.129).

The other possibility is $\psi \neq 0$. Since every exponential function that is not identically zero is nowhere zero, dividing (11.157) by

$$\psi(x + y) = \psi(x)\,\psi(y)$$

side by side, we obtain

$$\frac{\lambda\,g(x+y)}{\psi(x+y)} = \frac{\lambda\,g(x)}{\psi(x)} + \frac{\lambda\,g(y)}{\psi(y)} - \frac{(\lambda^2+1)}{\lambda^2}\left(\frac{\lambda\,g(x)}{\psi(x)}\right)\left(\frac{\lambda\,g(y)}{\psi(y)}\right). \quad (11.162)$$

When $\lambda^2 + 1 = 0$, we have

$$\frac{\lambda\,g(x+y)}{\psi(x+y)} = \frac{\lambda\,g(x)}{\psi(x)} + \frac{\lambda\,g(y)}{\psi(y)} \quad (11.163)$$

for all $x, y \in S$. Hence

$$\frac{\lambda\,g(x)}{\psi(x)} = A(x), \quad (11.164)$$

where $A : S \to \mathbb{F}$ is an additive function. Therefore

$$g(x) = \lambda^{-1}\,\psi(x)\,A(x), \quad (11.165)$$

and by (11.156) and (11.165), we get

$$f(x) = \psi(x) - \psi(x)\,A(x). \quad (11.166)$$

Letting $k = \lambda^{-1}$ from (11.165) and (11.166), we have the asserted solution (11.130). It is easy to check that the constant k satisfies $k^2 = -1$ because of $\lambda^2 + 1 = 0$.

When $\lambda^2 + 1 \neq 0$, (11.162) yields

$$\phi(x + y) = \phi(x)\,\phi(y), \quad (11.167)$$

where

$$\phi(x) = 1 - \frac{\lambda^2 + 1}{\lambda^2}\left(\frac{\lambda\,g(x)}{\psi(x)}\right). \quad (11.168)$$

Hence $\phi : S \to \mathbb{F}$ is an exponential function. Therefore

$$g(x) = b\,\psi(x) - b\,\phi(x)\,\psi(x), \quad (11.169)$$

and by (11.156) and (11.169), we get

$$f(x) = (1 - a)\,\psi(x) + a\,\phi(x)\,\psi(x), \quad (11.170)$$

where $b := \frac{\lambda}{\lambda^2+1}$ and $a := \frac{\lambda^2}{\lambda^2+1}$. It is easy to check that the constants satisfy $a^2 + b^2 = a$. Thus we have the asserted solution (11.131).

One can easily verify that $\psi(\sigma x) = \psi(x)$, $\phi(\sigma x) = \phi(x)$ and $E(\sigma x) = E(x)$ and the proof of the theorem is now complete. $\qquad\square$

11.8   Exercises

1. Let α be a nonzero real constant. Find all functions $f : \mathbb{R} \to \mathbb{R}$ that satisfy the functional equation

$$f(x + y + 2\alpha) + f(x - y + 2\alpha) = 2f(x)\,f(y)$$

for all $x, y \in \mathbb{R}$.

2. Let α be a nonzero real constant. Find all functions $f, g : \mathbb{R} \to \mathbb{R}$ that satisfy the functional equation

$$f(x + y + 2\alpha) + g(x - y + 2\alpha) = 2f(x)\,g(y)$$

for all $x, y \in \mathbb{R}$.

3. Let α be a nonzero real constant. Find all functions $f : \mathbb{R} \to \mathbb{C}$ that satisfy the functional equation

$$f(x + y + \alpha)\,f(x - y + \alpha) = f(x)^2 - f(y)^2$$

for all $x, y \in \mathbb{R}$.

4. Let α be a nonzero real constant. Find all functions $f : \mathbb{R} \to \mathbb{C}$ that satisfy the functional equation

$$f(x + y + \alpha)\,f(x - y + \alpha) = f(x)^2 + f(y)^2 - 1$$

for all $x, y \in \mathbb{R}$.

5. Let λ be a nonzero real constant. Find all functions $f, g : \mathbb{R} \to \mathbb{C}$ that satisfy the functional equation

$$f(x + y) - f(x - y) = \lambda\,f(x)\,g(y)$$

for all $x, y \in \mathbb{R}$.

6. Let λ be a nonzero real constant. Find all functions $f, g : \mathbb{R} \to \mathbb{C}$ that satisfy the functional equation

$$f(x + y) - g(x - y) = \lambda\,f(x)\,f(y)$$

for all $x, y \in \mathbb{R}$.

7. Let λ be a nonzero real constant. Find all functions $f, g : \mathbb{R} \to \mathbb{C}$ that satisfy the functional equation

$$f(x + y) - g(x - y) = \lambda\,g(x)\,g(y)$$

for all $x, y \in \mathbb{R}$.

8. Let λ be a nonzero real constant. Find all functions $f, g : \mathbb{R} \to \mathbb{C}$ that satisfy the functional equation

$$f(x+y) - g(x-y) = \lambda f(x) g(y)$$

for all $x, y \in \mathbb{R}$.

9. Find all functions $f, g : \mathbb{R} \to \mathbb{R}$ that satisfy the functional equation

$$g(x) f(x) = f\left(\frac{x+y}{2}\right)^2 - f\left(\frac{x-y}{2}\right)^2$$

for all $x, y \in \mathbb{R}$.

10. Find all functions $f, g : \mathbb{R} \to \mathbb{R}$ that satisfy the functional equation

$$g(x) g(x) = f\left(\frac{x+y}{2}\right)^2 - f\left(\frac{x-y}{2}\right)^2$$

for all $x, y \in \mathbb{R}$.

11. Find all functions $f : \mathbb{R} \to \mathbb{R}$ that satisfy the functional equation

$$f(x) + f(y) = f\left(\frac{x+y}{1-xy}\right)$$

for all $x, y \in \mathbb{R}$.

12. Find all functions $f, g, h : \mathbb{R} \to \mathbb{R}$ that satisfy the functional equation

$$f(x) + g(y) = h\left(\frac{x+y}{1-xy}\right)$$

for all $x, y \in \mathbb{R}$.

13. Find all functions $f, g : \mathbb{R} \to \mathbb{C}$ that satisfy the functional equation

$$f(x+y) + f(x-y) = f(x) f(y) + f(x) f(-y)$$

for all $x, y \in \mathbb{R}$.

14. Find all functions $f, g : \mathbb{R} \to \mathbb{C}$ that satisfy the functional equation

$$f(x+y) + f(x-y) = f(x) g(y) + f(x) g(-y)$$

for all $x, y \in \mathbb{R}$.

Chapter 12

Pompeiu Functional Equation

12.1 Introduction

Let $\mathbb{R}_\star$ be the set of real numbers except -1, that is $\mathbb{R}_\star = \mathbb{R} \setminus \{-1\}$. Then $(\mathbb{R}_\star, \circ)$ is an abelian group where the group operation is defined by

$$x \circ y = x + y + xy. \tag{12.1}$$

If ϕ is a homomorphism of $(\mathbb{R}_\star, \circ)$ into $(\mathbb{R}_\star, \circ)$, then ϕ satisfies

$$\phi(x \circ y) = \phi(x) \circ \phi(y)$$

for all $x, y \in \mathbb{R}_\star$. In view of (12.1), ϕ can be completely characterized by solving the functional equation

$$f(x + y + xy) = f(x) + f(y) + f(x)\,f(y), \tag{PE}$$

where $x, y \in \mathbb{R}_\star$. This functional equation is known as the Pompeiu functional equation.

In this chapter, we determine the solution of the Pompeiu functional equation, a generalized Pompeiu functional equation and the Pompeiu functional equation of Pexider type without any regularity assumption on unknown functions. The materials for this chapter are taken from Kannappan and Sahoo (1998).

12.2 Solution of the Pompeiu Functional Equation

This section is devoted to finding the solution of the Pompeiu functional equation (PE) without any regularity condition on the unknown function f.

Theorem 12.1. *If $f : \mathbb{R} \to \mathbb{R}$ satisfies the Pompeiu functional equation*

$$f(x + y + xy) = f(x) + f(y) + f(x) f(y) \qquad \text{(PE)}$$

for all $x, y \in \mathbb{R}$, then f is given by

$$f(x) = M(x + 1) - 1, \qquad (12.2)$$

where $M : \mathbb{R} \to \mathbb{R}$ is a multiplicative function.

Proof. Adding 1 to both sides of (PE), we get

$$1 + f(x + y + xy) = 1 + f(x) + f(y) + f(x) f(y)$$

which is

$$1 + f(x + y + xy) = (1 + f(x))(1 + f(y)) \qquad (12.3)$$

for all $x, y \in \mathbb{R}$. Defining $F : \mathbb{R} \to \mathbb{R}$ as

$$F(x) = 1 + f(x), \qquad (12.4)$$

we see that (12.3) reduces to

$$F(x + y + xy) = F(x) F(y) \qquad (12.5)$$

for all $x, y \in \mathbb{R}$. Replacing x by $x - 1$ and y by $y - 1$ in (12.5), we have

$$F(xy - 1) = F(x - 1) F(y - 1) \qquad (12.6)$$

for all $x, y \in \mathbb{R}$. If we define

$$M(x) = F(x - 1), \qquad (12.7)$$

then (12.6) reduces to

$$M(xy) = M(x) M(y) \qquad (12.8)$$

for all $x, y \in \mathbb{R}$. By (12.7) and (12.4), we get

$$\begin{aligned} f(x) &= F(x) - 1 \\ &= M(x + 1) - 1. \end{aligned}$$

This completes the proof of the theorem. $\qquad \square$

12.3   A Generalized Pompeiu Functional Equation

Now we consider the following generalized Pompeiu functional equation

$$f(ax + by + cxy) = f(x) + f(y) + f(x)\,f(y) \qquad \text{(FE)}$$

for all $x, y \in \mathbb{R}$. Here a, b, c are a priori chosen real parameters. If $a = b = c = 1$, then (FE) reduces to the Pompeiu functional equation. The following theorem was proved in Kannappan and Sahoo (1998).

Theorem 12.2. *The function $f : \mathbb{R} \to \mathbb{R}$ is a solution of the equation*

$$f(ax + by + cxy) = f(x) + f(y) + f(x)\,f(y) \qquad \text{(FE)}$$

if and only if f is given by

$$f(x) = \begin{cases} M(cx) - 1 & \text{if } a = 0 = b, \ c \neq 0 \\ E(x) - 1 & \text{if } a = 1 = b, \ c = 0 \\ M(cx + 1) - 1 & \text{if } a = 1 = b, \ c \neq 0 \\ k & \text{otherwise,} \end{cases}$$

where $M : \mathbb{R} \to \mathbb{R}$ is multiplicative, $E : \mathbb{R} \to \mathbb{R}$ is exponential, and k is a constant satisfying $k(k + 1) = 0$.

Proof. The only constant solutions of (FE) are $f \equiv 0$ and $f \equiv -1$. So we look for non-constant solutions of the functional equation (FE).

Substitution of $x = 0 = y$ in (FE) yields

$$f(0)[f(0) + 1] = 0.$$

Hence, either $f(0) = 0$ or $f(0) = -1$. Now we consider two cases.

Case 1.   Suppose $f(0) = -1$. Then $x = 0$ in (FE) gives

$$f(by) = f(0).$$

Hence when $b \neq 0$, the function $f \equiv -1$, which is not the case since we are looking for a non-constant solution. Similarly by putting $y = 0$ in (FE), we again get $f \equiv -1$ when $a \neq 0$. Likewise if $a = 0 = c$, $b \neq 0$ or $b = 0 = c$, $a \neq 0$, then we obtain $f \equiv -1$.

Suppose $a = 0 = b$. If c is also zero, then (FE) reduces to

$$(1 + f(x))\,(1 + f(y)) = 0$$

since $f(0) = -1$. That is, $f \equiv -1$, a constant function. So we assume that $c \neq 0$. Then replacing x by $\frac{x}{c}$ and y by $\frac{y}{c}$ in (FE), we obtain

$$M(xy) = M(x) M(y), \tag{12.9}$$

where

$$M(x) = 1 + f\left(\frac{x}{c}\right). \tag{12.10}$$

Therefore, we have

$$f(x) = M(cx) - 1, \tag{12.11}$$

where $M : \mathbb{R} \to \mathbb{R}$ is multiplicative. Therefore (12.11) is a solution of (FE) with $f(0) = -1$, $a = b = 0$ and $c \neq 0$.

Case 2. Suppose $f(0) = 0$. Let $a = 0$. Then $y = 0$ in (FE) gives $f \equiv 0$ which is not the case since we are looking for a non-constant solution. Assume that $a \neq 0$. Similarly $b \neq 0$. Setting $x = 0$ and $y = 0$ separately in (FE), we get

$$f(by) = f(y) \tag{12.12}$$

and

$$f(ax) = f(x) \tag{12.13}$$

so that (FE) becomes

$$f(ax + by + cxy) = f(ax) + f(by) + f(ax) f(by) \tag{12.14}$$

for all $x, y \in \mathbb{R}$.

Suppose $c = 0$. Then replacing x by $\frac{x}{a}$ and y by $\frac{y}{b}$ in (12.14), we have

$$E(x + y) = E(x) E(y), \tag{12.15}$$

where $E : \mathbb{R} \to \mathbb{R}$ given by

$$E(x) = 1 + f(x) \tag{12.16}$$

is an exponential map. Further, from (12.12), (12.13) and (12.16), we obtain

$$E(ax) = E(x) = E(bx). \tag{12.17}$$

Since $E(x)E(-x) = 1$, we have

$$E((a - b)x) = 1 = E((a - 1)x). \tag{12.18}$$

If $a \neq b$, then E is a constant map, and so f is also a constant function. If $a \neq 1$, then E, and hence f is a constant. Hence $a = 1 = b$. Then (12.16) yields

$$f(x) = E(x) - 1$$

which is a solution of (FE) for the case $a = b = 1$ and $c = 0$.

Finally, let $a \neq 0$, $b \neq 0$ and $c \neq 0$. Set $\alpha = \frac{c}{ab}$. Replacing x by $\frac{x}{a\alpha}$ and y by $\frac{y}{b\alpha}$ in (12.14), we obtain

$$F(x + y + xy) = F(x) F(y), \tag{12.19}$$

where

$$F(x) = 1 + f\left(\frac{x}{\alpha}\right). \tag{12.20}$$

Changing x to $x - 1$ and y to $y - 1$ in (12.19), we have

$$M(xy) = M(x) M(y),$$

where M is multiplicative and

$$M(x) = F(x - 1). \tag{12.21}$$

Thus by (12.20) and (12.21), we have

$$f(x) = F(\alpha x) - 1.$$

Hence by (12.21), we obtain

$$f(x) = M(1 + \alpha x) - 1. \tag{12.22}$$

If we use (12.22) in (12.12) and (12.13) and the fact that $\alpha = \frac{c}{ba}$, then we have

$$M\left(1 + \frac{c}{a}x\right) = M\left(1 + \frac{c}{b}x\right) = M\left(1 + \frac{c}{ab}x\right). \tag{12.23}$$

Recall that, since M is multiplicative,

$$M(x) M\left(\frac{1}{x}\right) = 1$$

(otherwise if $M(1) = 0$, then $M \equiv 0$ so that $f \equiv -1$). Changing separately x to $\frac{ax}{c}$ and x to $\frac{bx}{c}$ in (12.23), we get

$$M(1 + x) = M\left(1 + \frac{x}{b}\right) = M\left(1 + \frac{x}{a}\right). \tag{12.24}$$

Similarly, replacing x by $\frac{abx}{c}$ in (12.23), we have

$$M(1 + x) = M(1 + ax) = M(1 + bx). \tag{12.25}$$

Replacing x by $x - 1$ in (12.25), we obtain

$$M(x) = M(1 + a(x - 1))$$

which yields

$$M\left(\frac{1 - a + ax}{x}\right) = 1$$

if $x \neq 0$.

Suppose $a \neq 1$. Changing x to $(1 - a)x$, we have

$$M\left(a + \frac{1}{x}\right) = 1,$$

and thus again replacing x by $\frac{1}{x-a}$, we have

$$M(x) = 1$$

when $x \neq 0, a$. Similarly if $b \neq 1$, we get

$$M(x) = 1$$

when $x \neq 0, b$.

Hence, $M(x) = 1$ for all x which leads to f a constant. Therefore $a = 1 = b$. Then from (12.22), we obtain

$$f(x) = M(1 + cx) - 1, \tag{12.26}$$

where $M : \mathbb{R} \to \mathbb{R}$ is multiplicative.

Since no more cases are left, the proof of the theorm is complete. $\square$

12.4   Pexiderized Pompeiu Functional Equation

First we prove two lemmas that will be instrumental for establishing the pexiderized Pompeiu functional equation (12.58). In this section, the set of all nonzero real numbers will be denoted by $\mathbb{R}_o$.

Lemma 12.1. *Let $g, h : \mathbb{R}_o \to \mathbb{R}$ satisfy the functional equation*

$$g(xy) = g(y) + g(x)\,h(y) \tag{12.27}$$

for all $x, y \in \mathbb{R}_o$. Then for all $x, y \in \mathbb{R}_o$, $g(x)$ and $h(y)$ are given by

$$g(x) = 0, \qquad h(y) = arbitrary; \tag{12.28}$$

$$g(x) = L(x), \qquad h(y) = 1; \tag{12.29}$$

$$g(x) = \alpha \left[M(x) - 1 \right], \qquad h(y) = M(y), \qquad (12.30)$$

where $M : \mathbb{R}_o \to \mathbb{R}$ *is a multiplicative function not identically one,* $L : \mathbb{R}_o \to \mathbb{R}$ *is a logarithmic function not identically zero and* α *is an arbitrary nonzero constant.*

Proof. If $g \equiv 0$, then h is arbitrary and they satisfy the equation (12.27). Hence we have the solution (12.28). We assume hereafter that $g \not\equiv 0$.

Interchanging x with y in (12.27) and comparing the resulting equation to (12.27), we get

$$g(y) \left[h(x) - 1 \right] = g(x) \left[h(y) - 1 \right]. \qquad (12.31)$$

Suppose $h(x) = 1$ for all $x \in \mathbb{R}_o$. Then (12.27) yields $g(xy) = g(y) + g(x)$ and hence the function $g : \mathbb{R}_o \to \mathbb{R}$ is logarithmic. This yields the solution (12.29).

Finally, suppose $h(y) \neq 1$ for some y. Then from (12.31), we have

$$g(x) = \alpha \left[h(x) - 1 \right], \qquad (12.32)$$

where α is a nonzero constant, since $g \not\equiv 0$. Using (12.32) in (12.27) and simplifying, we obtain

$$h(xy) = h(x) \, h(y). \qquad (12.33)$$

Hence, $h : \mathbb{R}_o \to \mathbb{R}$ is a multiplicative function. This gives the asserted solution (12.30) and the proof of the lemma is now complete. $\square$

Lemma 12.2. *The general solutions* $f, g, h : \mathbb{R}_o \to \mathbb{R}$ *of the functional equation*

$$f(xy) = f(x) + f(y) + \alpha \, g(x) + \beta \, h(y) + g(x) \, h(y) \qquad (12.34)$$

for all $x, y \in \mathbb{R}_o$, *where* α *and* β *are a priori chosen constants, have values* $f(x), g(x)$ *and* $h(y)$ *given, for all* $x, y \in \mathbb{R}_o$, *by*

$$\left. \begin{array}{l} f(x) = L(x) + \alpha\beta \\ g(x) \text{ is arbitrary} \\ h(y) = -\alpha; \end{array} \right\} \qquad (12.35)$$

$$\left. \begin{array}{l} f(x) = L(x) + \alpha\beta \\ g(x) = -\beta \\ h(y) \text{ is arbitrary}; \end{array} \right\} \qquad (12.36)$$

$$\left. \begin{array}{l} f(x) = L_o(x) + \frac{1}{2} c \, L_1^2(x) + \alpha\beta \\ g(x) = c \, L_1(x) - \beta \\ h(y) = L_1(y) - \alpha; \end{array} \right\} \qquad (12.37)$$

$$\left.\begin{array}{l} f(x) = L(x) + \gamma\delta\left[M(x) - 1\right] + \alpha\beta \\ g(x) = \gamma\left[M(x) - 1\right] - \beta \\ h(y) = \delta\left[M(y) - 1\right] - \alpha, \end{array}\right\} \qquad (12.38)$$

where $M : \mathbb{R}_o \to \mathbb{R}$ is a multiplicative map not identically one; $L_o, L_1, L : \mathbb{R}_o \to \mathbb{R}$ are logarithmic functions with L_1 not identically zero; and c, δ, γ are arbitrary nonzero constants.

Proof. Interchanging x with y in (12.34) and comparing the resulting equation to (12.34), we obtain

$$[\alpha + h(y)][\beta + g(x)] = [\alpha + h(x)][\beta + g(y)]. \qquad (12.39)$$

Now we consider several cases.

Case 1. Suppose $h(y) = -\alpha$ for all $y \in \mathbb{R}_o$. Then (12.34) yields

$$f(xy) = f(x) + f(y) - \alpha\beta. \qquad (12.40)$$

Hence

$$f(x) = L(x) + \alpha\beta, \qquad (12.41)$$

where $L : \mathbb{R}_o \to \mathbb{R}$ is a logarithmic function. Hence we have the asserted solution (12.35).

Case 2. Suppose $g(x) = -\beta$ for all $x \in \mathbb{R}_o$. Then (12.34) yields

$$f(xy) = f(x) + f(y) - \alpha\beta.$$

Hence, as before,

$$f(x) = L(x) + \alpha\beta,$$

where $L : \mathbb{R}_o \to \mathbb{R}$ is a logarithmic function. Thus we have the asserted solution (12.36).

Case 3. Now we assume $h(x) \neq -\alpha$ for some $x \in \mathbb{R}_o$ and $g(x) \neq -\beta$ for some $x \in \mathbb{R}_o$. From (2.13), we get

$$\beta + g(y) = c[\alpha + h(y)], \qquad (12.42)$$

where c is a nonzero constant.

Using (12.34), we compute

$$\begin{aligned} f(x(yz)) = {}& f(x) + f(y) + f(z) + \alpha\, g(y) + \beta\, h(z) \\ & + g(y)h(z) + \alpha\, g(x) + \beta\, h(yz) + g(x)h(yz). \qquad (12.43) \end{aligned}$$

Again, using (12.34), we have

$$f((xy)z) = f(x) + f(y) + f(z) + \alpha g(x) + \beta h(y)$$
$$+ g(x)h(y) + \alpha g(xy) + \beta h(z) + g(xy)h(z). \qquad (12.44)$$

From (12.43) and (12.44), we obtain

$$[\alpha + h(z)] [g(y) - g(xy)] = [\beta + g(x)] [h(y) - h(yz)] \qquad (12.45)$$

for all $x, y \in \mathbb{R}_o$.

Since $g(x) \neq -\beta$ for some $x \in \mathbb{R}_o$, there exists a $x_o \in \mathbb{R}_o$ such that $g(x_o) + \beta \neq 0$. Letting $x = x_o$ in (12.45), we have

$$h(yz) = h(y) + [\alpha + h(z)] \, k(y), \qquad (12.46)$$

where

$$k(y) = \frac{g(yx_o) - g(y)}{g(x_o) + \beta}. \qquad (12.47)$$

The general solution of (12.46) can be obtained from Lemma 12.1 (add α to both sides). Hence, taking into consideration that $h(y) + \alpha \neq 0$, we have

$$h(y) = L_1(y) - \alpha \qquad (12.48)$$

or

$$h(y) = \delta \, [M(y) - 1] - \alpha, \qquad (12.49)$$

where L_1 is logarithmic not identically zero, M is multiplicative not identically one and δ is an arbitrary constant.

Now we consider two subcases.

Subcase 3.1. From (12.48) and (12.52), we have

$$g(y) = c \, L_1(y) - \beta. \qquad (12.50)$$

Using (12.48) and (12.50) in (12.34), we get

$$f(xy) = f(x) + f(y) + c \, L_1(x) L_1(y) - \alpha\beta. \qquad (12.51)$$

Defining

$$L_o(x) := f(x) - \frac{1}{2} c \, L_1^2(x) - \alpha\beta, \qquad (12.52)$$

we see that (12.51) reduces to

$$L_o(xy) = L_o(x) + L_o(y)$$

for all $x, y \in \mathbb{R}_o$, that is, L_o is logarithmic, and from (12.52), we have

$$f(x) = L_o(x) + \frac{1}{2} c L_1^2(x) + \alpha\beta. \tag{12.53}$$

Hence (12.53), (12.50) and (12.48) yield the asserted solution (12.37).

Subcase 3.2. Finally, from (12.49) and (12.42), we obtain

$$g(y) = \delta c \left[M(y) - 1 \right] - \beta. \tag{12.54}$$

With (12.49) and (12.54) in (12.34), we have

$$f(xy) = f(x) + f(y) - \alpha\beta + c\delta^2 \left[M(x) - 1 \right] \left[M(y) - 1 \right]. \tag{12.55}$$

Defining

$$L(x) := f(x) - c\delta^2 \left[M(x) - 1 \right] - \alpha\beta, \tag{12.56}$$

we see that (12.55) reduces to

$$L(xy) = L(x) + L(y)$$

for all $x, y \in \mathbb{R}_o$, that is, L is a logarithmic function. Using (12.56), we have

$$f(x) = L(x) + \gamma\delta \left[M(x) - 1 \right] + \alpha\beta, \tag{12.57}$$

where $\gamma = c\delta$. Hence (12.57), (12.54) and (12.49) yield the asserted solution (12.38). This completes the proof of the lemma. □

Using Lemma 12.2, we determine the general solution of the pexiderized Pompeiu functional equation following Kannappan and Sahoo (1998).

Theorem 12.3. *The functions $f, p, q, g, h : \mathbb{R}_\star \to \mathbb{R}$ satisfy the functional equation*

$$f(x + y + xy) = p(x) + q(y) + g(x) h(y) \tag{12.58}$$

for all $x, y \in \mathbb{R}_\star$ if and only if, for all $x, y \in \mathbb{R}_\star$,

$$\left.\begin{aligned}
&f(x) = L(x+1) + \alpha\beta + a + b \\
&p(x) = L(x+1) + b \\
&q(y) = L(y+1) + \alpha\beta + a + \beta h(y) \\
&g(x) = -\beta \\
&h(y) \text{ is arbitrary;}
\end{aligned}\right\} \tag{12.59}$$

$$\left.\begin{aligned}
&f(x) = L(x+1) + \alpha\beta + a + b \\
&p(x) = L(x+1) + +\alpha\beta + b + ag(x) \\
&q(y) = L(y+1) + a \\
&g(x) \text{ is arbitrary} \\
&h(y) = -\alpha;
\end{aligned}\right\} \tag{12.60}$$

$$\left.\begin{array}{l} f(x) = L(x+1) + \gamma\delta\left[M(x+1) - 1\right] + \alpha\beta + a + b \\ p(x) = L(x+1) + (\delta + \alpha)\gamma\left[M(x+1) - 1\right] + b \\ q(y) = L(y+1) + (\gamma + \beta)\delta\left[M(y+1) - 1\right] + a \\ g(x) = \gamma\left[M(x+1) - 1\right] - \beta \\ h(y) = \delta\left[M(y+1) - 1\right] - \alpha; \end{array}\right\} \quad (12.61)$$

$$\left.\begin{array}{l} f(x) = L_o(x+1) + \frac{1}{2}c\,L_1^2(x+1) + \alpha\beta + a + b \\ p(x) = L_o(x+1) + \frac{1}{2}c\,L_1^2(x+1) + \alpha c\,L_1(x+1) + b \\ q(y) = L_o(y+1) + \frac{1}{2}c\,L_1^2(y+1) + \beta\,L_1(y+1) + a \\ g(x) = c\,L_1(x+1) - \beta \\ h(y) = L_1(y+1) - \alpha, \end{array}\right\} \quad (12.62)$$

where $M : \mathbb{R}_o \to \mathbb{R}$ is a multiplicative function not identically one, $L_o, L_1, L : \mathbb{R}_o \to \mathbb{R}$ are logarithmic maps with L_1 not identically zero and $\alpha, \beta, \gamma, \delta, a, b, c$ are arbitrary real constants.

Proof. First, we substitute $y = 0$ in (12.58) and then we put $x = 0$ in (12.58) to obtain

$$p(x) = f(x) - a + \alpha\,g(x) \quad (12.63)$$

and

$$q(y) = f(y) - b + \beta\,h(y), \quad (12.64)$$

where $a := q(0)$, $b := p(0)$, $\alpha := -h(0)$, $\beta := -g(0)$. Using (3.5) and (3.6) in (12.58), we have

$$f(x+y+xy) = f(x) + f(y) - a - b + \alpha\,g(x) + \beta\,h(y) + g(x)\,h(y) \quad (12.65)$$

for $x, y \in \mathbb{R}_*$. Replacing x by $u - 1$ and y by $v - 1$ in (12.65) and then defining

$$F(u) := f(u-1) - a - b, \quad G(u) := g(u-1), \quad H(u) := h(u-1) \quad (12.66)$$

for all $u \in \mathbb{R}_o$, we obtain

$$F(uv) = F(u) + F(v) + \alpha\,G(u) + \beta\,H(v) + G(u)\,H(v) \quad (12.67)$$

for all $u, v \in \mathbb{R}_o$. The general solution of (12.67) can now be obtained from Lemma 12.2. The first two solutions of Lemma 12.2 (see (12.35) and (12.36)) together with (12.63) and (12.64) yield the solutions (12.59) and (12.60). The next two solutions of Lemma 12.2 (that is, solutions (12.37) and (12.38)) yield together with (12.63) and (12.64) the asserted solutions (12.61) and (12.62).

This completes the proof of the theorem. □

12.5   Concluding Remarks

Neagu (1984) reformulated the Pompeiu functional equation

$$f(x + y + xy) = f(x) + f(y) + f(x)f(y) \tag{PE}$$

in distribution and determined the distributional solution of (PE). From this distributional solution he then found the locally integrable solution of (PE). In 1985, he found the distributional solution of the following generalized Pompeiu functional equation (see Neagu (1985)):

$$f(x + y + xy) = g(x) + h(y) + h(x)\,g(y) \tag{12.68}$$

for all $x, y \in (-1, \infty)$.

Kannappan and Sahoo (1998) determined the general solution of the Pompeiu functional equation (PE) without any regularity condition on the unknown function $f : \mathbb{R} \to \mathbb{R}$. In the same paper, they also found the general solution of the pexiderized version of the Pompeiu functional equation, namely,

$$f(x + y + xy) = p(x) + q(y) + g(x)\,h(y) \tag{12.69}$$

for all $x, y \in \mathbb{R}$ without assuming any regularity assumption on the unknown functions $f, p, q, g, h : \mathbb{R}_\star \to \mathbb{R}$, where $\mathbb{R}_\star = \mathbb{R} \backslash \{-1\}$. Kannappan and Sahoo (1998) also determined the general solution of the functional equation

$$f(ax + by + cxy) = f(x) + f(y) + f(x)\,f(y) \tag{12.70}$$

without any regularity assumption of the unknown function $f : \mathbb{R} \to \mathbb{R}$. In the equation (12.70), a, b, c are real parameters.

Lee and Jun (2001) considered the pexiderized version of the functional equation (12.70), that is,

$$f(ax + by + cxy) = p(x) + q(y) + g(x)\,h(y) \tag{12.71}$$

for all $x, y \in \mathbb{R}$. They determined the general solution of the above functional equation (12.71) without any regularity assumption.

Chung (2001) reformulated the Pompeiu functional equation (PE) and its generalization as equations for Gevrey distributions and showed that every solution of each functional equation in the space of Gevrey distributions is Gevrey differentiable.

Chung, Chung and Kim (2006) studied the generalized Pompeiu functional equation (12.69) in the space of Schwartz distributions and then determined the locally integrable solutions of the functional equation (12.69) as an application.

12.6 Exercises

1. Find all functions $f : \mathbb{R} \to \mathbb{R}$ that satisfy the functional equation

$$f(x + y - xy) = f(x) + f(y) + f(x)\, f(y)$$

for all $x, y \in \mathbb{R}$.

2. Find all functions $f, g, h : \mathbb{R} \to \mathbb{R}$ that satisfy the functional equation

$$f(x + y - xy) = g(x) + h(y) + g(x)\, h(y) \; \cdot$$

for all $x, y \in \mathbb{R}$.

3. Find all functions $f : \mathbb{R} \to \mathbb{R}$ that satisfy the functional equation

$$f(x + y - xy) + f(x - y + xy) = 2\, f(x)$$

for all $x, y \in \mathbb{R}$.

4. Let a, b, c be a priori chosen real numbers. Find all functions $f, g, h, p, q : \mathbb{R} \to \mathbb{R}$ that satisfy the functional equation

$$f(ax + by + cxy) = p(x) + q(y) + g(x)\, h(y)$$

for all $x, y \in \mathbb{R}$.

5. Find all functions $f : \mathbb{R} \to \mathbb{R}$ that satisfy the functional equation

$$f(x + y + xy) = f(x) + f(y) + xf(y) + yf(x)$$

for all $x, y \in \mathbb{R}$ with $x \neq -1$.

6. Find all functions $f, g, h, k, \ell : \mathbb{R} \to \mathbb{R}$ that satisfy the functional equation
$$f(x + y + xy) = g(x) + h(y) + xk(y) + y\ell(x)$$

for all $x, y \in \mathbb{R}$.

7. Let a, b, c, n be a priori chosen real numbers such that $ab \neq 0$ and $n > 0$. Find all functions $f, g : \mathbb{R} \to \mathbb{R}$ that satisfy the functional equation

$$f(ax + by + nxy + c) = g(x) + g(y)$$

for all $x, y \in \mathbb{R}$.

8. Let a, b, c, n be a priori chosen real numbers such that $ab \neq 0$ and $n > 0$. Find all functions $f, g : \mathbb{R} \to \mathbb{R}$ that satisfy the functional equation

$$f(ax + by + nxy + c) = g(x)\, g(y)$$

for all $x, y \in \mathbb{R}$.

9. Find all functions $f : \mathbb{R}^2 \to \mathbb{R}$ that satisfy the functional equation

$$f(x + y - xy, u + v) + f(x + y - xy, u - v) = 2f(x, u)\, f(y, v)$$

for all $x, y, u, v \in \mathbb{R}$.

10. Find all functions $f : (-\infty, 1) \to \mathbb{R}$ that satisfy the functional equation

$$f(x + y - xy) + x\, f(y) + y\, f(x) = f(x) + f(y)$$

for all $x, y \in (-\infty, 1)$.

11. Let α and β be any two positive real numbers. Find all functions $f, g, h : \mathbb{R} \to \mathbb{R}$ that satisfy the functional equation

$$f(x + y - xy) = (1 - x)^\alpha\, g(y) + (1 - y)^\beta\, h(x)$$

for all $x, y \in \mathbb{R}$.

Chapter 13

Hosszú Functional Equation

13.1 Introduction

In lattice theory one considers functions f which satisfy

$$f(x \vee y) + f(x \wedge y) = f(x) + f(y)$$

for all x, y in some lattice (see Birkhoff (1967), p. 74). If one applies the standard substitutions

$$x \vee y = x + y - xy \quad \text{and} \quad x \wedge y = xy$$

to transform the lattice concepts to Boolean ring, one obtains the Hosszú's functional equation

$$f(x + y - xy) + f(xy) = f(x) + f(y). \tag{HE}$$

The goal of this chapter is to present the general solutions of the Hosszú functional equation. In this chapter, a generalization of the functional equation (HE) will be considered and its solution will be presented without any regularity assumption.

13.2 Hosszú Functional Equation

In this section, we determine the general solution of the Hosszú functional equation without any regularity assumptions on the unknown function f.

The following theorem was proved independently by Blanusa (1970) and Daróczy (1971). Although their papers were published in different years, they had submitted their papers to the same journal about the same time. The proof of the following theorem is based on the proof due to Daróczy.

Theorem 13.1. *Let $f : \mathbb{R} \to \mathbb{R}$ satisfy the Hosszú functional equation (HE), that is*

$$f(x + y - xy) + f(xy) = f(x) + f(y)$$

for all $x, y \in \mathbb{R}$. Then there exist an additive function $A : \mathbb{R} \to \mathbb{R}$ and a constant $a \in \mathbb{R}$ such that

$$f(x) = A(x) + a. \tag{13.1}$$

Proof. Defining $G : \mathbb{R}^2 \to \mathbb{R}$ by

$$G(x, y) = f(x) + f(y) - f(xy), \tag{13.2}$$

we see that G satisfies

$$G(uv, w) + G(u, v) = G(u, vw) + G(v, w) \tag{13.3}$$

for all $u, v, w \in \mathbb{R}$. Since

$$G(x, y) = f(x) + f(y) - f(xy),$$

by (HE), G is also equal to

$$G(x, y) = f(x + y - xy). \tag{13.4}$$

Inserting (13.4) into (13.3), we obtain

$$f(uv + w - uvw) + f(u + v - uv) = f(u + vw - uvw) + f(v + w - vw) \tag{13.5}$$

for all $u, v, w \in \mathbb{R}$. Setting $w = \frac{1}{v}$ in (13.5), we get

$$f\left(uv + \frac{1}{v} - u\right) + f(u + v - uv) = f(1) + f\left(v + \frac{1}{v} - 1\right) \tag{13.6}$$

for all $u \in \mathbb{R}$ and $v \in \mathbb{R} \backslash \{0\}$.

Choose arbitrarily $x, y \in \mathbb{R}$ such that $x + y \in \mathbb{R}_+$ (the set of all positive real numbers). We look for $u, v \in \mathbb{R}$ with $v \neq 0$, satisfying

$$uv + \frac{1}{v} - u = x + 1 \tag{13.7}$$

and

$$u + v - uv = y + 1. \tag{13.8}$$

Adding (13.7) and (13.8), we see that

$$v + \frac{1}{v} = x + y + 2 \tag{13.9}$$

which yields the following quadratic equation

$$v^2 - (x + y + 2)v + 1 = 0. \tag{13.10}$$

This quadratic equation has two real distinct nonzero roots since the constant term is 1 and the discriminant

$$(x + y + 2)^2 - 4 = (x + y)(x + y + 4) > 0.$$

Let v_0 be one of the two distinct roots of (13.10). Clearly $v_0 \neq 1$, since

$$v_0 + \frac{1}{v_0} = (x + y + 2) \neq 2$$

by (13.9) (also from (13.10)). Hence if

$$u_0 = \frac{y + 1 - v_0}{1 - v_0},$$

then u_0 and v_0 satisfy the system of equations (13.7)–(13.8).

For arbitrary $x, y \in \mathbb{R}$ with $x + y \in \mathbb{R}_+$, we let (13.7) and (13.8) into (13.6) to give

$$f(x + 1) + f(y + 1) = f(1) + f(x + y + 1). \tag{13.11}$$

We define

$$A(x) = f(x + 1) - f(1). \tag{13.12}$$

Then (13.12) in (13.11) yields

$$A(x) + A(y) = A(x + y) \tag{13.13}$$

for all $x, y \in \mathbb{R}$ with $x + y \in \mathbb{R}_+$.

Now we extend (13.13) to $\mathbb{R}^2$ (see Figure 13.1).

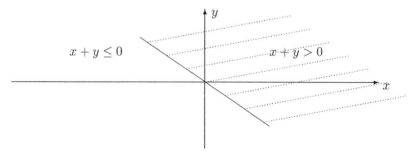

Figure 13.1. Illustration of the domain of equation (13.13).

If $x + y \leq 0$, then we can find a $t \in \mathbb{R}$ such that $t + x \in \mathbb{R}_+$ and $t + x + y \in \mathbb{R}_+$. Then (13.13) yields

$$
\begin{aligned}
A(t) + A(x + y) &= A(t + x + y) &&\text{(since } t + x + y > 0) \\
&= A(t + x) + A(y) &&\text{(by (13.13))} \\
&= A(t) + A(x) + A(y) &&\text{(since } t + x \in \mathbb{R}_+).
\end{aligned}
$$

Hence

$$
A(x + y) = A(x) + A(y)
$$

for all $x, y \in \mathbb{R}$. Therefore A is additive on $\mathbb{R}$.

By (13.12), we get

$$
\begin{aligned}
f(x + 1) - f(1) &= A(x) \\
f(x + 1) &= A(x) + f(1)
\end{aligned}
$$

or

$$
\begin{aligned}
f(x) &= A(x - 1) + f(1) \\
&= A(x) - A(1) + f(1) \\
&= A(x) + a.
\end{aligned}
$$

This completes the proof. $\qquad\square$

13.3   A Generalization of Hosszú Equation

In this section, we consider a generalization of the Hosszú functional equation due to Kannappan and Sahoo (1993). The following result is needed for finding the general solution of the generalized Hosszú functional equation.

Theorem 13.2. *Let $f, g, h, k : \mathbb{R} \to \mathbb{R}$ satisfy the functional equation*

$$
f(x + y) - g(x) - h(y) = k(xy) \tag{13.14}
$$

for all $x, y \in \mathbb{R}$. Then the general solution of (13.14) is given by

$$
\left.
\begin{aligned}
f(x) &= A_1(x^2) + A(x) + a \\
g(x) &= A_1(x^2) + A(x) + a - b_1 \\
h(x) &= A_1(x^2) + A(x) + a - b_2 \\
k(x) &= 2\,A_1(x) + b_1 + b_2 - a,
\end{aligned}
\right\} \tag{13.15}
$$

where $A, A_1 : \mathbb{R} \to \mathbb{R}$ *are additive functions and* a, b_1, b_2 *are arbitrary constants.*

Proof. Letting $x = 0$ and $y = 0$ in (13.14) separately, we obtain

$$h(y) = f(y) - g(0) - k(0) \tag{13.16}$$

and

$$g(x) = f(x) - h(0) - k(0), \tag{13.17}$$

respectively. Inserting (13.16) and (13.17) into (13.14), we get

$$f(x + y) - f(x) - f(y) = l(xy), \tag{13.18}$$

where

$$l(x) := k(x) - g(0) - h(0) - 2\,k(0). \tag{13.19}$$

Substituting $y = 1$ in (13.18), we obtain

$$l(x) = f(x + 1) - f(x) - f(1). \tag{13.20}$$

By (13.20), (13.18) can be rewritten as

$$f(x + y) + f(xy) = f(1 + xy) + f(x) + f(y) - f(1). \tag{13.21}$$

Defining

$$F(x) := f(x) - f(1) \tag{13.22}$$

the equation (13.21) can be written as

$$F(x + y) + F(xy) = F(1 + xy) + F(x) + F(y) \tag{13.23}$$

for all $x, y \in \mathbb{R}$.

Replacing y with $-y$ in (13.23), we obtain

$$F(x - y) + F(-xy) = F(1 - xy) + F(x) + F(-y). \tag{13.24}$$

Adding (13.24) to (13.23), we get

$$\begin{aligned}
F(x + y) &+ F(x - y) + F(xy) + F(-xy) \\
&= 2F(x) + F(y) + F(-y) + F(1 + xy) + F(1 - xy) \tag{13.25}
\end{aligned}$$

for all $x, y \in \mathbb{R}$. Interchanging x with y in (13.25), we have

$$\begin{aligned}
F(x + y) &+ F(y - x) + F(xy) + F(-xy) \\
&= 2F(y) + F(x) + F(-x) + F(1 + xy) + F(1 - xy). \tag{13.26}
\end{aligned}$$

Subtracting (13.26) from (13.25), we obtain

$$F(x - y) - F(y - x) = F(x) - F(-x) + F(-y) - F(y). \qquad (13.27)$$

With

$$\psi(x) := F(x) - F(-x) \qquad (13.28)$$

the above equation reduces to

$$\psi(x - y) = \psi(x) - \psi(y), \qquad x, y \in \mathbb{R}.$$

Thus ψ is additive, and we have

$$F(x) - F(-x) = A(x), \qquad (13.29)$$

where $A : \mathbb{R} \to \mathbb{R}$ is an additive function. Next, we subtract (13.24) from (13.23) to get

$$F(x + y) - F(x - y) + F(xy) - F(-xy)$$
$$= F(1 + xy) - F(1 - xy) + F(y) - F(-y). \qquad (13.30)$$

Using (13.29) in (13.30), we get

$$F(x + y) - F(x - y) + A(xy) = F(1 + xy) - F(1 - xy) + A(y). \qquad (13.31)$$

Define $w : \mathbb{R} \to \mathbb{R}$ by

$$w(x) := F(x) - \frac{1}{2} A(x) \qquad x \in \mathbb{R}. \qquad (13.32)$$

Then by (13.32), (13.31) reduces to

$$w(x + y) + w(1 - xy) = w(x - y) + w(1 + xy) \qquad (13.33)$$

for all $x, y \in \mathbb{R}$. Choose $2x = \sqrt{t + s} + \sqrt{t}$ and $2y = \sqrt{t + s} - \sqrt{t}$ so that

$$\left.\begin{array}{l} \sqrt{t} = x - y, \qquad t > 0 \\ s = 4xy. \end{array}\right\} \qquad (13.34)$$

Letting (13.34) in (13.33), we get

$$w(\sqrt{s + t}) + w\left(1 - \frac{s}{4}\right) = w(\sqrt{t}) + w\left(1 + \frac{s}{4}\right) \qquad (13.35)$$

for all $(s, t) \in D$, where

$$D := \{(s, t) \in \mathbb{R}^2 \,|\, t \in \mathbb{R}_+, \ s + t \in \mathbb{R}_+\}.$$

Defining

$$G(t) := w(\sqrt{t}), \qquad t \in \mathbb{R}_+ \tag{13.36}$$

and

$$H(t) := w\left(1 + \frac{t}{4}\right) - w\left(1 - \frac{t}{4}\right), \tag{13.37}$$

we see that (13.35) becomes

$$G(t + s) = G(t) + H(s), \qquad (s, t) \in D. \tag{13.38}$$

Now interchanging s and t in (13.38), we obtain

$$G(s + t) = G(s) + H(t)$$

and from which, using (13.38), we see that

$$H(s) = G(s) - \beta, \qquad s \in \mathbb{R}_+,$$

where β is a constant. Therefore (13.38) becomes

$$G(t + s) = G(t) + G(s) - \beta \qquad s, t > 0.$$

Hence, we obtain

$$G(t) = A_1(t) + \beta, \qquad t > 0,$$

where $A_1 : \mathbb{R} \to \mathbb{R}$ is an additive function. From (13.36), we see that

$$w(s) = A_1(s^2) + \beta \tag{13.39}$$

for all $s \in \mathbb{R}_+$.

Substitution of $x = 0$ in (13.33) gives $w(y) = w(-y)$. So (13.39) holds for $t < 0$. Further $x = y$ in (13.33) gives $w(0)$. Thus (13.39) holds for real numbers t.

From (13.22), (13.32) and (13.39), we get

$$f(x) = A_1(x^2) + \frac{1}{2} A(x) + \beta + f(1) \qquad x \in \mathbb{R}. \tag{13.40}$$

From (13.16) and (13.40), we get

$$h(x) = A_1(x^2) + \frac{1}{2} A(x) + \beta + f(1) - g(0) - k(0), \tag{13.41}$$

and from (13.17)(2.4) and (13.40), we obtain

$$g(x) = A_1(x^2) + \frac{1}{2} A(x) + \beta + f(1) - h(0) - k(0). \tag{13.42}$$

Finally from (13.19), (13.20) and (13.40), we get

$$
\begin{aligned}
k(x) &= l(x) + b \\
&= f(x+1) - f(x) + b - f(1) \\
&= A_1(1 + x^2 + 2x) - A_1(x) + \frac{1}{2} A(1+x) - \frac{1}{2} A(x) + b - f(1) \\
&= 2\,A_1(x) + A_1(1) + \frac{1}{2} A(1) + b - f(1),
\end{aligned}
$$

where
$$
b = g(0) + h(0) + 2\,k(0) = f(0) + k(0). \tag{13.43}
$$

From (13.39), (13.22) and (13.42), we see that

$$
A_1(1) + \frac{1}{2} A(1) + \beta = 0. \tag{13.44}
$$

Hence
$$
k(x) = 2\,A_1(x) - \beta + b - f(1). \tag{13.45}
$$

Letting

$$
a = f(1)+\beta, \quad b_1 = g(0)+k(0), \quad b_2 = h(0)+k(0), \quad b = b_1+b_2 \tag{13.46}
$$

into (13.40), (13.41), (13.42), (13.45) and replacing $\frac{1}{2} A(x)$ by $A(x)$, we obtain the asserted form (13.15). $\square$

Corollary 13.1. *Suppose $f : \mathbb{R} \to \mathbb{R}$ is such that the Cauchy difference $f(x+y) - f(x) - f(y)$ depends only on the product xy for all $x, y \in \mathbb{R}$. Then f is of the form*

$$
f(x) = A_1(x^2) + A(x) + a,
$$

where $A_1, A : \mathbb{R} \to \mathbb{R}$ are additive functions and a is an arbitrary constant.

Now we prove our main result.

Theorem 13.3. *Let $f, g, h, k : \mathbb{R} \to \mathbb{R}$ satisfy the functional equation*

$$
f(x + y + \alpha\,x\,y) + g(xy) = h(x) + k(y) \tag{FE}
$$

for all $x, y \in \mathbb{R}$ and $\alpha \in \mathbb{R}$. Then the general solution of (FE) is given by

$$
f(x) = \begin{cases} A_1(x^2) + A(x) + a & \text{if} \quad \alpha = 0 \\ A(\alpha\,x) + a & \text{if} \quad \alpha \neq 0 \end{cases} \tag{13.47}
$$

$$g(x) = \begin{cases} -2\,A_1(x) + a + b_1 + b_2 & if \quad \alpha = 0 \\ A(-\alpha^2\,x) + a + b_1 + b_2 & if \quad \alpha \neq 0 \end{cases} \tag{13.48}$$

$$h(x) = \begin{cases} A_1(x^2) + A(x) + a + b_1 & if \quad \alpha = 0 \\ A(\alpha\,x) + a + b_1 & if \quad \alpha \neq 0 \end{cases} \tag{13.49}$$

$$k(x) = \begin{cases} A_1(x^2) + A(x) + a + b_2 & if \quad \alpha = 0 \\ A(\alpha\,x) + a + b_2 & if \quad \alpha \neq 0, \end{cases} \tag{13.50}$$

where $A, A_1 : \mathbb{R} \to \mathbb{R}$ are additive functions and a, b_1, b_2 are arbitrary constants.

Proof. First, we treat the case $\alpha = 0$. In this case (FE) reduces to

$$f(x + y) + g(xy) = h(x) + k(y) \tag{13.51}$$

which is same as (13.14) and its solution can be obtained from Theorem 13.2. Thus, we have the first part of solution (13.47)–(13.50).

Next, we consider the case $\alpha \neq 0$. Letting $y = 0$ and $x = 0$ in (FE) separately, we see that

$$h(x) = f(x) + g(0) - k(0) \tag{13.52}$$

and

$$k(y) = f(y) + g(0) - h(0), \tag{13.53}$$

respectively. Use of (13.52) and (13.53) in (FE) yields

$$f(x + y + \alpha xy) + g(xy) = f(x) + f(y) + b_1 + b_2, \tag{13.54}$$

where

$$\left. \begin{array}{l} b_1 = g(0) - k(0), \\ b_2 = g(0) - h(0). \end{array} \right\} \tag{13.55}$$

Substitution of $y = -\frac{1}{\alpha}$ (recall $\alpha \neq 0$) in (13.54) gives

$$g(x) = f(-\alpha x) + b_1 + b_2. \tag{13.56}$$

Hence by (13.56), (13.54) becomes

$$f(x + y + \alpha xy) + f(-\alpha xy) = f(x) + f(y) \tag{13.57}$$

for all $x, y \in \mathbb{R}$. To determine the general solution of (13.57), we adopt a method similar to that of Daróczy (1971).

Defining $G : \mathbb{R}^2 \to \mathbb{R}$ by

$$G(x, y) := f(x) + f(y) - f(-\alpha xy), \tag{13.58}$$

we see that G satisfies

$$G\left(\alpha u, -\frac{v}{\alpha}\right) + G(\alpha uv, w) = G(\alpha u, vw) + G\left(-\frac{v}{\alpha}, w\right) \qquad (13.59)$$

for all $u, v, w \in \mathbb{R}$. By (13.58) and (13.57), (13.59) can be written as

$$f\left(\alpha u - \frac{v}{\alpha} - \alpha uv\right) + f(\alpha uv + w + \alpha^2 uvw)$$
$$= f(\alpha u + vw + \alpha^2 uvw) + f\left(-\frac{v}{\alpha} + w - vw\right). \qquad (13.60)$$

Setting $w = -\frac{1}{\alpha v}$ in (13.60), we obtain

$$f\left(\alpha u - \frac{v}{\alpha} - \alpha uv\right) + f\left(\alpha uv - \frac{1}{\alpha v} - \alpha u\right)$$
$$= f\left(-\frac{1}{\alpha}\right) + f\left(-\frac{v}{\alpha} - \frac{1}{\alpha v} + \frac{1}{\alpha}\right). \qquad (13.61)$$

for all $u \in \mathbb{R}$ and $v \in \mathbb{R} \setminus \{0\}$.

Choose arbitrarily $x, y \in \mathbb{R}$ such that $x + y \in \mathbb{R}_+$. We look for $u, v \in \mathbb{R}$ with $v \neq 0$, satisfying

$$\alpha u - \frac{v}{\alpha} - \alpha uv = \frac{y+1}{\alpha} \qquad (13.62)$$

and

$$\alpha uv - \frac{1}{\alpha v} - \alpha u = \frac{x+1}{\alpha}. \qquad (13.63)$$

Adding (13.62) and (13.63), we see that

$$-v - \frac{1}{v} = x + y + 2 \qquad (13.64)$$

which yields the following quadratic equation

$$v^2 + (x + y + 2)v + 1 = 0. \qquad (13.65)$$

This quadratic equation has two real distinct nonzero roots since the constant term is 1 and the discriminant is

$$(x + y + 2)^2 - 4 = (x + y)(x + y + 4) > 0.$$

Let v_o be one of the two distinct roots of (13.65). Clearly $v_o \neq 1$, since

$$v_o + \frac{1}{v_o} = -(x + y + 2) \neq 2$$

by (13.64). Hence if $u_o = \frac{y+1+v_o}{\alpha^2\,(1-v_o)}$, then u_o and v_o satisfy the system of equations (13.62)–(13.63).

For arbitrary $x, y \in \mathbb{R}$ with $x + y > 0$, we let (13.62) and (13.63) into (13.61) to obtain

$$f\left(\frac{x+1}{\alpha}\right) + f\left(\frac{y+1}{\alpha}\right) = f\left(-\frac{1}{\alpha}\right) + f\left(\frac{x+y+3}{\alpha}\right). \qquad (13.66)$$

We define

$$B(x) := f\left(\frac{x+1}{\alpha}\right) - f\left(-\frac{1}{\alpha}\right). \qquad (13.67)$$

Then by (13.67), the functional equation (13.66) reduces to

$$B(x + y + 2) = B(x) + B(y)$$

for all $x, y \in \mathbb{R}$ with $x + y \in \mathbb{R}_+$. For $x = 0$, $y > 0$ we see that $B(y+2) = B(0) + B(y)$ so that

$$B(x + y) + B(0) = B(x) + B(y)$$

for all $x, y \in \mathbb{R}$ with $x + y \in \mathbb{R}_+$. Now defining

$$A(x) := B(x) - B(0), \qquad x \in \mathbb{R},$$

we obtain

$$A(x + y) = A(x) + A(y) \qquad (13.68)$$

for all $x, y \in \mathbb{R}$ with $x + y \in \mathbb{R}_+$.

Now we extend (13.68) to $\mathbb{R}^2$. If $x + y \leq 0$, then we can find a $t \in \mathbb{R}$ such that $t + x \in \mathbb{R}_+$ and $t + x + y \in \mathbb{R}_+$. Then by (13.68), we get

$$\begin{aligned}
A(t) + A(x + y) \\
= A(t + x + y) \qquad &\text{(since } t + x + y \in \mathbb{R}_+\text{)} \\
= A(t + x) + A(y) \qquad &\text{(by (13.68))} \\
= A(t) + A(x) + A(y) \qquad &\text{(since } t + x \in \mathbb{R}_+\text{).}
\end{aligned}$$

Hence (13.68) holds for all x, y in $\mathbb{R}$ and A is an additive function on $\mathbb{R}$.

By (13.67), we get

$$f(x) = A(\alpha x) + A(-1) + f\left(-\frac{1}{\alpha}\right) + B(0). \qquad (13.69)$$

From (13.56) and (13.69), we get

$$g(x) = A(-\alpha^2 x) + A(-1) + f\left(-\frac{1}{\alpha}\right) + B(0) + b_1 + b_2, \qquad (13.70)$$

and from (13.52), (13.53), (13.55) and (13.69), we obtain

$$h(x) = A(\alpha x) + A(-1) + f\left(-\frac{1}{\alpha}\right) + B(0) + b_1 \qquad (13.71)$$

$$k(x) = A(\alpha x) + A(-1) + f\left(-\frac{1}{\alpha}\right) + B(0) + b_2. \qquad (13.72)$$

Letting $a = A(-1) + f(-\frac{1}{\alpha}) + B(0)$ into (13.69), (13.70), (13.71) and (13.72), we obtain the second part of solution (13.47)–(13.50).

It can be easily verified that (13.47)–(13.50) actually satisfy (FE). This completes the proof. $\qquad\qquad\square$

13.4 Concluding Remarks

The functional equation (HE) was mentioned for the first time by M. Hosszú at the International Symposium on Functional Equations (ISFE) held in Zakopane (Poland) in October 1967. Hosszú had solved (HE) under a differentiability assumption on the unknown function $f : \mathbb{R} \to \mathbb{R}$ (see Hosszú (1969)) and asked *what is the most general solution of the functional equation $f(x + y - xy) + f(xy) = f(x) + f(y)$ in the field of measurable functions?*

Swiatak (1968b) determined the solution of the Hosszú functional equation (HE) for all $x, y \in \mathbb{R}$ assuming that the unknown function $f : \mathbb{R} \to \mathbb{R}$ is continuous at the point $x = 0$ or at the point $x = 1$. She proved that the most general solution of the Hosszú functional equation satisfying one of these conditions has the form $f(x) = ax + b$, where a and b are arbitrary constants.

Fenyö (1969) found the solution of the Hosszú functional equation in the domain of distributions. Daróczy (1969) determined the general measurable solution of (HE) in the interval I, where I is either $(0, 1)$, $[0, 1]$ or $(-\infty, \infty)$. Assuming the integrability of f on the interval $(0, c)$, Swiatak (1968a) determined the solutions of the Hosszú functional equation. The same solution had been obtained earlier by Daróczy.

Blanusa (1970) and also independently Daróczy (1971) determined the general solution of the Hosszú functional equation. Let Γ and Ω be two arbitrary fields. A function $f : \Omega \to \Gamma$ is called additive if and only if $f(x + y) = f(x) + f(y)$ for all $x, y \in \Omega$. A function $f : \Omega \to \Gamma$ is called affine if and only if $f(x + y) + f(0) = f(x) + f(y)$ for all $x, y \in \Omega$. A function $f : \Omega \to \Gamma$ is called Jensen if and only if $2f(x+y) = f(2x) + f(2y)$ for all $x, y \in \Omega$. Let $\mathcal{B}(\Omega, \Gamma)$ denote the set of all functions f satisfying Hosszú functional equation with domain Ω and range Γ. Blanusa and Daróczy's result can be stated as follows:

Theorem 13.4. *Let Ω be a real field and Γ be a real or complex field. The function f belongs to $\mathcal{B}(\Omega, \Gamma)$ if and only if f is an affine function.*

Swiatak (1971) proved the following theorem concerning (HE).

Theorem 13.5. *Let Ω be a real field of characteristic different from 2 or 3, and Γ be an abelian group with no 2-torsion; and for every fixed $\gamma \in \Gamma$, there exists an additive map $A : \Omega \to \Gamma$ satisfying $A(1) = \gamma$. The function f belongs to $\mathcal{B}(\Omega, \Gamma)$ if and only if f is an affine function.*

Davison (1974a, 1974b), in two separate papers, studied the Hosszú functional equation in abstract structures. He proved the following two theorems.

Theorem 13.6. *Let Γ be an abelian group. Let Ω be either a ring of integers $\mathbb{Z}$, the ring of integers (mod k) $\mathbb{Z}_k$ or the field of rational numbers $\mathbb{Q}$. The function f belongs to $\mathcal{B}(\Omega, \Gamma)$ if and only if f is an affine function.*

Theorem 13.7. *Let Ω be a real field with more than 4 elements and Γ be an abelian group. The function f belongs to $\mathcal{B}(\Omega, \Gamma)$ if and only if f is an affine function.*

Glowacki and Kuczma (1979) considered the Hosszú functional equation when $\Omega = \mathbb{Z}$ and provided a new and didactic proof of Theorem 13.6 only when division by 6 is performable in the abelian group Γ. Davison and Redlin (1980) studied the equation (HE) for unknown functions f defined on a class of commutative rings more general than the class of fields.

Bagyinszki (1982) considered the following generalization of the Hosszú functional equation:

$$f(\tilde{x} + \tilde{y} - \tilde{x}\tilde{y}) + f(\tilde{x}\tilde{y}) = f(\tilde{x}) + f(\tilde{y}) \qquad (13.73)$$

for $\tilde{x}, \tilde{y} \in \Omega^n$, where $n \in \mathbb{N}$. Here $\tilde{x} = (x_1, x_2, ..., x_n)$, $\tilde{y} = (y_1, y_2, ..., y_n)$, $\tilde{x} + \tilde{y} = (x_1 + y_1, x_2 + y_2, ..., x_n + y_n)$ and $\tilde{x}\tilde{y} = (x_1 y_1, x_2 y_2, ..., x_n y_n)$.

Bagyinszki (1982) proved the following theorem.

Theorem 13.8. *Let $q = p^\alpha$, where p is a prime and $\alpha \geq 1$. Let $F_q = GF(q)$ be the corresponding finite field. The function f belongs to $\mathcal{B}(F_q, F_q)$ if and only if f is a polynomial of the form*

$$f(x) = \sum_{i=0}^{\alpha-1} a_i x^{p^i} + a_\alpha.$$

An analogous theorem also he proved for the function $f : F_q^n \to F_q$ satisfying the generalized Hosszú functional equation (13.73).

Lajkó (1973) determined the general solution of the following generalization of the Hosszú functional equation:

$$f(x) + f(y) - f(xy) = h(x + y - xy) \qquad (13.74)$$

for all $x, y \in \mathbb{R}$. He proved that the general solution of this equation is given by $f(x) = h(x) = A(x - 1) + f(1)$, where $A : \mathbb{R} \to \mathbb{R}$ is an additive function. He also studied the last functional equation when $f : \mathbb{R}_o \to \mathbb{R}$ and $h : \mathbb{R} \to \mathbb{R}$. Kannappan and Sahoo (1993) studied the pexiderized version of (HE), namely,

$$f(x + y + \alpha xy) + g(xy) = h(x) + k(y) \qquad (13.75)$$

for all $x, y \in \mathbb{R}$, where α is a real parameter. Section 2 of this chapter is based on Kannappan and Sahoo (1993). Koh (1994) determined the solution of (13.75) when $\alpha = 1$ in the space of distributions.

Daróczy (1999) considered the following Hosszú type functional equation:

$$f(x + y - x \circ y) + f(x \circ y) = f(x) + f(y) \qquad (13.76)$$

for all $x, y \in \mathbb{R}$, where $x \circ y = \ln(e^x + e^y)$. He found that the most general solution of the functional equation (13.76) is of the form $f(x) = A(x) + f(0)$, where $A : \mathbb{R} \to \mathbb{R}$ is an additive function.

13.5   Exercises

1. Find all functions $f : \mathbb{R} \to \mathbb{R}$ that satisfy the functional equation

$$f(x + y + xy) = f(x) + f(y) - f(xy)$$

for all $x, y \in \mathbb{R}$.

2. Find all functions $f : \mathbb{R} \to \mathbb{R}$ that satisfy the functional equation

$$f(xy + x) + f(y) = f(xy + y) + f(x)$$

for all $x, y \in \mathbb{R}$.

3. Find all functions $f : \mathbb{R} \to \mathbb{R}$ that satisfy the functional equation

$$f(x + y) + f(xy) = f(1 + xy) + f(x) + f(y)$$

for all $x, y \in \mathbb{R}$.

4. Find all functions $f : \mathbb{R} \to \mathbb{R}$ that satisfy the functional equation

$$f(x + y) = f(xy) + f(x) + f(y)$$

for all $x, y \in \mathbb{R}$.

5. Find all continuous functions $f : (0, 1) \to \mathbb{R}$ that satisfy the functional equation

$$f(x + y) = f(xy) + f(x) + f(y)$$

for all $x, y \in (0, 1)$.

6. Find all continuous functions $f : \mathbb{R} \to \mathbb{R}$ that satisfy the functional equation

$$f(x + y) + f(xy) = f(x) + f(y) + f(xy + 1)$$

for all $x, y \in \mathbb{R}$.

7. Find all functions $f, g, h, k : \mathbb{R} \to \mathbb{R}$ that satisfy the functional equation

$$f(x + y - xy) = g(x) + h(y) - k(xy)$$

for all $x, y \in \mathbb{R}$.

8. Let $a, b, c, d, \alpha, \beta$ be any six nonzero real numbers. Find all functions $f : \mathbb{R} \to \mathbb{R}$ that satisfy the functional equation

$$f(ax + by - \alpha xy) = cf(x) + df(y) - \beta f(xy)$$

for all $x, y \in \mathbb{R}$.

9. Let n be a given positive integer greater than or equal to two. Find all functions $f : \mathbb{R}^n \to \mathbb{R}$ that satisfy the functional equation

$$f(\tilde{x} + \tilde{y} - \tilde{x}\tilde{y}) = f(\tilde{x}) + f(\tilde{y}) - f(\tilde{x}\tilde{y})$$

for all $\tilde{x}, \tilde{y} \in \mathbb{R}^n$, where $\tilde{x} = (x_1, x_2, ..., x_n)$, $\tilde{y} = (y_1, y_2, ..., y_n)$, $\tilde{x} + \tilde{y} = (x_1 + y_1, x_2 + y_2, ..., x_n + y_n)$ and $\tilde{x}\tilde{y} = (x_1 y_1, x_2 y_2, ..., x_n y_n)$.

Chapter 14

Davison Functional Equation

14.1 Introduction

During the 17th International Symposium on Functional Equations, Davison (1980) introduced the functional equation

$$f(xy) + f(x + y) = f(xy + x) + f(y) \tag{14.1}$$

and asked for the general solution if the domain and the range of f are (commutative) fields. During the same meeting Benz (1980) gave the general continuous solution of (14.1) when f is an unknown function from the set of reals $\mathbb{R}$ to $\mathbb{R}$.

First, the continuous solution of Davison functional equation will be presented in this chapter and then the general solution will be given without any regularity assumption on the unknown function.

14.2 Continuous Solution of Davison Functional Equation

In this section, we present the continuous solution of Davison functional equation due to Benz (1980).

Theorem 14.1. *Every continuous solution of the equation*

$$f(xy) + f(x + y) = f(xy + x) + f(y) \tag{14.1}$$

for all $x, y \in \mathbb{R}$ is of the form

$$f(x) = ax + b, \tag{14.2}$$

where a and b are real constants.

Proof. First we write (14.1) as

$$f(xy + x) - f(xy) = f(x + y) - f(y) \qquad (14.3)$$

for all $x, y \in \mathbb{R}$. Letting xy for y in (14.3), we obtain

$$f(x^2y + x) - f(x^2y) = f(x + xy) - f(xy), \qquad (14.4)$$

for all $x, y \in \mathbb{R}$. Using the last equation and (14.3), we get

$$f(x^2y + x) - f(x^2y) = f(x + y) - f(y) \qquad (14.5)$$

for all $x, y \in \mathbb{R}$. Again substituting xy for y in (14.5), we obtain

$$f(x^3y + x) - f(x^3y) = f(x + xy) - f(xy)$$

for all $x, y \in \mathbb{R}$. Using (14.3) in the last equation, we obtain

$$f(x^3y + x) - f(x^3y) = f(x + y) - f(y).$$

Next, using induction, we obtain

$$f(x^ny + x) - f(x^ny) = f(x + y) - f(y) \qquad (14.6)$$

for all $x, y \in \mathbb{R}$ and for all $n \in \mathbb{N}$. Replacing y by $x^{-1}y$ in (14.3), it can be shown that

$$f\left(x^{-n}y + x\right) - f\left(x^{-n}y\right) = f(x + y) - f(y)$$

for all $n \in \mathbb{N}$. Hence

$$f(x^ny + x) - f(x^ny) = f(x + y) - f(y)$$

holds for all $n \in \mathbb{Z}$.

Suppose $0 \neq |x| < 1$. Then from (14.6), we get

$$f(x + y) - f(y) = \lim_{n \to \infty} [f(x^ny + x) - f(x^ny)].$$

Since f is continuous and $0 \neq |x| < 1$, we get

$$f(x + y) - f(y) = f\left(\lim_{n \to \infty} x^ny + x\right) - f\left(\lim_{n \to \infty} x^ny\right)$$
$$= f(x) - f(0).$$

Hence

$$f(x + y) = f(x) + f(y) - f(0)$$

for all $y \in \mathbb{R}$ and $0 \neq |x| < 1$.

Suppose $|x| > 1$. Then again taking $n \to -\infty$, we get

$$f(x + y) - f(y) = f\left(\lim_{n \to -\infty} x^n y + x\right) - f\left(\lim_{n \to -\infty} x^n y\right)$$
$$= f(x) - f(0).$$

Hence

$$f(x + y) = f(x) + f(y) - f(0)$$

for all $y \in \mathbb{R}$ and $|x| > 1$.

Therefore

$$f(x + y) = f(x) + f(y) - f(0) \tag{14.7}$$

for all $x \in \mathbb{R} \setminus \{0, 1, -1\}$ and $y \in \mathbb{R}$. Notice that (14.7) holds if $x = 0$. Taking $x \to 1$, we get

$$\lim_{x \to 1} f(x + y) = \lim_{x \to 1} f(x) + f(y) - f(0)$$

which is

$$f(1 + y) = f(1) + f(y) - f(0).$$

Similarly, letting $x \to -1$, we obtain

$$f(-1 + y) = f(-1) + f(y) - f(0).$$

Hence, we have

$$f(x + y) = f(x) + f(y) - f(0)$$

for all $x, y \in \mathbb{R}$. Defining

$$A(x) = f(x) - f(0),$$

we see that

$$A(x + y) = A(x) + A(y)$$

and

$$A(x) = ax,$$

where a is an arbitrary constant. Therefore

$$f(x) = ax + b,$$

where $b = f(0)$.

This completes the proof of the theorem. $\qquad\qquad \square$

14.3   General Solution of Davison Functional Equation

Next we give the general solution of the Davison equation without any regularity assumption. The general solution of (14.1) was given by Girgensohn and Lajkó in 2000.

Theorem 14.2. *The function* $f : \mathbb{R} \to \mathbb{R}$ *satisfies the functional equation* (14.1) *for all* $x, y \in \mathbb{R}$ *if and only if* f *is of the form*

$$f(x) = A(x) + b, \qquad (14.8)$$

where $A : \mathbb{R} \to \mathbb{R}$ *is an additive function and* $b \in \mathbb{R}$ *an arbitrary constant.*

Proof. Replacing y by $y + 1$ in (14.1), we obtain

$$f(xy + x) + f(x + y + 1) = f(xy + 2x) + f(y + 1) \qquad (14.9)$$

for all $x, y \in \mathbb{R}$. Adding (14.9) with (14.1), we get

$$f(xy) + f(x + y) + f(x + y + 1) = f(y) + f(xy + 2x) + f(y + 1) \quad (14.10)$$

for all $x, y \in \mathbb{R}$. Next, we replace x by $\frac{x}{2}$ and y by $2y$ in (14.10) to get

$$f(xy) + f\left(\tfrac{x}{2} + 2y\right) + f\left(\tfrac{x}{2} + 2y + 1\right)$$
$$= f(2y) + f(xy + x) + f(2y + 1) \qquad (14.11)$$

for all $x, y \in \mathbb{R}$. Subtracting (14.1) from (14.11), we obtain

$$f\left(\frac{x}{2} + 2y\right) + f\left(\frac{x}{2} + 2y + 1\right) - f(x + y)$$
$$= f(2y) + f(2y + 1) - f(y) \qquad (14.12)$$

for all $x, y \in \mathbb{R}$. Next, we replace x by $x - y$ in (14.12) to get

$$f\left(\frac{x}{2} + \frac{3y}{2}\right) + f\left(\frac{x}{2} + \frac{3y}{2} + 1\right) = f(2y) + f(2y + 1) - f(y) + f(x).$$

Replacing y by $\frac{y}{3}$ in the last equation, we obtain

$$f\left(\tfrac{x}{2} + \tfrac{y}{2}\right) + f\left(\tfrac{x}{2} + \tfrac{y}{2} + 1\right)$$
$$= f\left(\tfrac{2y}{3}\right) + f\left(\tfrac{2}{3}y + 1\right) - f\left(\tfrac{y}{3}\right) + f(x) \qquad (14.13)$$

for all $x, y \in \mathbb{R}$. Defining

$$\left.\begin{array}{l} A_1(t) = f\left(\frac{2t}{3}\right) + f\left(\frac{2t}{3} + 1\right) - f\left(\frac{t}{3}\right) \\ A_2(t) = f(t) \\ A_3(t) = f\left(\frac{t}{2}\right) + f\left(\frac{t}{2} + 1\right) \end{array}\right\} \tag{14.14}$$

and using (14.13), we see that

$$A_1(y) + A_2(x) = A_3(x + y) \tag{14.15}$$

for all $x, y \in \mathbb{R}$. The equation (14.15) is a Pexider equation and its solution can be found from Theorem 8.1 as

$$\left.\begin{array}{l} A_1(t) = A(t) + a \\ A_2(t) = A(t) + b \\ A_3(t) = A(t) + a + b, \end{array}\right\} \tag{14.16}$$

where $A : \mathbb{R} \rightarrow \mathbb{R}$ is an additive function and $a, b \in \mathbb{R}$ are arbitrary constants. Now using (14.16) and (14.14), we obtain

$$f(x) = A(x) + b$$

which is the asserted solution. $\qquad\square$

14.4 Concluding Remarks

Davison (1980), during the 17th ISFE, introduced the functional equation

$$f(xy) + f(x + y) = f(xy + x) + f(y)$$

and asked for the general solution if the domain and range of f are (commutative) fields. W. Benz (1980), during the same meeting proposed the following result: Every continuous solution $f : \mathbb{R} \rightarrow \mathbb{R}$ of (1) for all $x, y \in \mathbb{R}$ is of the form $f(x) = ax + b$, where a, b are real constants. The general real solution of the Davison functional equation was found in 2000 by Girgensohn and Lajkó (2000). They also determined the solution of the Pexiderized version

$$f(xy) + g(x + y) = h(xy + x) + k(y)$$

for $x, y \in \mathbb{R}$ and for $x, y \in \mathbb{R}_+$.

The proof of Girgensohn and Lajkó (2000) works if the domain of the function is replaced by a field of characteristics not equal to 2 and 3 and the codomain is replaced by an abelian group. Davison (2001) determined the solution of the Davison functional equation when the domain of the unknown function f is the set of natural numbers $\mathbb{N}$ or the ring of (rational) integers $\mathbb{Z}$.

14.5   Exercises

1. If a function $f : \mathbb{Z} \to \mathbb{R}$ satisfies the functional equation

$$f(x + y) + f(xy) = f(xy + x) + f(y)$$

for all $x, y \in \mathbb{Z}$, then show that

$$f(5) + f(0) = f(3) + f(2).$$

2. If a function $f : \mathbb{N} \to \mathbb{R}$ satisfies the functional equation

$$f(x + y) + f(xy) = f(xy + x) + f(y)$$

for all $x, y \in \mathbb{N}$, then show that for each $x \in \mathbb{N}$

$$f(2x) = f(x + 1) + f(x) - f(1)$$

and

$$f(2x + 1) = f(x + 3) + f(x + 2) - f(3) - f(2) + f(1).$$

3. Find all functions $f : \mathbb{Q} \to \mathbb{R}$ that satisfy the functional equation

$$f(x + y) + f(xy) = f(xy + x) + f(y)$$

for all $x, y \in \mathbb{Q}$.

4. Find all functions $f : [0, \infty) \to \mathbb{R}$ that satisfy the functional equation

$$f(xy) + f(x + y) = f(xy + x) + f(y)$$

for all $x, y \in [0, \infty)$.

5. Find all functions $f, g, h, k : \mathbb{R} \to \mathbb{R}$ that satisfy the functional equation

$$f(xy) + g(x + y) = h(xy + x) + k(y)$$

for all $x, y \in \mathbb{R}$.

6. Find all functions $f, g, h, k : [0, \infty) \to \mathbb{R}$ that satisfy the functional equation

$$f(xy) + g(x + y) = h(xy + x) + k(y)$$

for all $x, y \in [0, \infty)$.

7. Find all functions $f : (0, 1) \to \mathbb{R}$ that satisfy the functional equation

$$f(xy) + f(y - xy) = f(x) + f(y)$$

for all $x, y \in (0, 1)$.

8. Find all functions $f, g, h, k : (0, 1) \to \mathbb{R}$ that satisfy the functional equation

$$f(xy) + g(y - xy) = h(x) + k(y)$$

for all $x, y \in (0, 1)$.

9. Find all functions $f : \mathbb{R} \to \mathbb{R}$ that satisfy the functional equation

$$f(x + y + xy) = f(x + y) + f(xy)$$

for all $x, y \in \mathbb{R}$.

10. Find all functions $f : \mathbb{R} \to \mathbb{R}$ that satisfy the functional equation

$$f(x + y - xy) + f(x + xy) = 2f(x) + f(y)$$

for all $x, y \in \mathbb{R}$.

11. Find all functions $f : \mathbb{R} \to \mathbb{R}$ that satisfy the functional equation

$$f(x + y + xy) + f(x + y) = 2f(x) + 2f(y) + f(xy)$$

for all $x, y \in \mathbb{R}$.

12. Show that if the functions $f, g, h, k : \mathbb{R}_+ \to \mathbb{R}$ satisfy the functional equation

$$f(xy) + g(x + y) = h(xy + x) + k(y)$$

for all $x, y \in \mathbb{R}_+$, then

$$h(x) = f(x) + c \qquad \forall\, x \in \mathbb{R}_+$$

for some constant $c \in \mathbb{R}$.

13. Show that if the functions $f, g, h, k : \mathbb{R}_+ \to \mathbb{R}$ satisfy the functional equation

$$f(xy) + g(x + y) = h(xy + x) + k(y)$$

for all $x, y \in \mathbb{R}_+$, then

$$g(x) = A(x) + c \qquad \forall\, x \in \mathbb{R}_+,$$

where $A : \mathbb{R} \to \mathbb{R}$ is additive and $c \in \mathbb{R}$ is an arbitrary constant.

Chapter 15

Abel Functional Equation

15.1 Introduction

Niels Henrik Abel was one of the foremost mathematicians of the 19th century. He was born on August 5, 1802 in the small village of Findoe, Norway. At the age of 16, he began reading the classic mathematical works of Newton, Lagrange and Gauss. When Abel was 18 years old, his father died and the burden of supporting the family fell upon him. He took in private pupils and did odd jobs, while continuing to do mathematical research. At the age of 19, Abel solved a problem that had been open for hundreds of years. He proved that, unlike the situation for equations of degree 4 or less, there is no finite closed formula for the solution of the general fifth degree equation.

Abel's work laid the groundwork for abstract algebra. In addition to his work in the theory of equations, Abel made outstanding contributions to the theory of elliptic functions, elliptic integrals, Abelian integrals and infinite series. Abel died on April 6, 1829, at the age of 27. Camille Jordan (1838–1922) introduced the concept of Abelian groups in honor of Abel.

Abel dealt with several functional equations. Some of the functional equations he considered are the following:

$$f(\phi(x)) = f(x) + 1, \tag{A1}$$

$$f(x, f(y, z)) = f(f(x, y), z), \tag{A2}$$

$$f(x) + f(y) = f\left(\frac{x + y}{1 - xy}\right), \tag{A3}$$

$$f(x + y) = g(x)f(y) + f(x)g(y), \tag{A4}$$

$$f(x + y) = g(xy) + h(x - y), \tag{A5}$$

$$f(x) + f(y) = g(xf(y) + yf(x)), \tag{A6}$$

$$f(x + y)f(x - y) = f(x)^2 g(y)^2 - g(x)^2 f(y)^2, \tag{A7}$$

$$f(x + y)f(x - y) = f(x)^2 f(y)^2 - c^2 g(x)^2 g(y)^2 \tag{A8}$$

for all $x, y \in \mathbb{R}$. Abel presented a general method of solving functional equations by reducing them to differential equations.

In 1665, Newton discovered the binomial series. The binomial series is given by

$$(1 + x)^\alpha = 1 + \binom{\alpha}{1} y + \binom{\alpha}{2} y^2 + \cdots + \binom{\alpha}{n} y^n + \cdots ,$$

where α is a real number and

$$\binom{\alpha}{k} = \frac{\alpha(\alpha - 1)(\alpha - 2) \cdots (\alpha - k + 1)}{k!}.$$

The term $\binom{\alpha}{k}$ is called the generalized binomial coefficient.

Abel in 1826 justified "Newton's binomial series" by using the functional equation

$$f(z + w) = f(z)f(w)$$

for all $z, w \in \mathbb{C}$.

The main goal of this chapter is to determine the general solution of the Abel functional equation (A5) without any regularity assumption on the unknown functions.

15.2 General Solution of the Abel Functional Equation

In this section, we study the functional equation

$$\psi(x + y) = g(xy) + h(x - y)$$

for all $x, y \in \mathbb{R}$.

Theorem 15.1. *If the functions* $\psi, g, h : \mathbb{R} \to \mathbb{R}$ *satisfy the functional equation*

$$\psi(x + y) = g(xy) + h(x - y) \tag{FE}$$

for all $x, y \in \mathbb{R}$, *then*

$$\psi(x) = A\left(\frac{x^2}{4}\right) + \alpha + \beta \tag{15.1}$$

$$g(x) = A(x) + \alpha \tag{15.2}$$

$$h(x) = A\left(\frac{x^2}{4}\right) + \beta, \qquad (15.3)$$

where α, β are arbitrary constants and $A : \mathbb{R} \to \mathbb{R}$ is an additive function.

Proof. Substituting $y = 0$ in (FE), we obtain

$$\psi(x) = g(0) + h(x). \qquad (15.4)$$

Hence we see that $g(0) = \psi(0) - h(0)$. Letting $x = y = \frac{t}{2}$ in (FE), we have

$$\psi(t) = g\left(\frac{t^2}{4}\right) + h(0). \qquad (15.5)$$

Similarly, letting $x = -y = \frac{t}{2}$, we obtain

$$\psi(0) = g\left(-\frac{t^2}{4}\right) + h(t). \qquad (15.6)$$

Using (15.5) and (15.6) on (FE), we get

$$g\left(\frac{(x+y)^2}{4}\right) + h(0) = g(xy) + \psi(0) - g\left(-\frac{(x-y)^2}{4}\right)$$

for all $x, y \in \mathbb{R}$.

By transformations

$$u = xy \qquad \text{and} \qquad v = \frac{(x-y)^2}{4},$$

we obtain

$$g(u+v) = g(u) - g(-v) + g(0), \qquad (15.7)$$

where

$$g(0) = \psi(0) - h(0).$$

Note that $u + v \geq 0$, since

$$4(u+v) = (x+y)^2 \geq 0.$$

Then (15.7) holds for all $(u, v) \in D := \{(u, v) \mid v \geq 0, \ u+v \geq 0\}$. Letting $u = 0$ in (15.7), we get

$$g(v) + g(-v) = 2g(0). \qquad (15.8)$$

Hence

$$g(-v) = -g(v) + 2g(0). \qquad (15.9)$$

By (15.9), (15.7) yields

$$g(u + v) = g(u) + g(v) - g(0). \qquad (15.10)$$

Define

$$A(x) := g(x) - g(0). \qquad (15.11)$$

Then (15.10) yields

$$A(u + v) = A(u) + A(v) \qquad (15.12)$$

for all $(u, v) \in D$. Because of symmetry in u and v, we obtain from (15.12) that (15.12) holds on $D^* := \{(u, v) \in \mathbb{R}^2 \,|\, u + v \geq 0\}$. Now we extend (15.12) from D^* (see Figure 15.1) to all of $\mathbb{R}^2$.

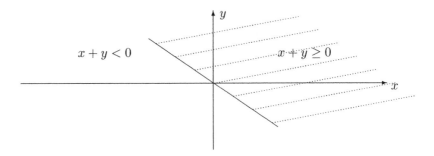

Figure 15.1. Illustration of the domain D^*.

Choose $u, v \in \mathbb{R}$ such that $u + v < 0$. Then there exists a real number $x \geq 0$ such that

$$x + u \geq 0$$

and

$$x + u + v \geq 0.$$

Consider

$$
\begin{aligned}
A(x) + A(u + v) &= A(x + u + v) && \text{(by (15.12))} \\
&= A(x + u) + A(v) && \text{(by (15.12))} \\
&= A(x) + A(u) + A(v) && \text{(by (15.12))}.
\end{aligned}
$$

Hence, we have

$$A(u + v) = A(u) + A(v)$$

for all $u, v \in \mathbb{R}$ with $u + v < 0$. Thus A is additive on $\mathbb{R}$. Therefore by (15.11), we obtain

$$g(x) = A(x) + \alpha, \qquad (15.13)$$

where $\alpha = g(0)$. Using (15.13) and (15.5), we get

$$\psi(x) = A\left(\frac{x^2}{4}\right) + \alpha + \beta, \qquad (15.14)$$

where $\beta = h(0)$. Using (15.13) and (15.14) on (15.6), we get

$$h(x) = A\left(\frac{x^2}{4}\right) + \beta. \qquad (15.15)$$

Now the proof of the theorem is complete. □

15.3   Concluding Remarks

In 1823, Abel studied the functional equation

$$\psi(x + y) = g(xy) + h(x - y)$$

for all $x, y \in \mathbb{R}$ which is, today, known as the Abel functional equation. Roşcău (1960) determined the solution of the Abel equation assuming the unknown functions to be continuous. However his proof was not quite correct. The general measurable solution of the Abel equation was found by Stamate (1964). Lajkó (1987) determined the general solution of the Abel equation without any regularity assumption. Independently, Aczél (1989) also determined the solution of the Abel equation by reducing it to the additive Cauchy functional equation. Lajkó (1994) determined the general solution of the Abel equation by an elementary method. Smajdor (1999) studied the Abel equation when the unknown functions are defined on the interval $[0, \infty)$ with range on an abelian semigroup with zero and satisfying cancellation law. Chung, Ebanks, Ng, Sahoo and Zeng (1994) determined the general solution of the Abel functional equation without any regularity assumption when the unknown functions are defined on certain type of fields and their range is in an abelian group.

The functional equation

$$f(x + y)\, h(x - y) = g(xy) \qquad \forall\, x, y \in \mathbb{R} \qquad (15.16)$$

is a multiplicative form of the Abel functional equation. The general measurable solution of this equation was obtained by Filipescu (1969). Lajkó (1994) gave the general solution of this functional equation without any regularity assumptions on the unknown functions. In particular he proved the following result.

Theorem 15.2. *If the functions $f, g, h : \mathbb{R} \to \mathbb{R}$ satisfy the functional equation (15.16), then either*

$$f(x) = a\, e^A \left(\frac{x^2}{4} \right), \quad h(x) = b\, e^{-A\left(\frac{x^2}{4} \right)}, \quad g(x) = ab\, e^A(x),$$

or

$$f(x) = 0, \quad h(x) = \text{arbitrary}, \quad g(x) = 0$$

or

$$f(x) = \text{arbitrary}, \quad h(x) = 0, \quad g(x) = 0,$$

where $A : \mathbb{R} \to \mathbb{R}$ is an additive function and a, b are nonzero real arbitrary constants.

15.4   Exercises

1. Find all functions $f, g, h : \mathbb{R} \to \mathbb{R}$ that satisfy the functional equation

$$f(x + y) = g(xy) + h(x) - h(y)$$

for all $x, y \in \mathbb{R}$.

2. Without using Theorem 15.1, find the general solutions $f. g : \mathbb{R} \to \mathbb{R}$ of the functional equation

$$f(x + y) = g(xy) + f(x - y)$$

for all $x, y \in \mathbb{R}$.

3. Find all functions $f, g, h : (0, \infty) \to \mathbb{R}$ that satisfy the functional equation

$$f(x + y) + g(xy) = h(x) + h(y)$$

for all $x, y \in (0, \infty)$.

4. Find all functions $f, g : (0, \infty) \to \mathbb{R}$ that satisfy the functional equation

$$g(xy) = f(x + y) - f(x) - f(y)$$

for all $x, y \in (0, \infty)$.

5. Find all functions $f, g : \mathbb{R} \to \mathbb{R}$ that satisfy the functional equation

$$f(x + y) = f(x)\, f(y)\, g(xy)$$

for all $x, y \in \mathbb{R}$.

6. Find all functions $f, g : \mathbb{R} \rightarrow \mathbb{R}$ that satisfy the functional equation

$$f(x - y) = f(x) f(y) g(xy)$$

for all $x, y \in \mathbb{R}$.

7. Find all functions $f, g, h : \mathbb{R} \rightarrow \mathbb{R}$ that satisfy the functional equation

$$f(xy) = g(x + y) + h\left(\frac{x}{y}\right)$$

for all $x, y \in \mathbb{R} \smallsetminus \{0\}$.

8. Find all functions $f : \mathbb{R} \rightarrow \mathbb{R}$ that satisfy the functional equation

$$f\left(\frac{x + y}{2}\right) + f\left(\frac{2xy}{x + y}\right) = f(x) + f(y)$$

for all $x, y \in \mathbb{R}$.

9. Find all functions $f, g : \mathbb{R} \rightarrow \mathbb{R}$ that satisfy the functional equation

$$f(x + y + xy) = g(xy) + f(x + y - xy)$$

for all $x, y \in \mathbb{R}$.

10. Let a, b, c, d be nonzero integers. Find all functions $f, g, h : \mathbb{R} \rightarrow \mathbb{R}$ that satisfy the functional equation

$$f(ax + by) = g(xy) + h(cx - dy)$$

for all $x, y \in \mathbb{R}$.

Chapter 16

Mean Value Type Functional Equations

16.1 Introduction

In this chapter, we will examine some functional equations that arise from the Lagrange mean value theorem. We call these functional equations the mean value type of functional equations. The study of this type of functional equations was started by Aczél (1985), Haruki (1979) and Kuczma (1991a, 1991b). A detailed account on mean value type functional equations can be found in the book *Mean Value Theorems and Functional Equations* by Sahoo and Riedel (1998).

16.2 The Mean Value Theorem

The mean value theorem (MVT) whose proof can be in the book *Mean Value Theorems and Functional Equations* by Sahoo and Riedel (1998) (see pages 25–27) is the following.

Theorem 16.1. *Suppose the real-valued function f is continuous on the closed interval $[a, b]$ and is differentiable on the open interval (a, b). Then there exists $\eta \in (a, b)$ such that*

$$f'(\eta) = \frac{f(b) - f(a)}{b - a}. \tag{16.1}$$

The equation of the tangent line at the point $(\eta, f(\eta))$ is given by

$$y = (x - \eta)f'(\eta) + f(\eta). \tag{16.2}$$

The equation of the secant line joining the points $(a, f(a))$ and $(b, f(b))$

is given by

$$y = (x - a)\left(\frac{f(b) - f(a)}{b - a}\right) + f(a). \tag{16.3}$$

If the secant line is parallel to the tangent line at η (see Figure 16.1), then

$$f'(\eta) = \frac{f(b) - f(a)}{b - a}. \tag{16.4}$$

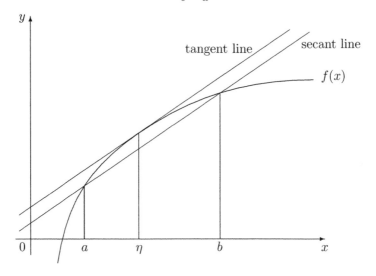

Figure 16.1. Geometrical interpretation of MVT.

This Lagrange's mean value theorem is an important theorem in differentiable calculus. This theorem was first discovered by Joseph Louis Lagrange (1736–1813), but the idea of applying the Rolle's theorem to a suitably contrived auxiliary function was given by Ossian Bonnet (1819–1892). However, the first statement of the theorem appears in a paper by renowned physicist André-Marie Ampére (1775–1836).

If $f : \mathbb{R} \to \mathbb{R}$ is a differentiable function and $[x, y]$ is any interval, then according to the mean-value theorem there exists a mean value $\eta \in (x, y)$ such that

$$\frac{f(x) - f(y)}{x - y} = f'(\eta(x, y)). \tag{16.5}$$

Obviously, $\eta(x, y)$ is a function of x and y. One may ask for what f does the mean value $\eta(x, y)$ depend on x and y in a given manner. From this point of view, (16.5) appears as a functional equation with unknown function f and given $\eta(x,y)$. The function $\eta(x,y)$ could, in general, be any linear or nonlinear combination of x,y such as

$$\eta(x,y) = \frac{1}{2}(x+y), \tag{16.6}$$

$$\eta(x,y) = \sqrt{xy} \qquad \text{or} \tag{16.7}$$

$$\eta(x,y) = \sqrt[p]{\frac{x^p + y^p}{2}}. \tag{16.8}$$

In (16.5), the derivative of f can be replaced by another unknown function, say, h. Then (16.5) becomes

$$\frac{f(x) - f(y)}{x - y} = h(\eta(x,y)). \tag{16.9}$$

With

$$\eta(x,y) = \frac{x+y}{2}$$

in (16.9), we get the functional equation

$$f(x) - f(y) = (x-y)h\left(\frac{x+y}{2}\right) \tag{16.10}$$

for all $x,y \in \mathbb{R}$.

16.3   A Mean Value Type Functional Equation

Haruki (1979) and independently also Aczél (1985) found the general solution of (16.10). Aczél proved the following theorem without any differentiability or other regularity conditions in 1963 but forgot about it; however the result and the proof were preserved in Aczél (1963).

Theorem 16.2. *Let* $f : \mathbb{R} \to \mathbb{R}$ *satisfy*

$$f(x) - f(y) = (x-y)h\left(\frac{x+y}{2}\right) \tag{16.10}$$

for all $x,y \in \mathbb{R}$. *Then there exist constants* a,b,c *such that*

$$\left.\begin{array}{l} f(x) = ax^2 + bx + c \\ h(x) = 2ax + b. \end{array}\right\} \tag{16.11}$$

Proof. Note that if f is a solution of (16.10), then so also is $f + c$, where c is a constant. Hence, without loss of generality, we may assume

$$f(0) = 0. \tag{16.12}$$

Inserting $y = 0$ in (16.10), we get

$$f(x) = xh\left(\frac{x}{2}\right). \tag{16.13}$$

Letting (16.13) into (16.10), we have

$$xh\left(\frac{x}{2}\right) - yh\left(\frac{y}{2}\right) = (x - y)h\left(\frac{x + y}{2}\right) \tag{16.14}$$

for all $x, y \in \mathbb{R}$. If h is a solution of (16.14), then $h + b$ is a solution of (16.14), where b is a constant. So, without loss of generality, we also assume

$$h(0) = 0. \tag{16.15}$$

Letting $x = -y$ in (16.14) yields

$$-yh\left(-\frac{y}{2}\right) = yh\left(\frac{y}{2}\right) \tag{16.16}$$

for all $y \in \mathbb{R}$. Thus replacing y by $2y$, we see that

$$-2yh(-y) = 2yh(y)$$

which is

$$h(-y) = -h(y)$$

for all $y \in \mathbb{R}\setminus\{0\}$. Since $h(0) = 0$, we see that

$$-h(-y) = h(y) \tag{16.17}$$

for all $y \in \mathbb{R}$. Thus h is an odd function.

Next, replacing y by $-y$ in (16.14), we get

$$xh\left(\frac{x}{2}\right) + yh\left(-\frac{y}{2}\right) = (x + y)h\left(\frac{x - y}{2}\right).$$

Since h is odd, the above equation yields

$$xh\left(\frac{x}{2}\right) - yh\left(\frac{y}{2}\right) = (x + y)h\left(\frac{x - y}{2}\right). \tag{16.18}$$

Subtracting (16.18) from (16.14), we see that

$$(x - y)h\left(\frac{x + y}{2}\right) = (x + y)h\left(\frac{x - y}{2}\right). \tag{16.19}$$

Substituting

$$2u = x + y \atop 2v = x - y \Bigg\}$$ (16.20)

in (16.19), we obtain

$$vh(u) = uh(v);$$ (16.21)

thus

$$h(u) = 2au,$$

where a is a constant. Removing the assumption $h(0) = 0$, we get

$$h(u) = 2au + b.$$ (16.22)

Letting (16.22) into (16.13), we have

$$f(x) = x(ax + b)$$

or

$$f(x) = ax^2 + bx.$$

As before, removing the $f(0) = 0$ assumption, we arrive at

$$f(x) = ax^2 + bx + c.$$ (16.23)

This completes the proof of the theorem.   □

16.4   Generalizations of Mean Value Type Equation

The functional equation

$$f(x) - f(y) = (x - y)h\left(\frac{x+y}{2}\right)$$ (16.10)

can be written as

$$\frac{f(x) - f(y)}{x - y} = g(x + y),$$ (16.24)

where

$$g(x) := h\left(\frac{x}{2}\right).$$ (16.25)

One disadvantage in the representation (16.24) is that it holds for all $x, y \in \mathbb{R}$ with $x \neq y$. The advantage in (16.24) is that it can be represented as

$$f[x, y] = g(x + y)$$ (16.26)

for all $x, y \in \mathbb{R}$ with $x \neq y$. Here

$$f[x, y] = \frac{f(x) - f(y)}{x - y}$$

represents the 2-point divided difference. The representation (16.26) allows us to generalize (16.26) to

$$f[x, y, z] = g(x + y + z) \tag{16.27}$$

or even to

$$f[x_1, x_2, \ldots, x_n] = g\left(\sum_{k=1}^{n} x_k\right). \tag{16.28}$$

Definition 16.1. *Given n distinct points, $x_1, x_2, \ldots, x_n$ the n-point divided difference of $f : \mathbb{R} \to \mathbb{R}$ is recursively defined as*

$$f[x_1] = f(x_1)$$

and

$$f[x_1, x_2, \ldots, x_n] = \frac{f[x_1, x_2, \ldots, x_{n-1}] - f[x_2, \ldots, x_n]}{x_1 - x_n}.$$

If $n = 2$, then the above yields

$$f[x, y] = \frac{f(x) - f(y)}{x - y}.$$

Similarly, the three-point-divided difference of f is deduced from the definition as

$$f[x, y, z] = \frac{f[x, y] - f[y, z]}{x - z},$$

that is,

$$f[x, y, z] = \frac{1}{(x - z)} \left[\frac{f(x) - f(y)}{x - y} - \frac{f(y) - f(z)}{y - z} \right]$$

which is

$$f[x, y, z] = \frac{(y - z)f(x) + (z - x)f(y) + (x - y)f(z)}{(x - z)(x - y)(y - z)}.$$

Next we present a few basic facts.

Proposition 16.1. *If $f(x) = bx + c$, then*

$$f[x, y, z] = 0.$$

Proposition 16.2. *If $f(x) = ax^2 + bx + c$, then*

$$f[x, y, z] = a.$$

Proposition 16.3. *If $f(x) = ax^3 + bx^2 + cx + d$, then*

$$f[x, y, z] = a(x + y + z).$$

Proposition 16.4.

$$f[x_1, x_2, \ldots, x_n] = \sum_{i=1}^{n} \frac{f(x_i)}{\prod_{\substack{j \neq i}} (x_i - x_j)}.$$

The functional equation (16.27) was, for the first time, studied by Bailey (1992). He gave the differentiable solution of (16.27) and asked to determine the general solution of (16.27) without the differentiability assumption. In 1993, Chung and Sahoo gave the locally integrable solution of this functional equation. In 1995, Kannappan and Sahoo determined the general solution of (16.28) without any regularity assumption.

Theorem 16.3. *Let $f, h : \mathbb{R} \to \mathbb{R}$ satisfy the functional equation*

$$f[x_1, x_2, x_3] = h(x_1 + x_2 + x_3) \qquad \text{(BE)}$$

for all $x_1, x_2, x_3 \in \mathbb{R}$ with $x_1 \neq x_2$, $x_2 \neq x_3$, $x_3 \neq x_1$. The f is a polynomial of degree at most three and h is linear.

Proof. If $f(x)$ is a solution of (BE), so also is

$$f(x) + a_0 + a_1 x,$$

where a_0 and a_1 are arbitrary real constants. This can be verified by direct substitution into (BE), that is,

$$(x_2 - x_3)f(x_1) + (x_3 - x_1)f(x_2) + (x_1 - x_2)f(x_3)$$
$$= (x_1 - x_3)(x_1 - x_2)(x_2 - x_3)h(x_1 + x_2 + x_3). \qquad (16.29)$$

Letting

$$f(x_i) + a_0 + a_1 x_i$$

for $i = 1, 2, 3$ for $f(x_i)$ in (16.29), we see that

$$(x_2 - x_3)[f(x_1) + a_0 + a_1 x_1] + (x_3 - x_1)[f(x_2) + a_0 + a_1 x_2]$$
$$+ (x_1 - x_2)[f(x_3) + a_0 + a_1 x_3]$$
$$= (x_2 - x_3)f(x_1) + (x_3 - x_1)f(x_2) + (x_1 - x_2)f(x_3)$$

$$= (x_1 - x_3)(x_1 - x_2)(x_2 - x_3)h(x_1 + x_2 + x_3),$$

which is (BE).

Let

$$g(x) = f(x) + a_0 + a_1 x. \tag{16.30}$$

Then $x = 0$ yields

$$f(0) = g(0) - a_0.$$

Picking $g(0) = a_0$, we get

$$f(0) = 0. \tag{16.31}$$

In other words, by a suitable choice for a_0, we may assume $f(0) = 0$ without loss of generality.

Now letting $x = \alpha$ in (16.24), we get

$$f(\alpha) = g(\alpha) - a_0 - a_1 \alpha. \tag{16.32}$$

Letting $a_0 + a_1 \alpha = g(\alpha)$, we get $f(\alpha) = 0$. Hence we may assume, without loss of generality, that $f(\alpha) = 0$.

Now substitute $(x, 0, \alpha)$ for (x_1, x_2, x_3) in (16.29) to get

$$f(x) = x(x - \alpha)h(x + \alpha). \tag{16.33}$$

Again substituting $(x, 0, y)$ for (x_1, x_2, x_3) in (16.29), we have

$$-yf(x) + xf(y) = -xy(x - y)h(x + y). \tag{16.34}$$

Hence

$$\frac{f(x)}{x} - \frac{f(y)}{y} = (x - y)h(x + y) \tag{16.35}$$

for all $x, y \in \mathbb{R}$, $x, y \neq 0$.

Define

$$g(x) = \frac{f(x)}{x}, \quad x \neq 0. \tag{16.36}$$

Then (16.35) yields

$$g(x) - g(y) = (x - y)h(x + y) \tag{16.37}$$

for all $x, y \in \mathbb{R} \backslash \{0\}$.

Substitution of $y = -x$ into (16.37) yields

$$g(x) - g(-x) = 2xh(0) \tag{16.38}$$

for $x \neq 0$. Replacing y by $-y$ into (16.37), we get

$$g(x) - g(-y) = (x + y)h(x - y) \tag{16.39}$$

for $x, y \neq 0$. Subtracting (16.37) from (16.39), we have

$$g(y) - g(-y) = (x + y)h(x - y) - (x - y)h(x + y).$$

Using (16.38) in the above equation, we obtain

$$(x + y)[h(x - y) - h(0)] = (x - y)[h(x + y) - h(0)]. \tag{16.40}$$

Fix a nonzero $u \in \mathbb{R}$. Choose a real number $v \in \mathbb{R}$ such that $(u+v)/2 \neq 0$ and $(u - v)/2 \neq 0$. There are plenty of choices for such v. Let

$$x = \frac{u + v}{2}, \qquad y = \frac{u - v}{2}$$

so that

$$\left. \begin{array}{l} u = x + y \\ v = x - y \end{array} \right\}. \tag{16.41}$$

Letting (16.41) into (16.40), we obtain

$$u[h(v) - h(0)] = v[h(u) - h(0)] \tag{16.42}$$

for all $v \in \mathbb{R} \setminus \{-u, u\}$. Hence, for a fixed $u = u_1$, we have from (16.42)

$$h(v) = \frac{h(u_1) - h(0)}{u_1} v + h(0) \tag{16.43}$$

or

$$h(v) = a_1 v + b_1 \tag{16.44}$$

for all $v \in \mathbb{R} \setminus \{-u_1, u_1\}$. Again letting $u = u_2$ in (16.42), we get

$$h(v) = a_2 v + b_2 \tag{16.45}$$

for all $v \in \mathbb{R} \setminus \{-v_2, v_2\}$.

Since the sets $\{u_1, -u_1\}$ and $\{u_2, -u_2\}$ are disjoint, by (16.44) and (16.45), we obtain

$$h(v) = av + b \tag{16.46}$$

for all $v \in \mathbb{R}$. Now inserting (16.46) into (16.33), we get

$$\begin{aligned} f(x) &= (x^2 - x\alpha)h(x + \alpha) \\ &= (x^2 - x\alpha)[a(x + \alpha) + b] \\ &= ax^3 + bx^2 + cx, \end{aligned} \tag{16.47}$$

where $c = -a\alpha^2 - b\alpha$. If we remove the assumption of $f(0) = 0$, we get

$$f(x) = ax^3 + bx^2 + cx + d$$

for all $x \neq 0, \alpha$. This now yields that f is a polynomial of at most degree 3 and h is linear. □

In 1989, Walter Rudin raised the following problem in the elementary problem section of *American Mathematical Monthly*. Let s and t be two given real numbers. Find all real functions f that satisfy

$$f[x, y] = f'(sx + ty)$$

for all real numbers x, y with $x \neq y$.

Kannappan, Sahoo and Jacobson (1995), among others, investigated the general solution of

$$f[x, y] = h(sx + ty) \tag{16.48}$$

for all $x, y \in \mathbb{R}$ with $x \neq y$.

Theorem 16.4. *Suppose s and t are two real parameters. The functions $f, h : \mathbb{R} \to \mathbb{R}$ satisfy (16.48), that is*

$$f[x, y] = h(sx + ty)$$

for all $x, y \in \mathbb{R}$ with $x \neq y$ if and only if

$$f(x) = \begin{cases} ax + b & \text{if } s = 0 = t \\ ax + b & \text{if } s = 0, \ t \neq 0 \\ ax + b & \text{if } s \neq 0, \ t = 0 \\ atx^2 + ax + b & \text{if } s = t \neq 0 \\ \frac{A(tx)}{t} + b & \text{if } s = -t \neq 0 \\ \beta x + b & \text{if } s^2 \neq t^2 \end{cases}$$

$$h(y) = \begin{cases} \text{arbitrary} & \text{if } s = 0 = t \\ a & \text{if } s = 0, \ t \neq 0 \\ a & \text{if } s \neq 0, \ t = 0 \\ \alpha y + a & \text{if } s = t \neq 0 \\ \frac{A(y)}{y} & \text{if } s = -t \neq 0 \\ \beta & \text{if } s^2 \neq t^2, \end{cases}$$

where $A : \mathbb{R} \to \mathbb{R}$ is an additive function and a, b, α, β are arbitrary constants.

Proof. Rewriting (16.48), we obtain

$$\frac{f(x) - f(y)}{x - y} = h(sx + ty). \tag{16.49}$$

Case 1. Suppose $s = 0 = t$. Then (16.49) yields

$$f(x) - f(y) = (x - y)h(0)$$

which is

$$f(x) - xh(0) = f(y) - yh(0).$$

Hence

$$f(x) - ax = f(y) - ay, \tag{16.50}$$

where $h(0) = a$. Since the left side depends on x and the right side depends on y, each side is a constant, say, b. Thus, we have

$$f(x) = ax + b \tag{16.51}$$

and h is arbitrary with $h(0) = a$.

Case 2. Suppose $s = 0$ and $t \neq 0$. Then (16.49) implies

$$f(x) - f(y) = (x - y)h(ty) \tag{16.52}$$

for all $x, y \in \mathbb{R}$. Letting $y = 0$ in (16.52), we obtain

$$f(x) = xh(0) + f(0)$$

which is

$$f(x) = ax + b, \quad x \neq 0, \tag{16.53}$$

where $a = h(0)$ and $b = f(0)$. So, (16.53) holds for $x = 0$ also. Substitution of (16.53) into (16.52) yields

$$a(x - y) = (x - y)h(ty)$$

which is

$$h(ty) = a, \quad y \in \mathbb{R}.$$

This yields

$$h(y) = a, \quad \forall \, y \in \mathbb{R}. \tag{16.54}$$

Case 3. Suppose $s \neq 0$ and $t = 0$. Then (16.49) implies

$$f(x) - f(y) = (x - y)h(sx). \tag{16.55}$$

Similar to the solution of (16.52), we obtain the solution

$$\left. \begin{array}{l} f(x) = ax + b \\ h(y) = b. \end{array} \right\}$$

Case 4.   Suppose $s \neq 0 \neq t$. Letting $x = 0$ in (16.49), we get

$$\frac{f(0) - f(y)}{0 - y} = h(ty)$$

or

$$f(y) = yh(ty) + f(0)$$

which is

$$f(y) = yh(ty) + b, \tag{16.56}$$

where $b = f(0)$. Similarly letting $y = 0$ in (16.49), we get

$$f(x) = xh(sx) + b. \tag{16.57}$$

Letting (16.56) and (16.57) into (16.49), we see that

$$xh(xs) - yh(ty) = (x - y)h(sx + ty) \tag{16.58}$$

for all $x, y \in \mathbb{R}\backslash\{0\}$ with $x \neq y$. Replacing x by $\frac{x}{s}$ and y by $\frac{y}{t}$ in (16.58), we have

$$\frac{x}{s}h(x) - \frac{y}{t}h(y) = \left(\frac{x}{s} - \frac{y}{t} \right) h(x + y) \tag{16.59}$$

for all $x, y \in \mathbb{R}\backslash\{0\}$ with $x \neq y$.

Case 4 gives rise to 3 subcases.

Subcase (i).   Suppose $s = t$. Then (16.59) yields

$$xh(x) - yh(y) = (x - y)h(x + y) \tag{16.60}$$

for all $x, y \in \mathbb{R}\backslash\{0\}$ with $x \neq y$. Letting $y = -x$ in (16.60), we get

$$xh(x) + xh(-x) = 2xh(0). \tag{16.61}$$

Next, replacing y by $-y$ in (16.60), we see that

$$xh(x) + yh(-y) = (x + y)h(x - y). \tag{16.62}$$

Subtracting (16.60) from (16.62) and using (16.61), we obtain

$$2yh(0) = (x + y)h(x - y) - (x - y)h(x + y).$$

Letting $u = x + y$ and $v = x - y$ into the above equation, we have

$$(u - v)h(0) = uh(v) - vh(u)$$

which is

$$v[h(u) - h(0)] = u[h(v) - h(0)] \tag{16.63}$$

for all $u, v, u - v, u + v \in \mathbb{R}\backslash\{0\}$. Thus

$$h(u) = \alpha u + a, \tag{16.64}$$

where α is a constant and $a := h(0)$. Thus (16.64) holds for $u = 0$ also. Letting (16.64) into (16.49), we see that

$$f(x) - f(y) = (x - y)[\alpha tx + \alpha ty + a],$$

that is,

$$f(x) - \alpha tx^2 - ax = f(y) - \alpha ty^2 - ay.$$

Hence

$$\left. \begin{array}{l} f(x) = \alpha tx^2 + ax + b \\ h(y) = \alpha y + b. \end{array} \right\} \tag{16.65}$$

Subcase (ii). Suppose $s = -t$. Then (16.59) gives

$$xh(x) + yh(y) = (x + y)h(x + y) \tag{16.66}$$

for all $x, y \in \mathbb{R} \setminus \{0\}$ with $x + y \neq 0$. We define $A : \mathbb{R} \to \mathbb{R}$ as follows:

$$A(x) = \begin{cases} xh(x) & \text{if } x \neq 0 \\ 0 & \text{if } x = 0. \end{cases}$$

Then (16.66) yields

$$A(x) + A(y) = A(x + y) \tag{16.67}$$

for all $x, y, x + y \in \mathbb{R}\backslash\{0\}$. Now we show that A is additive on $\mathbb{R}$. In view of (16.67) it is enough to show that

$$A(x) + A(-x) = 0 \tag{16.68}$$

for all $x \in \mathbb{R}$.

Interchanging y with $-y$ in (16.66), we obtain

$$xh(x) - yh(-y) = (x - y)h(x - y). \tag{16.69}$$

Subtracting (16.69) from (16.66), we see that

$$yh(y) + yh(-y) = (x + y)h(x + y) - (x - y)h(x - y). \tag{16.70}$$

By the definition of A, (16.70) reduces to

$$A(y) - A(-y) = A(x + y) - A(x - y). \tag{16.71}$$

Replacing x with $-x$ in (16.71), we obtain

$$A(y) - A(-y) = A(-x + y) - A(-x - y). \tag{16.72}$$

Hence, by (16.71) and (16.72), we get

$$A(x + y) + A(-(x + y)) = A(x - y) + A(-(x - y)). \tag{16.73}$$

Letting $u = x + y$, $v = x - y$ in (16.73), we have

$$A(u) + A(-u) = A(v) + A(-v). \tag{16.74}$$

Hence

$$A(u) + A(-u) = \gamma \tag{16.75}$$

where γ is a constant. Hence from the definition of A, we have

$$xh(x) - xh(-x) = \gamma \tag{16.76}$$

for all $x \in \mathbb{R}\backslash\{0\}$. Letting $s = -t$ in (16.49), we have

$$f(x) - f(y) = (x - y)h(-t(x - y)). \tag{16.77}$$

Interchanging x and y in (16.77), we obtain

$$f(y) - f(x) = (y - x)h(-t(y - x)). \tag{16.78}$$

Adding (16.77) and (16.78), we get

$$0 = (x - y)h(-t(x - y)) - (x - y)h(t(x - y))$$

which is

$$0 = t(x - y)h(t(x - y)) - t(x - y)h(-t(x - y)).$$

By (16.76), the above yields

$$0 = \gamma.$$

Hence $\gamma = 0$ and (16.75) reduces to

$$A(x) + A(-x) = 0.$$

Hence $A : \mathbb{R} \to \mathbb{R}$ is additive on $\mathbb{R}$ (as $A(0) = 0$).

Now by (16.56) and definition of A, we obtain

$$f(y) = \frac{A(ty)}{t} + b. \tag{16.79}$$

Thus in this case, we have

$$\left.\begin{array}{l} f(x) = \dfrac{A(tx)}{t} + b \\[2mm] h(x) = \dfrac{A(x)}{x}, \quad x \neq 0. \end{array}\right\} \tag{16.80}$$

Subcase (iii). Suppose $s^2 \neq t^2$. Interchanging x with y in (16.59), we obtain

$$\frac{y}{s}h(y) - \frac{x}{t}h(x) = \left(\frac{y}{s} - \frac{x}{t}\right)h(x+y). \tag{16.81}$$

Subtracting (16.81) from (16.59), we have

$$\left(\frac{1}{s} + \frac{1}{t}\right)[xh(x) - yh(y)] = \left(\frac{1}{s} + \frac{1}{t}\right)(x-y)h(x+y)$$

which is, since $s^2 \neq t^2$,

$$xh(x) - yh(y) = (x-y)h(x+y). \tag{16.82}$$

This is in fact (16.60) and thus we have

$$h(x) = \alpha x + b. \tag{16.83}$$

Letting (16.83) into (16.81), we have

$$\frac{y}{s}(\alpha y + b) - \frac{x}{t}(\alpha x + b) = \left(\frac{y}{s} - \frac{x}{t}\right)(\alpha x + \alpha y + b).$$

Hence

$$\alpha xy \left(\frac{1}{t} - \frac{1}{s}\right) = 0.$$

Therefore

$$\alpha = 0.$$

Hence

$$h(x) = b. \tag{16.84}$$

Therefore (16.49) yields

$$f(x) - f(y) = b(x-y).$$

That is,

$$f(x) = bx + c.$$

This completes the proof. $\qquad\qquad\qquad\qquad\qquad\qquad\qquad\square$

Using this theorem, now we give the differentiable solution of the functional equation

$$f[x, y, z] = h(ux + vy + wz)$$

for $x, y, z \in \mathbb{R}$ with $x \neq y$, $y \neq z$, $z \neq x$.

Theorem 16.5. *Let $u, v, w \in \mathbb{R}$ be parameters. Let $f, g : \mathbb{R} \to \mathbb{R}$ with f differentiable satisfy the functional equation*

$$f[x, y, z] = h(ux + vy + wz) \tag{16.85}$$

for all distinct $x, y, z \in \mathbb{R}$. Then f, h are given by

$$f(x) = \begin{cases} ax^2 + bx + c & \text{if } u = v = w = 0 \\ dux^3 + ax^2 + bx + c & \text{if } u = v = w \neq 0 \\ ax^2 + bx + c & \text{otherwise} \end{cases}$$

and

$$h(x) = \begin{cases} \text{arbitrary} & \text{if } u = v = w = 0 \\ dx + a & \text{if } u = v = w \neq 0 \\ a & \text{otherwise,} \end{cases}$$

where a, b, c, d are arbitrary constants. The converse is also true.

Proof. It is easy to see that f and h enumerated in the theorem satisfy the equation (16.85). Now we prove the converse.

Without loss of generality, we may assume

$$f(0) = 0.$$

Letting $z = 0$ in (16.85), we get

$$q(x) - q(y) = (x - y)h(ux + vy), \tag{16.86}$$

where

$$q(x) = \frac{f(x)}{x}, \quad x \neq 0. \tag{16.87}$$

Since f is differentiable, q can be extended to $\mathbb{R}$ by defining

$$q(x) = \begin{cases} \frac{f(x)}{x}, & \text{if } x \neq 0 \\ f'(0), & \text{if } x = 0. \end{cases}$$

Hence (16.86) holds for all $x, y \in \mathbb{R}$. The solutions of (16.86) can be obtained from Theorem 16.4. We consider several cases based on the solutions of Theorem 16.4.

Case 1. Suppose $u = v = 0$. Apply Theorem 16.4 to get

$$\left.\begin{array}{l} q(x) = ax + b \\ h \text{ arbitrary, } h(0) = b. \end{array}\right\} \tag{16.88}$$

Hence

$$\left.\begin{array}{l} f(x) = ax^2 + bx \\ h \text{ arbitrary, } h(0) = b. \end{array}\right\}$$

Now removing the assumption $f(0) = 0$, we have

$$\left.\begin{array}{l} f(x) = ax^2 + bx + c \\ h \text{ arbitrary, } h(0) = b. \end{array}\right\} \tag{16.89}$$

Using (16.89) in (16.85), we get

$$a = h(wz). \tag{16.90}$$

If $w = 0$, then $h(0) = a$ and (16.89) holds. If $w \neq 0$, then we replace z by $\frac{x}{w}$ to get

$$h(x) = a$$

for all $x \in \mathbb{R}$. So, we have

$$\left.\begin{array}{l} f(x) = ax^2 + bx + c \\ h(x) = a. \end{array}\right\} \tag{16.91}$$

Case 2. Suppose $u \neq 0$ and $v = 0$. Then (16.86) gives

$$q(x) - q(y) = (x - y)h(ux). \tag{16.92}$$

Again from Theorem 16.4, we get

$$\left.\begin{array}{l} q(x) = ax + b \\ h(x) = a. \end{array}\right\} \tag{16.93}$$

Hence using the definition of q and removing the assumption $f(0) = 0$, we get

$$\left.\begin{array}{l} f(x) = ax^2 + bx + c \\ h(x) = a. \end{array}\right\} \tag{16.94}$$

Case 3. Suppose $u = v \neq 0$. Theorem 16.4 yields

$$\left.\begin{array}{l} q(x) = dux^2 + ax + b \\ h(x) = dx + a, \end{array}\right\} \tag{16.95}$$

from which after removing the assumption $f(0) = 0$, we get

$$\left.\begin{array}{l} f(x) = dux^3 + ax^2 + bx + c \\ h(x) = dx + a. \end{array}\right\} \tag{16.96}$$

Inserting (16.96) into (16.85) and then substituting $x = 0$, we have

$$du(y + z) = d(uy + wz) + a. \tag{16.97}$$

Hence, we have

$$d(w - u)z + a = 0 \tag{16.98}$$

for all $z \in \mathbb{R}$. If $w = u$, then $a = 0$, and f and g given by (16.96) is a solution of (16.85). If $w \neq u$, then (16.98) gives

$$d = 0 \quad \text{and} \quad a = 0$$

and (16.96) reduces to

$$\left.\begin{array}{l} f(x) = bx + c \\ h(x) = 0. \end{array}\right\} \tag{16.99}$$

Case 4. Suppose $u = -v \neq 0$. By Theorem 16.4, we get

$$\left.\begin{array}{l} q(x) = \dfrac{A(ux)}{u} + b \\[2mm] h(x) = \dfrac{A(x)}{x}, \quad x \neq 0. \end{array}\right\} \tag{16.100}$$

Since f is differentiable, q is continuous and hence $A : \mathbb{R} \to \mathbb{R}$ is continuous. Therefore

$$A(x) = ax, \tag{16.101}$$

where a is a constant. Hence (16.100) and (16.101) yield

$$\left.\begin{array}{l} q(x) = ax + b \\[2mm] h(x) = \dfrac{ax}{x}, \quad x \neq 0, \end{array}\right\} \tag{16.102}$$

that is,

$$\left.\begin{array}{l} f(x) = ax^2 + bx + c \\ h(x) = a. \end{array}\right\} \tag{16.103}$$

Case 5. Suppose $u^2 \neq v^2$. By Theorem 16.4 and previous arguments, we get

$$\left.\begin{array}{l} f(x) = ax^2 + bx + c \\ h(x) = a. \end{array}\right\} \tag{16.104}$$

Since no more cases are left, the proof of Theorem 16.5 is complete. $\quad\square$

Remark 16.1. *Note that the definition*

$$q(x) = \begin{cases} \frac{f(x)}{x} & \text{if } x \neq 0 \\ f'(0) & \text{if } x = 0 \end{cases}$$

is key to reduce (16.85) *to*

$$q(x) - q(y) = (x - y)h(ux + vy)$$

for all $x, y \in \mathbb{R}$ *with* $x \neq y$. *Without the differentiability of* f, *the function* q *cannot be defined at* $x = 0$. *Hence*

$$q(x) - q(y) = (x - y)h(ux + vy)$$

has to be solved for all $x, y \in \mathbb{R} \setminus \{0\}$ *with* $x \neq y$.

16.5 Concluding Remarks

Mean value type functional equations such as

$$f(x) - f(y) = (x - y) f'(\eta(x, y)) \tag{16.105}$$

with $\eta(x, y) = \frac{x+y}{2}$ and for all distinct $x, y \in \mathbb{R}$ were first time studied by the Romanian mathematician Pompeiu (1930). The functional equation (16.105) when $\eta(x, y) = \frac{x+y}{2}$ is a special case of the equation

$$f(x) - f(y) = (x - y) h(x + y) \tag{16.106}$$

for all $x, y \in \mathbb{R}$. The general solution of the functional equation (16.106) without any regularity assumption on f and h was determined by Aczél (1963). Without knowing the result of Aczél, Haruki (1979) considered the following more general functional equation

$$f(x) - g(y) = (x - y) h(x + y) \tag{16.107}$$

for all $x, y \in \mathbb{R}$. He proved, using Jensen functional equation, that the general solution $f, g, h : \mathbb{R} \to \mathbb{R}$ of (16.107) is given by

$$f(x) = g(x) = ax^2 + bx + c \qquad \text{and} \qquad h(x) = ax + b,$$

where a, b, c are arbitrary real constants. Aczél (1985) determined the general solution $f, g, h : \mathbb{F} \to \mathbb{F}$ of (16.107) when $\mathbb{F}$ is a field of characteristic different from 2. Sablik (1992) extended the result of Aczél (1985) to the case where variables and values of the unknown functions are in some abelian groups.

Using the notion of 3-point divided difference, Bailey (1992) generalized (16.106) to

$$f[x, y, z] = h(x + y + z) \qquad (16.108)$$

for all distinct $x, y, z \in \mathbb{R}$. He proved that if $f : \mathbb{R} \to \mathbb{R}$ is a differentiable function satisfying (16.108), the f is a polynomial of degree at most 3. The differentiability condition on f played an important role in determining the solution of (16.108). Bailey (1992) concluded his paper with the following: *One is led to wonder if*

$$f[x_1, x_2, ..., x_n] = h(x_1 + x_2 + \cdots + x_n] \qquad (16.109)$$

and f is continuous (or perhaps differentiable) will imply that f is a polynomial of degree no more than n. At this point we have no answer.

Chung and Sahoo (1993) showed that if a Lebesgue-integrable function $f : \mathbb{R} \to \mathbb{R}$ satisfies the functional equation (16.109) for distinct $x_1, x_2, ..., x_n \in \mathbb{R}$, then f is a polynomial of degree at most n. This answers the question raised by Bailey (1992). In 1993, Kannappan and Sahoo (1995) determined the general solutions $f, h : \mathbb{R} \to \mathbb{R}$ of the functional equation (16.109) without any regularity assumption on the unknown functions f and h. Without knowing the work of Kannappan and Sahoo, Anderson (1996) found the solution of (16.109) without any regularity assumption. Schwaiger (1994) found the general solution of (16.109) in a field $\mathbb{F}$ of characteristic different from 2 provided $\mathbb{F}$ has enough points in it. Davies and Rousseau (1998) determined the solution of (16.109) in arbitrary field $\mathbb{F}$ including the field with characteristic 2. They proved that f is a polynomial of degree at most n if $\mathbb{F}$ is a field of characteristic different from 2. This result also holds when $\mathbb{F} = GF(2)$ or $\mathbb{F} = GF(4)$. This result fails for all other fields of characteristic different from 2 with $|\mathbb{F}| > 4$.

Aczél and Kuczma (1989) generalized the functional equation (16.105) to

$$f(x) - f(y) = (x - y) h(\eta(x, y)) \qquad (16.110)$$

for all x, y in some real interval $I \subset \mathbb{R}$. They determined the solution of (16.110) when $\eta(x, y)$ is an arithmetic, geometric or harmonic means of x and y. They also considered the case where $\eta(x, y)$ is a quasiarithmetic mean with the additional assumption that 0 is in I.

In 1989, Walter Rudin raised the following problem in the elementary problem section of the *American Mathematical Monthly*. Let s and t be two given real numbers. Find all real functions f that satisfy

$$f(x) - f(y) = (x - y) f'(sx + ty) \qquad (16.111)$$

for all numbers $x, y \in \mathbb{R}$. Kannappan, Sahoo and Jacobson (1995), investigated the general solution of

$$f(x) - f(y) = (x - y) h(sx + ty) \qquad (16.112)$$

for all $x, y \in \mathbb{R}$ which is a generalization of (16.106). They determined the general solution of (16.112) without any regularity assumption on the unknown functions. Kannappan (2003) determined the general solution $f : G \to \mathbb{F}$, $h : G \to Hom(G, \mathbb{F})$ of the functional equation (16.112) holding for all $x, y \in G$, where G is an abelian 2-divisible groups and $\mathbb{F}$ is a field of characteristic zero.

Another generalization of (16.112) is

$$f(x) - g(y) = (x - y) [h(sx + ty) + \psi(x) + \phi(y)] \qquad (16.113)$$

for all $x, y \in \mathbb{R}$. Here $f, g, h, \psi, \phi : \mathbb{R} \to \mathbb{R}$ are unknown functions to be determined. Kannappan, Riedel and Sahoo (1998) determined the general solution of the functional equation (16.113) without any regularity assumption on the unknown functions. The following functional equation

$$f(x) - g(y) = (x - y) [h(sx + ty) + k(tx + sy) + \psi(x) + \phi(y)] \quad (16.114)$$

for all $x, y \in \mathbb{R}$, where $f, g, h, k, \psi, \phi : \mathbb{R} \to \mathbb{R}$ are unknown functions to be determined, is a generalization of (16.113). The general solution of this functional equation was determined by Sahoo (2007) assuming f, g, ψ, ϕ to be two time differentiable and h, k to be four time differentiable. The general solution of (16.114) without any regularity assumption on the unknown functions is not known.

A special case of the functional equation (16.113) is

$$f(x) - f(y) = (x - y) \left[h \left(\frac{x + y}{2} \right) - g(x) - g(y) \right] \qquad (16.115)$$

holding for all $x, y \in I$, where $I \subset \mathbb{R}$ is an arbitrary interval on the real line with positive length. Ger (2002) obtained the solution $f, h, g : I \to \mathbb{R}$ of the equation (16.115) without any regularity assumption.

A mean value theorem for functions in two variables yields the following functional equation (see Kannappan and Sahoo (1997b)):

$$f(u, v) - f(x, y) = (u - x) g_1(x + u, y + v) + (v - y) g_2(x + u, y + v) \quad (16.116)$$

for all $x, y \in \mathbb{R}$ with $(u-x)^2+(v-y)^2 \neq 0$. Kannappan and Sahoo (1997b) proved the following result: For some $g_1, g_2 : \mathbb{R}^2 \to \mathbb{R}$, if $f : \mathbb{R}^2 \to \mathbb{R}$ satisfies (16.116) for all $x, y, u, v \in \mathbb{R}$ with $(u - x)^2 + (v - y)^2 \neq 0$, then $f : \mathbb{R}^2 \to \mathbb{R}$ is of the form

$$f(x, y) = B(x, y) + ax^2 + bx + cy^2 + dy + \alpha,$$

where $B : \mathbb{R}^2 \to \mathbb{R}$ is a biadditive function and a, b, c, d, α are arbitrary constants. For the a priori chosen real parameters s and t, Riedel and Sahoo (1997) determined the solution of the functional equation

$$\begin{aligned} f(u, v) &- f(x, y) \\ &= (u - x)\, g_1(sx + tu, sy + tv) + (v - y)\, g_2(sx + tu, sy + tv) \end{aligned}$$

for all $x, y \in \mathbb{R}$ with $(u - x)^2 + (v - y)^2 \neq 0$ without any regularity assumption on the unknown functions $f, g_1, g_2 : \mathbb{R}^2 \to \mathbb{R}$. The following functional equation is the multivariable analogue of the latter equation

$$f(x) - f(y) = \langle (y - x),\, g_i(sx + ty) \rangle \tag{16.117}$$

for all $x, y \in \mathbb{R}^n$. Here $f, g_i : \mathbb{R}^n \to \mathbb{R}$ ($i = 1, 2, ..., n$) are unknown functions, $< \cdot, \cdot >$ is the inner product in $\mathbb{R}^n$, $|| \cdot ||$ is the norm in $\mathbb{R}^n$, and s and t are a priori chosen real parameters. The general solution of this functional equation is also determined by Riedel and Sahoo (1997) without any regularity assumption.

The functional equation (16.105) was originated from the Lagrange mean value theorem. There are other mean value theorems such as the Pompeiu mean value theorem and the Flett mean value theorem. These mean value theorems also give rise to other mean value type functional equations. The Pompeiu mean value theorem states that for every real valued function f differentiable on an interval $[a, b]$ not containing 0 and for all pairs $x_1 \neq x_2$ in $[a, b]$, there exists ξ in $]a, b[$ such that

$$\frac{x_1 f(x_2) - x_2 f(x_1)}{x_1 - x_2} = f(\xi) - \xi f'(\xi).$$

From the Pompeiu mean value theorem, one obtains the following mean value type functional equation

$$xf(y) - yf(x) = (x - y)\, h(\xi(x, y)) \tag{16.118}$$

for all $x, y \in \mathbb{R}$ with $x \neq y$. Aczél and Kuczma (1989) solved the equation (16.118) for all x, y in some real interval $I \subset \mathbb{R}$. They determined the solution of (16.118) when $\xi(x, y)$ is arithmetic, geometric or harmonic means of x and y. They also considered the case where $\xi(x, y)$ is a quasiarithmetic mean with the additional assumption that 0 is in I.

Kannappan, Riedel and Sahoo (1998) determined the general solution $f, h : \mathbb{R} \to \mathbb{R}$ of the functional equation

$$xf(y) - yf(x) = (x - y)[h(x + y) - h(x) - h(y)] \qquad (16.119)$$

for all $x, y \in \mathbb{R}$ without assuming any regularity assumption on the unknown functions. They proved that $f(x) = 3ax^3 + 2bx^2 + cx + d$ and $g(x) = -ax^3 - bx^2 - A(x) - d$, where $A : \mathbb{R} \to \mathbb{R}$ is additive and a, b, c, d are arbitrary constants. Ger (2002) studied the functional equation

$$xf(y) - yf(x) = (x - y)[h(x + y) - g(x) - g(y)] \qquad (16.120)$$

for all $x, y \in I$, where $I \subset \mathbb{R}$ is an arbitrary interval on the real line with positive length. She determined the general solution of (16.120) without any regularity assumption on the unknown functions.

For a priori chosen real parameters s and t, the functional equation

$$xf(y) - yf(x) = (x - y)[h(sx + ty) + k(tx + sy) + \psi(x) + \phi(y)] \quad (16.121)$$

for all $x, y \in \mathbb{R}$ has not been studied. We raise the following open problem: Determine the general solution $f, h, k, \psi, \phi : \mathbb{R} \to \mathbb{R}$ of the functional equation (16.121) holding without any regularity assumption on the unknown functions.

T.M. Flett (1958) proved that if $f : [x, y] \to \mathbb{R}$ is differentiable on $[x, y]$ and satisfies $f'(x) = f'(y)$, then there exists η in the open interval (x, y) such that

$$f(\eta) - f(x) = (\eta - x)f'(\eta(x, y)). \qquad (16.122)$$

Flett's conclusion implies that the tangent at $(\eta, f(\eta))$ passes through the point $(x, f(x))$. Davitt, Powers, Riedel and Sahoo (1999) extended the Flett mean value theorem that does not depend on the hypothesis $f'(x) = f'(y)$, but reduces to Flett mean value theorem when this is the case. They proved that *if $f : [x, y] \to \mathbb{R}$ is a differentiable function, then there exists a point $\eta \in (x, y)$ such that*

$$f(\eta) - f(x) = (\eta - x) f'(\eta) - \frac{1}{2} \frac{f'(y) - f'(x)}{y - x} (\eta - x)^2. \qquad (16.123)$$

Letting $\eta(x, y) = sx + ty$ for $s + t = 1$ and $0 < s, t < 1$, one obtains the following functional equation:

$$f(sx + ty) - f(x)$$
$$= (sx + ty - x) h(sx + ty) - \frac{1}{2} \frac{h(y) - h(x)}{y - x} (sx + ty - x)^2. \qquad (16.124)$$

for all $x, y \in \mathbb{R}$. Riedel and Sablik (2000) determined the general solution of (16.124) without any regularity assumption on f and h.

For details on the various mean value type functional equations, the interested reader is referred to the book *Mean Value Theorems and Functional Equations* by Sahoo and Riedel (1998).

16.6   Exercises

1. Find all functions $f, g : \mathbb{R} \to \mathbb{R}$ that satisfy the functional equation

$$f(x) - f(y) = (x - y) \left[\frac{g(x) + g(y)}{2} \right]$$

for all $x, y \in \mathbb{R}$.

2. Find all functions $f, g : \mathbb{R} \to \mathbb{R}$ that satisfy the functional equation

$$x f(y) - y f(x) = (x - y) g(x + y)$$

for all $x, y \in \mathbb{R}$.

3. Find all functions $f, g : (0, \infty) \to \mathbb{R}$ that satisfy the functional equation

$$f(x) - f(y) = (x - y) g\left(\sqrt{xy} \right)$$

for all $x, y \in (0, \infty)$.

4. Find all functions $f, g : (0, \infty) \to \mathbb{R}$ that satisfy the functional equation

$$f(x) - f(y) = (x - y) g\left(\frac{2xy}{x + y} \right)$$

for all $x, y \in (0, \infty)$.

5. Find all functions $f, g : (0, \infty) \to \mathbb{R}$ that satisfy the functional equation

$$x f(y) - y f(x) = (x - y) g\left(\sqrt{xy} \right)$$

for all $x, y \in (0, \infty)$.

6. Find all functions $f, g : (0, \infty) \to \mathbb{R}$ that satisfy the functional equation

$$x f(y) - y f(x) = (x - y) g\left(\frac{2xy}{x + y} \right)$$

for all $x, y \in (0, \infty)$.

7. Let s, t be any two a priori chosen real numbers. Find all functions $f, g : \mathbb{R} \to \mathbb{R}$ that satisfy the functional equation

$$xf(y) - yf(x) = (x - y)\, g(sx + ty)$$

for all $x, y \in \mathbb{R}$.

8. Find all functions $f, g : \mathbb{R} \to \mathbb{R}$ that satisfy the functional equation

$$g(x) - g(y) = \frac{1}{6}\,(x - y)\left[f(x) + 4f\left(\frac{x + y}{2}\right) + f(y)\right]$$

for all $x, y \in \mathbb{R}$.

9. Find all functions $f, g : \mathbb{R} \to \mathbb{R}$ that satisfy the functional equation

$$g(x) - g(y) = (x - y)\, f(x + y) + (x + y)\, f(x - y)$$

for all $x, y \in \mathbb{R}$.

10. Find all functions $f, g : \mathbb{R} \to \mathbb{R}$ that satisfy the functional equation

$$xf(y) - yf(x) = (x - y)\,[g(x + y) - g(x) - g(y)]$$

for all $x, y \in \mathbb{R}$.

11. Find all functions $f, g, h, k : \mathbb{R} \to \mathbb{R}$ that satisfy the functional equation

$$f(x) - g(y) = (x - y)\,[h(x + y) + k(x) + k(y)]$$

for all $x, y \in \mathbb{R}$.

12. Find all functions $f, g, h, \psi, \phi : \mathbb{R} \to \mathbb{R}$ that satisfy the functional equation

$$f(x) - g(y) = (x - y)\,[h(x + y) + \psi(x) + \phi(y)]$$

for all $x, y \in \mathbb{R}$.

13. Let s, t be any two a priori chosen real numbers. Find all functions $f, g, h, k : \mathbb{R} \to \mathbb{R}$ that satisfy the Sahoo-Riedel type functional equation

$$f(x) - g(y) = (x - y)\,[h(sx + ty) + k(x) + k(y)]$$

for all $x, y \in \mathbb{R}$.

14. Find all differentiable functions $f, g : \mathbb{R} \to \mathbb{R}$ that satisfy the functional equation

$$f(y + x)g(y + x) + f(y - x)g(y - x) = f(y)[g(y + x) + g(y - x)]$$

for all $x, y \in \mathbb{R}$.

15. Find all functions $f, g : \mathbb{R}^2 \to \mathbb{R}$ that satisfy the functional equation

$$f(u, v) - f(x, y) = (u - x)g(u + x, v + y) + (v - y)h(u + x, v + y)$$

for all $x, y, u, v \in \mathbb{R}$ with $(u - x)^2 + (v - y)^2 \neq 0$.

16. Find all functions $f, g : \mathbb{R}^2 \to \mathbb{R}$ that satisfy the functional equation

$$f(u, v) - f(x, y) = (u - x)g(su + tx, sv + ty) + (v - y)h(su + tx, sv + ty)$$

for all $x, y, u, v \in \mathbb{R}$ with $(u - x)^2 + (v - y)^2 \neq 0$.

17. Find all functions $f : \mathbb{R}^2 \to \mathbb{R}$ that satisfy the functional equation

$$f(u + x, v + y) + f(u - x, v) + f(u, v - y)$$
$$= f(u - x, v - y) + f(u + x, v) + f(u, v + y)$$

for all $x, y, u, v \in \mathbb{R}$.

18. Find all functions $f, g : \mathbb{R} \to \mathbb{R}$ that satisfy the functional equation

$$f[x_1, x_2, ..., x_n] = g(x_1 + x_2 + \cdots + x_n)$$

for all $x_1, x_2, ..., x_n \in \mathbb{R}$ with $x_i \neq x_j$ for $i \neq j$.

19. If $f(x) = ax^3 + bx^2 + cx + d$, then show that $f[x, y, z] = a(x + y + z)$, where a, b, c, d are real constants.

20. Prove that

$$f[x_1, x_2, \ldots, x_n] = \sum_{i=1}^{n} \frac{f(x_i)}{\prod_{j \neq i} (x_i - x_j)},$$

where $x_1, x_2, ..., x_n$ are distinct points in $\mathbb{R}$.

Chapter 17

Functional Equations for Distance Measures

17.1 Introduction

Let I denote the open unit interval $(0, 1)$ and J denote the open-closed unit interval $(0, 1]$. Let $\mathbb{R}$ denote the set of real numbers. Let $\mathbb{R}_+ = \{x \in \mathbb{R} \mid x > 0\}$ and $\mathbb{R}_1 = \{x \in \mathbb{R}_+ \mid x \neq 1\}$. Let

$$\Gamma_n^o = \left\{ P = (p_1, p_2, ..., p_n) \mid 0 < p_k < 1, \sum_{k=1}^{n} p_k = 1 \right\}$$

denote the set of all n-ary discrete complete probability distributions (without zero probabilities); that is, Γ_n^o is the class of discrete distributions on a finite set Ω of cardinality n with $n \geq 2$. Over the years, many distance measures between discrete probability distributions have been proposed.

Most similarity, affinity or distance measures $\mu_n : \Gamma_n^o \times \Gamma_n^o \to \mathbb{R}_+$ that have been proposed between two discrete probability distributions can be represented in the *sum form*

$$\mu_n(P, Q) = \sum_{k=1}^{n} \phi(p_k, q_k), \tag{17.1}$$

where $\phi : I \times I \to \mathbb{R}$ is a real-valued function on unit square or a monotonic transformation of the right side of (17.1), that is,

$$\mu_n(P, Q) = \psi \left(\sum_{k=1}^{n} \phi(p_k, q_k) \right), \tag{17.2}$$

where $\psi : \mathbb{R} \to \mathbb{R}_+$ is an increasing function on $\mathbb{R}$. The function ϕ is called a *generating function*. It is also referred to as the *kernel* of $\mu_n(P, Q)$. Some examples of sum form distance measures between two discrete probability distributions P and Q in Γ_n^o are (see Kannappan and Sahoo (1997a)):

(a) Directed divergence

$$\phi(x, y) = x(\log x - \log y),$$

$$D_n(P, Q) = \sum_{k=1}^{n} p_k \log\left(\frac{p_k}{q_k}\right);$$

(b) Symmetric J-divergence

$$\phi(x, y) = (x - y)(\log x - \log y),$$

$$J_n(P, Q) = \sum_{k=1}^{n} (p_k - q_k) \log\left(\frac{p_k}{q_k}\right);$$

(c) Hellinger coefficient

$$\phi(x, y) = \sqrt{xy},$$

$$H_n(P, Q) = \sum_{k=1}^{n} \sqrt{p_k q_k};$$

(d) Jeffreys distance

$$\phi(x, y) = \left(\sqrt{x} - \sqrt{y}\right)^2,$$

$$K_n(P, Q) = \sum_{k=1}^{n} \left(\sqrt{p_k} - \sqrt{q_k}\right)^2;$$

(e) Chernoff coefficient

$$\phi(x, y) = x^\alpha y^{1-\alpha},$$

$$C_{n,\alpha}(P, Q) = \sum_{k=1}^{n} p_k^\alpha q_k^{1-\alpha} \qquad \alpha \in I;$$

(f) Variational distance

$$\phi(x, y) = |x - y|,$$

$$V_n(P, Q) = \sum_{k=1}^{n} |p_k - q_k|;$$

(g) Proportional distance

$$\phi(x, y) = \min\{x, y\},$$

$$X_n(P, Q) = \sum_{k=1}^{n} \min\{p_k, q_k\};$$

(h) Kagan affinity measure

$$\phi(x, y) = \frac{(y - x)^2}{y},$$

$$A_n(P, Q) = \sum_{k=1}^{n} q_k \left[1 - \frac{p_k}{q_k}\right]^2;$$

(i) Vajda affinity measure

$$\phi(x, y) = y \left|\frac{x}{y} - 1\right|^\alpha,$$

$$A_{n,\alpha}(P, Q) = \sum_{k=1}^{n} q_k \left|\frac{p_k}{q_k} - 1\right|^\alpha, \qquad \alpha \geq 1;$$

(j) Matusita distance

$$\phi(x, y) = |x^\alpha - y^\alpha|^{\frac{1}{\alpha}},$$

$$M_{n,\alpha}(P, Q) = \sum_{k=1}^{n} |p_k^\alpha - q_k^\alpha|^{\frac{1}{\alpha}}, \qquad 0 < \alpha \leq 1;$$

(k) Divergence measure of degree α

$$\phi(x, y) = \frac{1}{2^{\alpha-1} - 1} \left[x^\alpha y^{1-\alpha} - x\right],$$

$$B_{n,\alpha}(P, Q) = \frac{1}{2^{\alpha-1} - 1} \left[\sum_{k=1}^{n} \left(p_k^\alpha q_k^{1-\alpha} - p_k\right)\right], \qquad \alpha \neq 1;$$

(l) Cosine α-divergence measure

$$\phi(x, y) = \frac{1}{2} \left[x - \sqrt{xy} \cos\left(\alpha \log\left(\frac{x}{y}\right)\right)\right],$$

$$N_{n,\alpha}(P, Q) = \frac{1}{2} \left[1 - \sum_{k=1}^{n} \sqrt{p_k q_k} \cos\left(\alpha \log\frac{p_k}{q_k}\right)\right];$$

(m) Divergence measure of Higashi and Klir

$$\phi(x, y) = x \log\frac{2x}{x + y} + y \log\frac{2y}{x + y},$$

$$I_n(P,Q) = \sum_{k=1}^{n} \left[p_k \log \left(\frac{2p_k}{p_k + q_k} \right) + q_k \log \left(\frac{2q_k}{p_k + q_k} \right) \right];$$

(n) Csiszar f-divergence measure

$$\phi(x,y) = xf\left(\frac{x}{y} \right),$$

$$Z_{n,\alpha}(P,Q) = \sum_{k=1}^{n} p_k f\left(\frac{p_k}{q_k} \right);$$

(o) Kullback-Leibler type f-distance measure

$$\phi(x,y) = x[f(x) - f(y)],$$

$$L_{n,\alpha}(P,Q) = \sum_{k=1}^{n} p_k[f(p_k) - f(q_k)].$$

The following

(p) Renyi's divergence measure

$$\phi(x,y) = x^\alpha y^{1-\alpha}, \qquad\qquad \psi(x) = \frac{1}{\alpha - 1} \log x$$

$$R_{n,\alpha}(P\|Q) = \frac{1}{\alpha - 1} \log \left(\sum_{k=1}^{n} p_k^\alpha q_k^{1-\alpha} \right), \qquad \alpha \neq 1,$$

is a monotonic transformation of sum form distance measures. Renyi's divergence measure is the logarithm of the so-called exponential entropy

$$E_{n,\alpha}(P,Q) = \left(\sum_{k=1}^{n} p_k^\alpha q_k^{1-\alpha} \right)^{\frac{1}{\alpha-1}}.$$

In order to derive axiomatically the principle of minimum divergence, Shore and Johnson (1980) formulated a set of four axioms, namely, *uniqueness, invariance, system independence* and *subset independence*. They proved that if a functional $\mu_n : \Gamma_n^o \times \Gamma_n^o \to \mathbb{R}_+$ satisfies the axioms of uniqueness, invariance and subset independence, then there exists a generating function (or kernel) $\phi : I \times I \to \mathbb{R}$ such that

$$\mu_n(P,Q) = \sum_{k=1}^{n} \phi(p_k, q_k)$$

for all $P, Q \in \Gamma_n^o$. In view of this result one can conclude that the above sum form representation is not artificial. In most applications involving distance measures between probability distributions, one encounters the minimization of $\mu_n(P,Q)$ and the sum form representation makes problems tractable.

A sequence of measures $\{\mu_n\}$ is said to be *symmetrically compositive* if for some $\lambda \in \mathbb{R}$,

$$\mu_{nm}(P \star R, Q \star S) + \mu_{nm}(P \star S, Q \star R)$$
$$= 2\mu_n(P,Q) + 2\mu_m(R,S) + \lambda\mu_n(P,Q)\mu_m(R,S)$$

for all $P, Q \in \Gamma_n^o$, $S, R \in \Gamma_m^o$, where

$$P * R = (p_1r_1, p_1r_2, ..., p_1r_m, p_2r_1, ..., p_2r_m, ..., p_nr_m).$$

If $\lambda = 0$, then $\{\mu_n\}$ is said to be *symmetrically additive*.

The functional equations

$$f(pr, qs) + f(ps, qr) = (r + s)f(p, q) + (p + q)f(r, s) \qquad (17.3)$$

and

$$f(pr, qs) + f(ps, qr) = f(p, q)\, f(r, s) \qquad (17.4)$$

holding for all $p, q, r, s \in I$ are instrumental in the characterization of symmetrically compositive sum form distance measures with a measurable generating function.

In this chapter, we present the general solution of the functional equations (17.3) and (17.4), and the functional equation

$$f_1(pr, qs) + f_2(ps, qr) = g(p, q) + h(r, s), \qquad (17.5)$$

for all $p, q, r, s \in I$.

17.2   Solution of Two Functional Equations

In this section, following Chung, Kannappan, Ng and Sahoo (1989) we present the general solution of the functional equations (17.3) and (17.4) without any regularity assumption.

A map $L : \mathbb{R}_+ \to \mathbb{R}$ is called *logarithmic* if and only if

$$L(xy) = L(x) + L(y)$$

for all $x, y \in \mathbb{R}_+$. A function $\ell : \mathbb{R}_+^2 \to \mathbb{R}$ is called *bilogarithmic* if and only if it is logarithmic in each variable. The capital letter L along with its subscripts is used exclusively for a logarithmic map.

The following theorem concerns the general solution of the functional equation (17.3).

Theorem 17.1. *A function $f : I^2 \to \mathbb{R}$ satisfies the functional equation* (17.3), *that is,*

$$f(pr, qs) + f(ps, qr) = (r + s)f(p, q) + (p + q)f(r, s),$$

for all $p, q, r, s \in I$ if and only if

$$f(p, q) = p\left[L_1(q) - L_2(p)\right] + q\left[L_1(p) - L_2(q)\right], \qquad (17.6)$$

where $L_1, L_2 : \mathbb{R}_+ \to \mathbb{R}$ are logarithmic functions.

Proof. It is easy to verify that f given by (17.6) satisfies (17.3). Obviously, $f = 0$ is a solution of (17.3) and is of the form (17.6). We now suppose $f \neq 0$. First, we will show that f satisfying (17.3) for all $p, q, r, s \in I$ can be extended (uniquely) from I^2 to $\mathbb{R}_+^2$. Setting $r = s = \lambda$ in (17.3), we get

$$f(\lambda p, \lambda q) = \lambda f(p, q) + \lambda(p + q)\ell(\lambda), \qquad (17.7)$$

where

$$\ell(\lambda) := (2\lambda)^{-1} f(\lambda, \lambda).$$

Next, we show that ℓ is logarithmic. Replacing λ by $\lambda_1 \lambda_2$ in (17.7), we get

$$f(\lambda_1 \lambda_2 p, \lambda_1 \lambda_2 q) = \lambda_1 \lambda_2 f(p, q) + \lambda_1 \lambda_2 (p + q)\ell(\lambda_1 \lambda_2). \qquad (17.8)$$

Using (17.7) twice, $f(\lambda_1 \lambda_2 p, \lambda_1 \lambda_2 q)$ can be written as

$$
\begin{aligned}
f(\lambda_1 \lambda_2 p, \lambda_1 \lambda_2 q) &= \lambda_1 f(\lambda_2 p, \lambda_2 q) + \lambda_1 \lambda_2 (p + q)\ell(\lambda_1) \\
&= \lambda_1 \left[\lambda_2 f(p, q) + \lambda_2)p + q)\ell(\lambda_2)\right] \\
&\qquad + \lambda_1 \lambda_2 (p + q)\ell(\lambda_1). \qquad (17.9)
\end{aligned}
$$

Comparing (17.8) and (17.9), we see that

$$\ell(\lambda_1 \lambda_2) = \ell(\lambda_1) + \ell(\lambda_2),$$

that is, ℓ is logarithmic.

Now, we extend f to $\bar{f}$ as follows. For any $p, q \in \mathbb{R}_+$, choose a positive λ sufficiently small such that $\lambda, \lambda p, \lambda q \in I$. Define

$$\bar{f}(p, q) = \frac{1}{\lambda} f(\lambda p, \lambda q) - (p + q)\ell(\lambda). \qquad (17.10)$$

From

$$f(\lambda\mu p, \lambda\mu q) = f(\lambda \cdot \mu p, \lambda \cdot \mu q) = f(\mu \cdot \lambda p, \mu \cdot \lambda q),$$

using (17.7), it follows that the right side of (17.10) is independent of λ and thus $\bar{f}$ is well defined. For $p, q, r, s \in \mathbb{R}_+$, choose $\lambda \in I$ such that $\lambda p, \lambda q, \lambda r, \lambda s \in I$. Then

$$\bar{f}(pr, qs) + \bar{f}(ps, qr)$$

$$= \frac{1}{\lambda^2} \left\{ f(\lambda^2 pr, \lambda^2 qs) + f(\lambda^2 ps, \lambda^2 qr) \right\} - (pr + qs)\ell(\lambda^2) - (ps + qr)\ell(\lambda^2)$$

$$= \frac{1}{\lambda^2} \left\{ f(\lambda p\, \lambda r, \lambda q\, \lambda s) + f(\lambda p\, \lambda s, \lambda q\, \lambda r) \right\} - (p + q)(r + s)\ell(\lambda^2)$$

$$= \frac{1}{\lambda} \left\{ (r + s)f(\lambda p, \lambda q) + (p + q)f(\lambda r, \lambda s) \right\}$$

$$\qquad - 2(p + q)(r + s)\,\ell(\lambda) \quad \text{(by using (17.3) } \ell \text{ logarithmic)}$$

$$= (r + s)\,\bar{f}(p, q) + (p + q)\,\bar{f}(r, s).$$

Thus $\bar{f}$ satisfies (17.3) on $\mathbb{R}_+$. From here on, let us simply assume that f satisfies (17.3) for all $p, q, r, s \in \mathbb{R}_+$. Set $p = q$, $r = s$ in (17.3) to get

$$f(pr, pr) = rf(p, p) + pf(r, r). \tag{17.11}$$

From this it follows that

$$L(p) := \frac{1}{p} f(p, p) \tag{17.12}$$

is logarithmic on $\mathbb{R}_+$. Setting $q = s = 1$ in (17.3), we get

$$f(p, r) = (1 + r)g(p) + (1 + p)g(r) - g(pr), \tag{17.13}$$

where $g(p) := f(1, p)$. Note that (17.3) implies $f(1, 1) = 0$ and f is symmetric. Now (17.12) and (17.13) give

$$g(p^2) = 2(1 + p)\,g(p) - L(p), \qquad p \in \mathbb{R}_+. \tag{17.14}$$

With $p = r$, $q = s$ in (17.3) yields

$$f(p^2, q^2) + f(pq, pq) = 2(p + q)\,f(p, q). \tag{17.15}$$

Putting (17.12), (17.13) and (17.14) into (17.15), we have

$$(1 - p)(1 - q)[2g(pq) - L(pq)]$$

$$= (1 - q)(1 - pq)[2g(q) - L(q)] + (1 - q)(1 - pq)[2g(p) - L(p)].$$

Defining

$$2L_2(x) = \begin{cases} (1-x)^{-1}\left[(2g(x) - L(x)\right], & x \neq 1 \\ 0, & x = 1, \end{cases}$$

we get from this definition and $g(1) = 0$ that

$$g(p) = \frac{1}{2}L(p) + (1-p)L_2(p), \qquad p \in \mathbb{R}_+,$$

and from the above equation, $L_2(pq) = L_2(p) + L_2(q)$ follows whenever $p \neq 1$, $q \neq 1$, and $pq \neq 1$. The function L_2 evidently satisfies $L_2(pq) = L_2(p) + L_2(q)$ when $p = 1$ or $q = 1$. To check that this equation is also true for the case $p \neq 1$ but $pq = 1$, we have to show that $L_2(p^{-1}) = -L_2(p)$ for $p \neq 1$. The latter is equivalent to

$$\left(1 + \frac{1}{p}\right)f(p,p) + 2pf\left(1, \frac{1}{q}\right) = 2f(1,p),$$

which can be obtained by putting $q = p$, $r = 1$, and $s = \frac{1}{p}$ in (17.3). Thus, L_2 is logarithmic on $\mathbb{R}_+$. Now using (17.13), we get (17.6), where $L_1 = \frac{1}{2}L + L_2$. This proves the theorem.   □

The next theorem gives the general solution of the equation (17.4).

Theorem 17.2. *Suppose $f : I^2 \to \mathbb{R}$ satisfies the functional equation (17.4), that is*

$$f(pr, qs) + f(ps, qr) = f(p,q)\,f(r,s)$$

for all $p, q, r, s \in I$. Then

$$f(p,q) = M_1(p)\,M_2(q) + M_1(q)\,M_2(p), \tag{17.16}$$

where $M_1, M_2 : \mathbb{R} \to \mathbb{C}$ are multiplicative functions. Further, either M_1 and M_2 are both real or M_2 is the complex conjugate of M_1. The converse is also true.

Proof. Without loss of generality, let us assume that $f \neq 0$. Observe first that with $r = s = \lambda$ in (17.4), we get the M-homogeneity law

$$f(\lambda p, \lambda q) = \frac{1}{2}f(\lambda, \lambda)f(p,q) = M(\lambda)f(p,q),$$

where $M : I \to \mathbb{R}$, $M(\lambda) := \frac{1}{2}f(\lambda, \lambda)$. Since $f \neq 0$, M is multiplicative and can be uniquely extended to a multiplicative $\bar{M}$ ($\neq 0$) on $\mathbb{R}_+$ (see

Theorem 4.3 in Chapter 4 or Aczél, Baker, Djoković, Kannappan and Rado (1971)). As in Theorem 17.1, it is easy to show that

$$\bar{f}(p,q) = \bar{M}\left(\frac{1}{\lambda}\right) f(\lambda p, \lambda q) \tag{17.17}$$

for $p, q \in \mathbb{R}_+$ (λ small such that $\lambda p, \lambda q \in I$) extends f and (17.4) to $\mathbb{R}_+^2$. We assume from here on that f has this extended meaning.

We then fix $s = q = 1$ in (17.4) and define $g(p) = f(p, 1)$ to get

$$f(p,q) = g(p)\,g(q) - g(pq). \tag{17.18}$$

The assumption $f \neq 0$ implies that g is not multiplicative, and $f(1,1) = 2$. Setting $s = 1$ in (17.4) and using (17.18), we obtain

$$g(pr)g(q) - 2g(pqr) = g(p)g(q)g(r) - g(p)g(qr) - g(r)g(pq). \tag{17.19}$$

We fix q in (17.19) to get

$$F(pr) = F(p)\frac{g(r)}{2} + F(r)\frac{g(p)}{2}, \tag{17.20}$$

where $F(p) := g(p)g(q) - 2g(pq)$.

If F is identically 0, then $M_1 := 2^{-1}g$ is multiplicative and $f(p,q) = 2M_1(pq)$, and f thus is of the form (17.16) with $M_2 = M_1$. If $F \neq 0$, then from (17.20) (see Chung, Kannappan and Ng (1985), Lemma 5, Remark 4) and since g is not multiplicative, we have $g = M_1 + M_2$, where $M_1, M_2 : \mathbb{R}_+ \to \mathbb{C}$ are multiplicative; that is, f is of the form (17.16) where M_1, M_2 are complex valued. Suppose $M_k(p) = a_k(p) + ib_k(p)$ with real a_k, b_k ($k = 1, 2$). Since f is real valued, letting $q = 1$ in (17.18), we get $b_1(p) = -b_2(p)$. Again using f real, that is, the imaginary part of $f = 0$, we get

$$[a_2(p) - a_1(p)]b(q) + [a_2(q) - a_1(q)]b(p) = 0. \tag{17.21}$$

Suppose there is a p_o such that $a_1(p_o) \neq a_2(p_o)$. Then from (17.21), we have $b(q) = c\,[a_2(q) - a_1(q)]$, that is,

$$c\,[a_2(p) - a_1(p)][a_2(q) - a_1(q)] = 0$$

for all p, q. Then $c = 0$, so $b(q) = 0$ and M_1 and M_2 are real. Otherwise $a_1 = a_2$ and M_2 equals the conjugate of M_1 as desired. The converse part is straightforward. This completes the proof of the theorem. $\square$

17.3   Some Auxiliary Results

The following auxiliary results are needed to determine the general solution of the functional equation (17.5). This section is adapted from Sahoo (1999).

Lemma 17.1. *The function $f : I^2 \to \mathbb{R}$ satisfies the functional equation*

$$f(pr, qs) = f(p, q) + f(r, s) \tag{17.22}$$

for all $p, q, r, s \in I$ if and only if

$$f(p, q) = L_1(p) + L_2(q), \tag{17.23}$$

where $L_1, L_2 : \mathbb{R}_+ \to \mathbb{R}$ are logarithmic.

Proof. Let $a \in I$ be a fixed element and consider

$$
\begin{aligned}
f(p, q) &= f(p, q) + 2f(a, a) - 2f(a, a) \\
&= f(paa, qaa) - 2f(a, a) \\
&= f(pa \cdot a, a \cdot qa) - 2f(a, a) \\
&= f(pa, a) + f(a, qa) - 2f(a, a) \\
&= L_1(p) + L_2(q),
\end{aligned}
$$

where

$$
\begin{aligned}
L_1(p) &= f(pa, a) - f(a, a) \\
L_2(q) &= f(a, qa) - f(a, a).
\end{aligned}
$$

Next, we show that L_1 and L_2 are logarithmic functions on $\mathbb{R}_+$. Observe that

$$
\begin{aligned}
L_1(pq) &= f(pqa, a) - f(a, a) \\
&= f(pqa, a) + f(a, a) - 2f(a, a) \\
&= f(pqaa, aa) - 2f(a, a) \\
&= f(pa \cdot qa, a \cdot a) - 2f(a, a) \\
&= f(pa, a) + f(qa, a) - 2f(a, a) \\
&= L_1(p) + L_1(q)
\end{aligned}
$$

for all $p, q \in I$. Hence L_1 is logarithmic. It is well known that L_1 can be extended to $\mathbb{R}_+$ from I. Similarly, it can be shown that L_2 is a logarithmic map on $\mathbb{R}_+$. This completes the proof of the lemma. $\square$

Lemma 17.2. *The functions* $f, g, h : I^2 \to \mathbb{R}$ *satisfy the functional equation*

$$f(pr, qs) = g(p, q) + h(r, s) \tag{17.24}$$

for all $p, q, r, s \in I$ *if and only if*

$$f(p, q) = L_1(p) + L_2(q) + \alpha + \beta, \tag{17.25}$$
$$g(p, q) = L_1(p) + L_2(q) + \alpha, \tag{17.26}$$
$$h(r, s) = L_1(r) + L_2(s) + \beta, \tag{17.27}$$

where $L_1, L_2 : \mathbb{R}_+ \to \mathbb{R}$ *are logarithmic and* α, β *are arbitrary constants.*

Proof. Let a and b be any two fixed elements in I. Inserting $r = a$ and $s = b$ in (17.24), we obtain

$$g(p, q) = f(pa, qb) - h(a, b). \tag{17.28}$$

Now again letting $p = b$ and $q = a$ in (17.24), we have

$$h(r, s) = f(rb, sa) - g(b, a). \tag{17.29}$$

By (17.28) and (17.29), (17.24) yields

$$f(pr, qs) = f(pa, qb) + f(rb, sa) + k, \tag{17.30}$$

where $k = -h(a, b) - g(b, a)$. Replacing p by bp, q by aq, r by ar, and s by bs in (17.30), we obtain

$$f(abpr, abqs) = f(abp, abq) + f(abr, abs) + k. \tag{17.31}$$

Defining

$$F(p, q) = f(abp, abq) + k, \tag{17.32}$$

we see that the last equation transforms into

$$F(pr, qs) = F(p, q) + F(r, s) \tag{17.33}$$

for all $p, q, r, s \in I$. By Lemma 3.1, we obtain

$$F(p, q) = L_1(p) + L_2(q), \tag{17.34}$$

where $L_1, L_2 : \mathbb{R}_+ \to \mathbb{R}$ are logarithmic. Therefore

$$f(abp, abq) = L_1(p) + L_2(q) - k \tag{17.35}$$

which is

$$f(p, q) = L_1(p) + L_2(q) + \gamma, \tag{17.36}$$

where γ is a constant. Using (17.36) in (17.28) and (17.29), we obtain (17.26) and (17.27), respectively. Letting (17.36), (17.26) and (17.27) into (17.24), we see that $\gamma = \alpha + \beta$ and thus we have (17.25). This completes the proof. □

Lemma 17.3. *The function $f : I^2 \to \mathbb{R}$ satisfies the functional equation*

$$f(pr, qs) + f(ps, qr) = 2f(p, q) + 2f(r, s) \qquad (17.37)$$

for all $p, q, r, s \in I$ if and only if

$$f(p, q) = L(p) + L(q) + \ell\left(\frac{p}{q}, \frac{p}{q}\right), \qquad (17.38)$$

where $L : \mathbb{R}_+ \to \mathbb{R}$ is logarithmic and $\ell : \mathbb{R}_+^2 \to \mathbb{R}$ is bilogarithmic.

Proof. Setting $r = \lambda = s$ in (17.37), we obtain

$$f(\lambda p, \lambda q) = f(p, q) + L(\lambda), \qquad (17.39)$$

where

$$L(\lambda) := f(\lambda, \lambda). \qquad (17.40)$$

It is easy to show that L is logarithmic. Consider

$$f(\lambda_1 \lambda_2 p, \lambda_1 \lambda_2 q) = f(p, q) + L(\lambda_1 \lambda_2). \qquad (17.41)$$

Also we get

$$
\begin{aligned}
f(\lambda_1 \lambda_2 p, \lambda_1 \lambda_2 q) &= f(\lambda_2 p, \lambda_2 q) + L(\lambda_1) \\
&= f(p, q) + L(\lambda_2) + L(\lambda_1).
\end{aligned} \qquad (17.42)
$$

Thus from (17.41) and (17.42), we see that $L(\lambda_1 \lambda_2) = L(\lambda_1) + L(\lambda_2)$ for all $\lambda_1, \lambda_2 \in I$. Hence L is logarithmic on I and it can be extended uniquely to $\mathbb{R}_+$.

Now we extend f to $\bar{f}$ from I^2 to $\mathbb{R}_+^2$ as follows: For $p, q \in \mathbb{R}_+$, choose $\lambda \in \mathbb{R}_+$ sufficiently small such that $\lambda, \lambda p, \lambda q \in I$. Define

$$\bar{f}(p, q) = f(\lambda p, \lambda q) - L(\lambda). \qquad (17.43)$$

It is easy to show that $\bar{f}$ in (17.43) is well defined, that is, independent of the choice of λ. To show this, using (17.39) we write $f(\lambda \mu p, \lambda \mu q)$ in two different ways:

$$f(\lambda \mu p, \lambda \mu q) = f(\lambda p, \lambda q) + L(\mu)$$

and also

$$f(\lambda \mu p, \lambda \mu q) = f(\mu p, \mu q) + L(\lambda).$$

Hence from the last two equations, we get

$$f(\lambda p, \lambda q) - L(\lambda) = f(\mu p, \mu q) - L(\mu). \qquad (17.44)$$

Thus $\bar{f}$ is independent of the choice of λ.

Next we establish that $\bar{f}$ satisfies the functional equation (17.37). Choose $p, q, r, s \in \mathbb{R}_+$ and $\lambda \in I$ such that $\lambda p, \lambda q, \lambda r, \lambda s \in I$. Next we compute

$$
\begin{aligned}
\bar{f}(pr, qs) + \bar{f}(ps, qr) &= f(\lambda^2 pr, \lambda^2 qs) + f(\lambda^2 ps, \lambda^2 qr) - 2L(\lambda^2) \\
&= f(\lambda p \lambda r, \lambda q \lambda s) + f(\lambda p \lambda s, \lambda q \lambda r) - 4L(\lambda) \\
&= 2f(\lambda p, \lambda q) + 2f(\lambda r, \lambda s) - 4L(\lambda) \\
&= 2\bar{f}(p, q) + 2\bar{f}(r, s).
\end{aligned}
$$

Hence $\bar{f}$ satisfies (17.37) for all $p, q, r, s \in \mathbb{R}_+$. Here after, we simply assume that f satisfies (17.37) for all $p, q, r, s \in \mathbb{R}_+$.

A substitution of $p = q = r = s = 1$ in (17.37) yields $f(1, 1) = 0$. Further, substituting $p = q = 1$ in (17.37), we see that $f(r, s) = f(s, r)$ for all $r, s \in \mathbb{R}_+$. Letting $q = s = 1$ in (17.37), we have

$$
f(p, r) = 2g(p) + 2g(r) - g(pr), \tag{17.45}
$$

where

$$
g(p) := f(p, 1). \tag{17.46}
$$

Note that $g(1) = 0$ in view of $f(1, 1) = 0$. Letting $s = 1$ in (17.37) and then using (17.45) in the resulting equation, we obtain

$$
g(pqr) + g(p) + g(q) + g(r) = g(pr) + g(qr) + g(pq) \tag{17.47}
$$

for $p, q, r \in \mathbb{R}_1$. Defining

$$
2\ell(p, r) = g(pr) - g(p) - g(r), \tag{17.48}
$$

we see that (17.47) reduces to

$$
\ell(pq, r) = \ell(p, r) + \ell(q, r) \tag{17.49}
$$

for all $p, q, r \in \mathbb{R}_1 := \{x \in \mathbb{R}_+ | x \neq 1\}$. Hence ℓ is logarithmic on $\mathbb{R}_1^2$ on the first variable. Since the right side of (17.48) is symmetric with respect to p and r, so also is the left side. Thus

$$
\ell(p, r) = \ell(r, p);
$$

that is, ℓ is a real-valued bilogarithmic function on $\mathbb{R}_1^2$.

Again, defining

$$
G(p) = g(p) - \ell(p, p) \tag{17.50}
$$

and using (17.50) in (17.48) and the symmetry of ℓ, we obtain

$$G(pr) = G(p) + G(r). \tag{17.51}$$

Thus

$$G(p) = L(p), \qquad p \in \mathbb{R}_1, \tag{17.52}$$

where $L : \mathbb{R}_1 \to \mathbb{R}$ is an arbitrary logarithmic function. Now using (17.52) and (17.50), we obtain

$$g(p) = L(p) + \ell(p, p) \tag{17.53}$$

for all $p \in \mathbb{R}_1$. The equation (17.53) in (17.45) yields

$$f(p, q) = L(p) + L(q) + \ell\left(\frac{p}{q}, \frac{p}{q}\right) \tag{17.54}$$

for all $p, q \in \mathbb{R}_1$. By (17.46) and (17.53), we see that

$$f(p, 1) = L(p) + \ell(p, p)$$

for all $p \in \mathbb{R}_1$, and since $f(1, 1) = 0$, this extends to all $\mathbb{R}_+$. Using the fact that $f(p, 1) = L(p) + \ell(p, p)$ for all $p \in \mathbb{R}_+$, and the symmetry of f, we see that (17.54) holds for all $p, q \in \mathbb{R}_+$. This completes the proof. $\square$

Lemma 17.4. *The function $f : I^2 \to \mathbb{R}$ satisfies the functional equation*

$$f(pr, qs) + f(ps, qr) = 2f(p, q) + f(r, s) + f(s, r) \tag{17.55}$$

for all $p, q, r, s \in I$ if and only if

$$f(p, q) = L_0(q) - L_0(p) + L_1(p) + L_1(q) + \ell\left(\frac{p}{q}, \frac{p}{q}\right), \tag{17.56}$$

where $L_0, L_1 : \mathbb{R}_+ \to \mathbb{R}$ are logarithmic and $\ell : \mathbb{R}_+^2 \to \mathbb{R}$ is bilogarithmic.

Proof. As in the proof of the previous lemma, we define $\bar{f} : \mathbb{R}_+^2 \to \mathbb{R}$ as

$$\bar{f}(p, q) = f(\lambda p, \lambda q) - L(\lambda),$$

where $L(\lambda) := f(\lambda, \lambda)$. Then L is logarithmic and $\bar{f}$ satisfies the functional equation (17.55) for all $p, q, r, s \in \mathbb{R}_+$. Hence from here on, we simply assume that f satisfies (17.55) for all $p, q, r, s \in \mathbb{R}_+$.

Interchanging p with r and q with s in (17.55), we get

$$f(pr, qs) + f(qr, ps) = 2f(r, s) + f(p, q) + f(q, p). \tag{17.57}$$

Subtracting (17.57) from (17.55), we obtain

$$\phi(ps, qr) = \phi(p, q) + \phi(s, r),\tag{17.58}$$

where

$$\phi(p, q) := f(p, q) - f(q, p).\tag{17.59}$$

From Lemma 3.1, we have $\phi(p, q) = 2L_2(p) + 2L_0(q)$, where $L_0, L_2 : \mathbb{R}_+ \to \mathbb{R}$ are logarithmic. Since ϕ is skew symmetric, we see that $L_2 = -L_0$ and hence

$$f(q, p) = f(p, q) + 2L_0(p) - 2L_0(q).\tag{17.60}$$

Letting (17.60) into (17.55), we see that

$$f(pr, qs) + f(ps, qr) = 2f(p, q) + 2f(r, s) + 2[L_0(r) - L_0(s)].\tag{17.61}$$

Defining

$$F(p, q) = f(p, q) + [L_0(p) - L_0(q)],\tag{17.62}$$

we obtain from (17.61)

$$F(pr, qs) + F(ps, qr) = 2F(p, q) + 2F(r, s)$$

for all $p, q, r, s \in \mathbb{R}_+$. Hence by Lemma 17.3, we have

$$F(p, q) = L_1(p) + L_1(q) + \ell\left(\frac{p}{q}, \frac{p}{q}\right),\tag{17.63}$$

where $L_1 : \mathbb{R}_+ \to \mathbb{R}$ is logarithmic and $\ell : \mathbb{R}_+^2 \to \mathbb{R}$ is bilogarithmic. From (17.62) and (17.63), we get the asserted solution (17.56). This completes the proof. □

Lemma 17.5. *The functions $f, g, h : I^2 \to \mathbb{R}$ satisfy the functional equation*

$$f(pr, qs) + f(ps, qr) = g(p, q) + h(r, s)\tag{17.64}$$

for all $p, q, r, s \in I$ if and only if

$$f(p, q) = L_0(q) - L_0(p) + L_1(p) + L_1(q) + \ell\left(\frac{p}{q}, \frac{p}{q}\right) + \alpha + \beta,\tag{17.65}$$

$$g(p, q) = 2\left[L_0(q) - L_0(p) + L_1(p) + L_1(q) + \ell\left(\frac{p}{q}, \frac{p}{q}\right)\right] + 2\alpha,\tag{17.66}$$

$$h(r, s) = 2\left[L_1(r) + L_1(s) + \ell\left(\frac{r}{s}, \frac{r}{s}\right)\right] + 2\beta,\tag{17.67}$$

where $L_0, L_1 : \mathbb{R}_+ \to \mathbb{R}$ are logarithmic and $\ell : \mathbb{R}_+^2 \to \mathbb{R}$ is bilogarithmic, and α, β are arbitrary real constants.

Proof. Let $a \in I$ be a fixed element. Substituting $r = a = s$ in (17.64), we get

$$g(p, q) = 2f(pa, qa) - h(a, a). \tag{17.68}$$

Similarly, letting $p = a = q$ in (17.64), we have

$$h(r, s) = f(ra, sa) + f(sa, ra) - g(a, a). \tag{17.69}$$

Letting (17.68) and (17.69) into (17.64), we obtain

$$f(pr, qs) + f(ps, qr) = 2f(pa, qa) + f(ra, sa) + f(sa, ra) + 2\alpha_0. \tag{17.70}$$

Replacing p by pa, r by ra, q by qa, and s by sa, we obtain

$$F(pr, qs) + F(ps, qr) = 2F(p, q) + F(r, s) + F(s, r), \tag{17.71}$$

where

$$F(p, q) := f\left(pa^2, qa^2\right) + \alpha_0. \tag{17.72}$$

By Lemma 17.4 and (17.72), we obtain

$$f(p, q) = L_0(q) - L_0(p) + L_1(p) + L_1(q) + \ell\left(\frac{p}{q}, \frac{p}{q}\right) + \gamma, \tag{17.73}$$

where γ is a constant. From (17.73) and (17.68), we obtain (17.66). Similarly, (17.73) and (17.69), we obtain (17.67). Letting (17.66), (17.67) and (17.73) into (17.64), we have $\gamma = \alpha + \beta$. Hence, with this and (17.73), we have the asserted solution (17.65). $\qquad\square$

Lemma 17.6. *The functions $f, g : I^2 \to \mathbb{R}$ satisfy the functional equation*

$$f(pr, qs) - f(ps, qr) = g(r, s) \tag{17.74}$$

for all $p, q, r, s \in I$ if and only if

$$f(p, q) = \phi(pq) + L(q), \tag{17.75}$$
$$g(p, q) = L(q) - L(p), \tag{17.76}$$

where $L : \mathbb{R}_+ \to \mathbb{R}$ is logarithmic and $\phi : \mathbb{R}_+ \to \mathbb{R}$ is arbitrary.

Proof. First, substituting $r = a = q$ in (17.74), we obtain

$$f(pa, sa) - f(ps, a^2) = g(a, s). \tag{17.77}$$

Replacing p by pa and s by sa in (17.77) and rearranging, we get

$$f(a^2 p, a^2 s) = f(a^2 ps, a^2) + g(a, as). \tag{17.78}$$

Similarly, replacing p by pa, r by ra, s by sa, and q by qa in (17.74), we have

$$f(a^2 pr, a^2 qs) - f(a^2 ps, a^2 qr) = g(ar, as). \qquad (17.79)$$

Using (17.78) in (17.79), we obtain

$$g(a, aqs) - g(a, aqr) = g(ar, as). \qquad (17.80)$$

Again letting $p = a = q$ in (17.74), we get

$$f(ar, as) - f(as, ar) = g(r, s). \qquad (17.81)$$

Hence, by (17.77), the equation (17.81) yields

$$g(r, s) = g(a, s) - g(a, r). \qquad (17.82)$$

As before, by replacing r by ar and s by as, we have

$$g(ar, as) = g(a, as) - g(a, ar). \qquad (17.83)$$

Thus from (17.83) and (17.82), we see that

$$g(a, aqs) - g(a, aqr) = g(a, as) - g(a, ar). \qquad (17.84)$$

Defining $\psi : I^2 \to \mathbb{R}$ as

$$\psi(x) = g(a, ax), \qquad (17.85)$$

we have from (17.84)

$$\psi(qs) - \psi(qr) = \psi(s) - \psi(r) \qquad (17.86)$$

for all $q, s, r \in I$. From (17.86), we obtain

$$\psi(qs) - \psi(s) = \psi(qr) - \psi(r) \qquad (17.87)$$

for all $q, s, r \in I$. The last equation yields

$$\psi(qs) - \psi(s) = \delta(q), \qquad (17.88)$$

where $\delta : I \to \mathbb{R}$. Interchanging q with s in (17.88), we see that

$$\psi(sq) - \psi(q) = \delta(s). \qquad (17.89)$$

From (17.88) and (17.89), we have

$$\delta(q) - \psi(q) = \delta(s) - \psi(s) = k,$$

where k is a constant. Hence

$$\delta(q) = \psi(q) + k. \qquad (17.90)$$

Using (17.90) in (17.89), we have

$$\psi(sq) - \psi(q) + \psi(s) + k. \tag{17.91}$$

Thus

$$\psi(x) = L(x) + k, \tag{17.92}$$

where $L : \mathbb{R}_+ \to \mathbb{R}$ is logarithmic. From (17.85), (17.82) and (17.92), we obtain

$$g(r, s) = L(s) - L(r). \tag{17.93}$$

Using (17.77) and (17.93), we see that

$$f(pa, sa) = \sigma(ps) + g(a, s), \tag{17.94}$$

where

$$\sigma(ps) := f(ps, a^2). \tag{17.95}$$

Now (17.93) and (17.93) give

$$f(pa, sa) = \sigma(ps) + L(s) + \alpha, \tag{17.96}$$

where α is a constant. Letting (17.96) into (17.74), we observe that σ is an arbitrary function. Hence

$$f(p, s) = \phi(ps) + L(s), \tag{17.97}$$

where $\phi = \sigma + \alpha$. This completes the proof of the lemma. □

17.4   Solution of a Generalized Functional Equation

Now we are ready to determine the general solution of (17.5) without any regularity assumptions on the unknown functions.

Theorem 17.3. *The functions $f_1, f_2, g, h : I^2 \to \mathbb{R}$ satisfy the functional equation (17.5), that is,*

$$f_1(pr, qs) + f_2(ps, qr) = g(p, q) + h(r, s),$$

for all $p, q, r, s \in I$ if and only if

$$f_1(p, q) = \mathcal{L}(p, q) + L_2(q) + \ell\left(\frac{p}{q}, \frac{p}{q}\right) + \phi(pq) - \alpha, \tag{17.98}$$

$$f_2(p, q) = \mathcal{L}(p, q) - L_2(q) + \ell\left(\frac{p}{q}, \frac{p}{q}\right) - \phi(pq) - \alpha, \tag{17.99}$$

$$g(p,q) = 2\mathcal{L}(p,q) + 2\ell\left(\frac{p}{q},\frac{p}{q}\right) + 2\beta - 2\alpha, \tag{17.100}$$

$$h(r,s) = 2L_1(r) + 2L_1(s) + L_2(s) - L_2(r) + 2\ell\left(\frac{r}{s},\frac{r}{s}\right) - 2\beta, \tag{17.101}$$

where $\mathcal{L}(p,q) := L_0(q) - L_0(p) + L_1(p) + L_1(q)$, $L_0, L_1, L_2 : \mathbb{R}_+ \to \mathbb{R}$ *are logarithmic,* $\ell : \mathbb{R}_+^2 \to \mathbb{R}$ *is bilogarithmic,* $\phi : \mathbb{R}_+ \to \mathbb{R}$ *is arbitrary, and* α, β *are arbitrary real constants.*

Proof. It is easy to check that solution (17.98)–(17.101) enumerated in the theorem satisfies (17.5).

Interchanging r with s in (17.5), we get

$$f_1(ps,qr) + f_2(pr,qs) = g(p,q) + h(s,r). \tag{17.102}$$

Adding (17.102) to (17.5), we obtain

$$F(pr,qs) + F(ps,qr) = 2g(p,q) + 2H(r,s), \tag{17.103}$$

where

$$F(p,q) = f_1(p,q) + f_2(p,q) \tag{17.104}$$
$$2H(r,s) = h(r,s) + h(s,r). \tag{17.105}$$

Further, subtracting (17.102) from (17.5), we obtain

$$f(pr,qs) - f(ps,qr) = k(r,s), \tag{17.106}$$

where

$$f(p,q) = f_1(p,q) - f_2(p,q) \tag{17.107}$$
$$2H(r,s) = h(r,s) - h(s,r). \tag{17.108}$$

Using (17.104), (17.105), (17.107) and (17.108), we obtain

$$f_1 = \frac{1}{2}\left(F + f\right), \qquad f_2 = \frac{1}{2}\left(F - f\right), \qquad h = \frac{1}{2}\left(2H + k\right). \tag{17.109}$$

From Lemma 17.6, Lemma 17.5 and (17.109) the asserted solution follows. This completes the proof of the theorem. $\qquad\square$

17.5 Concluding Remarks

In statistical estimation problems various measures between probability distributions play significant roles. Hellinger coeeficient, Jeffreys distance, Chernoff coefficient, directed divergence, and its symmetrization J-divergence are examples of such measures. Chung, Kannappan, Ng and Sahoo (1989) characterized these and like measures through a composition law and the sum form they possess. The functional equations

$$f(pr, qs) + f(ps, qr) = (r + s)f(p, q) + (p + q)f(r, s) \qquad (17.110)$$

and

$$f(pr, qs) + f(ps, qr) = f(p, q)\, f(r, s) \qquad (17.111)$$

holding for all $p, q, r, s \in I$ were instrumental in their deduction. These and like functional equations are referred to as functional equations for stochastic distance measures. The general solution $f : I \to \mathbb{R}$ of each of the functional equations (17.110) and (17.111) was determined by Chung, Kannappan, Ng and Sahoo (1989).

Kannappan, Sahoo and Chung (1993) generalized the functional equation (17.110) to the following:

$$f_1(pr, qs) + f_2(ps, qr) = (r + s)g(p, q) + (p + q)h(r, s). \qquad (17.112)$$

They determined the general solution $f_1, f_2, g, h : J \to \mathbb{R}$ of the functional equation (17.112) without any regularity assumption on the unknown function. We present their result in the next theorem.

Theorem 17.4. *The functions $f_1, f_2, g, h : J^2 \to \mathbb{R}$ satisfy the functional equation*

$$f_1(pr, qs) + f_2(ps, qr) = (r + s)g(p, q) + (p + q)h(r, s)$$

for all $p, q, r, s \in J$ if and only if

$$\left. \begin{array}{l} f_1(p, q) = \mathcal{L}(p, q) + (c - \beta)\,(p - q) + \phi(pq) \\ f_2(p, q) = \mathcal{L}(p, q) - (c + \beta)\,(p - q) - \phi(pq) \\ g(p, q) = \mathcal{L}(p, q) - (\beta + \alpha)\,p + (\beta - \alpha)\,q \\ h(p, q) = \mathcal{L}(p, q) - (\alpha + c)\,p + (\alpha - c)\,q, \end{array} \right\}$$

where

$$\mathcal{L}(p, q) := p\,[L_1(q) - L_2(p)] + q\,[L_1(p) - L_2(q)],$$

α, β, c are arbitrary constants, $\phi(p)$ is an arbitrary function on $(0, 1]$, and L_1, L_2 are logarithmic functions on J.

The functional equation (17.111) was generalized by Kannappan, Sahoo and Chung (1994) to

$$f_1(pr, qs) + f_2(ps, qr) = g(p, q)\, h(r, s) \qquad (17.113)$$

for all $p, q, r, s \in J$. They determined the general complex-valued solution $f_1, f_2, g, h : J^2 \to \mathbb{C}$ of this functional equation without any regularity condition. They found 11 sets of solutions of the equation (17.113).

Riedel and Sahoo (1995) considered another generalization of (17.110), namely,

$$f(pr, qs) + f(ps, qr) = g(r, s) f(p, q) + g(p, q) f(r, s) \qquad (17.114)$$

for all $p, q, r, s \in J$. They proved the following theorem concerning this functional equation.

Theorem 17.5. *The functions* $f, g : J^2 \to \mathbb{C}$ *satisfy the functional equation*

$$f(pr, qs) + f(ps, qr) = f(p, q) g(r, s) + f(r, s) g(p, q)$$

for all $p, q, r, s \in J$ *if and only if*

$$\left. \begin{array}{l} f(p, q) = 0 \\ g(p, q) \text{ arbitrary}; \end{array} \right\}$$

$$\left. \begin{array}{l} f(p, q) = M(p) M(q) \left[L(p) + L(q) + l\left(\frac{p}{q}, \frac{p}{q} \right) \right] \\ g(r, s) = 2 M(r) M(s); \end{array} \right\}$$

$$\left. \begin{array}{l} f(p, q) = \frac{a}{2} \left[M_3(p) M_4(q) + M_3(q) M_4(p) \right] \\ g(r, s) = \frac{1}{2} \left[M_3(r) M_4(s) + M_3(s) M_4(r) \right]; \end{array} \right\}$$

$$\left. \begin{array}{l} f(p, q) = M_1(p) M_2(q) \left[L_1(p) + L_2(q) \right] + M_1(q) M_2(p) \left[L_1(q) + L_2(p) \right] \\ g(r, s) = M_1(r) M_2(s) + M_1(s) M_2(r), \end{array} \right\}$$

where $M, M_1, M_2, M_3, M_4 : J \to \mathbb{C}$ *are multiplicative functions,* $L, L_1, L_2 : J \to \mathbb{C}$ *are logarithmic functions, and* a *is an arbitrary nonzero complex constant.*

The functional equation

$$f(pr, qs) + f(ps, qr) = g(r, s) h(p, q) + g(p, q) h(r, s) \qquad (17.115)$$

for all $p, q, r, s \in J$ is a generalization of the functional equation (17.114). The complex-valued solution $f, g, h : J^2 \to \mathbb{C}$ of the functional equation (17.115) was determined by Riedel and Sahoo (1997) without any regularity condition on the unknown functions.

The functional equations (17.110) and (17.111) can be simultaneously generalized to

$$f_1(pr, qs) + f_2(ps, qr) = g_1(r, s) h_1(p, q) + g_2(p, q) h_2(r, s) \qquad (17.116)$$

for all $p, q, r, s \in I$. The general solution this functional equation is not known (see Kannappan and Sahoo (1997a)). Next we restate their open problem. Find all functions $f_1, f_2, g_1, g_2, h_1, h_2 : I^2 \to \mathbb{C}$ satisfying the functional equation (17.116) for all $p, q, r, s \in I$. To solve this equation (17.116), one needs a series of functional equations which are the specializations of (17.116). These equations are the following:

$$f(pr, qs) + f(ps, qr) = 2\, f(p, q) + 2\, f(r, s), \tag{17.117}$$

$$f_1(pr, qs) + f_2(ps, qr) = (r + s)\, f_3(p, q) + (p + q)\, f_4(r, s), \tag{17.118}$$

$$f_1(pr, qs) + f_2(ps, qr) = f_3(p, q)\, f_4(r, s), \tag{17.119}$$

$$f_1(pr, qs) + f_2(ps, qr) = f_3(p, q) + f_4(r, s), \tag{17.120}$$

$$f(pr, qs) + f(ps, qr) = g(r, s)\, f(p, q) + g(p, q)\, f(r, s), \tag{17.121}$$

$$f(pr, qs) + f(ps, qr) = g(r, s)\, h(p, q) + g(p, q)\, h(r, s), \tag{17.122}$$

$$f(pr, qs) + f(ps, qr) = g_1(r, s)\, h_1(p, q) + g_2(p, q)\, h_2(r, s), \tag{17.123}$$

$$f(pr, qs) - f(ps, qr) = g_1(r, s)\, h_1(p, q) + g_2(p, q)\, h_2(r, s) \tag{17.124}$$

for all $p, q, r, s \in I$. Solutions of these functional equations are unknown when the domain of the functions is I^2. In the previous works by Kannappan, Sahoo and Chung (1993, 1994) and Riedel and Sahoo (1995, 1997), the process of obtaining the solutions of the functional equations they studied depended heavily on the boundary point 1. Without this point, the technique employed in finding the solution would not work.

17.6   Exercises

1. Without using Theorem 17.1, find all functions $f : \mathbb{R}^2_+ \to \mathbb{R}$ that satisfy the functional equation

$$f(pr, qs) + f(ps, qr) = (r + s)f(p, q) + (p + q)f(r, s)$$

for all $p, q, r, s \in \mathbb{R}^2_+$.

2. Without using Lemma 17.3, find all functions $f : \mathbb{R}^2_+ \to \mathbb{R}$ that satisfy the functional equation

$$f(pr, qs) + f(ps, qr) = 2f(p, q) + 2f(r, s)$$

for all $p, q, r, s \in \mathbb{R}^2_+$.

3. Find all functions $f, g : \mathbb{R}_+^2 \to \mathbb{R}$ that satisfy the functional equation

$$f(pr, qs) + f(ps, qr) = g(r, s)f(p, q) + g(p, q)f(r, s)$$

for all $p, q, r, s \in \mathbb{R}_+^2$.

4. Find all functions $f_1, f_2, g, h : \mathbb{R}_+^2 \to \mathbb{R}$ that satisfy the functional equation

$$f_1(pr, qs) + f_2(ps, qr) = (r + s)g(p, q) + (p + q)h(r, s)$$

for all $p, q, r, s \in \mathbb{R}_+^2$.

5. Find all functions $f_1, f_2, g, h : \mathbb{R}_+^2 \to \mathbb{R}$ that satisfy the functional equation

$$f_1(pr, qs) + f_2(ps, qr) = g(p, q) h(r, s)$$

for all $p, q, r, s \in \mathbb{R}_+^2$.

6. Find all functions $f, g, h : \mathbb{R}_+^2 \to \mathbb{R}$ that satisfy the functional equation

$$f(pr, qs) + f(ps, qr) = g(r, s)h(p, q) + g(p, q)h(r, s)$$

for all $p, q, r, s \in \mathbb{R}_+^2$.

7. Find all functions $f, g, h : \mathbb{R}_+^2 \to \mathbb{R}$ that satisfy the functional equation

$$f(pr, qs) - f(ps, qr) = g(r, s)h(p, q) + g(p, q)h(r, s)$$

for all $p, q, r, s \in \mathbb{R}_+^2$.

8. Find all functions $f_1, f_2, g, h : \mathbb{R}_+^2 \to \mathbb{R}$ that satisfy the functional equation

$$f_1(pr, qs) + f_2(ps, qr) = g(p, q)h(r, s)$$

for all $p, q, r, s \in \mathbb{R}_+^2$.

9. Find all functions $f_1, f_2, g, h : \mathbb{R}_+^2 \to \mathbb{R}$ that satisfy the functional equation

$$f_1(pr, qs) + f_2(ps, qr) = g(p, q) + h(r, s)$$

for all $p, q, r, s \in \mathbb{R}_+^2$.

10. Find all functions $f_1, f_2, g_1, g_2, h_1, h_2 : \mathbb{R}_+^2 \to \mathbb{R}$ that satisfy the functional equation

$$f_1(pr, qs) + f_2(ps, qr) = g_1(r, s)h_1(p, q) + g_2(p, q)h_2(r, s)$$

for all $p, q, r, s \in \mathbb{R}_+^2$.

Chapter 18

Stability of Additive Cauchy Equation

18.1 Introduction

A certain formula or equation is applicable to model a physical process if a small change in the formula or equation gives rise to a small change in the corresponding result. When this happens, we say the formula or equation is stable. In an application, a functional equation such as the additive Cauchy functional equation $f(x + y) - f(x) - f(y) = 0$ may not be true for all $x, y \in \mathbb{R}$ but it may be true approximately, that is,

$$f(x + y) - f(x) - f(y) \approx 0$$

for all $x, y \in \mathbb{R}$. This can be stated mathematically as

$$|f(x + y) - f(x) - f(y)| \leq \varepsilon \tag{18.1}$$

for some small positive ε and for all $x, y \in \mathbb{R}$. We would such as to know when small changes in a particular equation like additive Cauchy functional equation have only small effects on its solutions. This is the essence of the stability theory.

In 1940, S.M. Ulam (see Ulam (1960)) asked the following question: Given a group G, a metric group H with metric $d(\cdot, \cdot)$ and a positive number ϵ, does there exist a $\delta > 0$ such that if $f : G \to H$ satisfies

$$d(f(xy), f(x)f(y)) \leq \delta$$

for all $x, y \in G$, then a homomorphism $\phi : G \to H$ exists with

$$d(f(x), \phi(x)) \leq \varepsilon$$

for all $x \in G$? These kinds of questions form the material of the stability theory. For Banach spaces, the above problem was solved by D.H. Hyers (1941) with $\delta = \epsilon$ and

$$\phi(x) = \lim_{n \to \infty} \frac{f(2^n x)}{2^n}.$$

In 1968, S.M. Ulam proposed the following general problem: *When is it true that by changing a little in the hypotheses of a theorem one can still assert that the thesis of the theorem remains true or approximately true?* According to Gruber (1978), this kind of stability problem is of particular interest in probability theory.

In this chapter, we present Hyers' result along with a theorem due to Th.M. Rassias that generalizes the result of Hyers. We also point out some other generalizations related to the stability of the additive Cauchy equation.

18.2   Cauchy Sequence and Geometric Series

In this section, we briefly review Cauchy sequence and the sum of a geometric series. The notion of Cauchy sequence and the sum of a geometric series will be used in the proof of Hyers' theorem.

Definition 18.1. *A sequence $\{x_n\}$ of real numbers is said to be a Cauchy sequence if for every $\epsilon > 0$ there exists a natural number N such that for all natural numbers $n, m \geq N$, the terms x_n and x_m satisfy*

$$|x_n - x_m| < \epsilon.$$

The proof of the following theorem can be found in any elementary book on real analysis.

Theorem 18.1. *A sequence of real numbers is convergent if and only if it is a Cauchy sequence.*

Example 18.1. *The following are Cauchy sequences:*

(a) $\left\{ \frac{n+1}{n} \right\}_{n=1}^{\infty}$,

(b) $\left\{ 1 + \frac{1}{2!} + \frac{1}{3!} + \cdots + \frac{1}{n!} \right\}_{n=1}^{\infty}$.

Example 18.2. *The following are not Cauchy sequences:*

(a) $\{(-1)^n\}_{n=1}^{\infty}$,

(b) $\left\{ n + \frac{(-1)^n}{n} \right\}_{n=1}^{\infty}$.

Now we consider the sum of a geometric series.

Theorem 18.2. *Let $r \in [0, 1)$. Then*

$$S_n = 1 + r + r^2 + \cdots + r^{n-1} = \frac{1 - r^n}{1 - r}$$

and

$$S = 1 + r + r^2 + \cdots + r^n + \cdots = \frac{1}{1 - r}.$$

Further

$$S_n = \sum_{k=0}^{n-1} r^k < \sum_{k=0}^{\infty} r^k = S.$$

Proof. To prove the first equality write

$$S_n = 1 + r + r^2 + \cdots + r^{n-1}. \tag{18.2}$$

Then

$$r S_n = r + r^2 + \cdots + r^n. \tag{18.3}$$

Subtract (18.3) from (18.2) to obtain

$$S_n - r S_n = 1 - r^n. \tag{18.4}$$

Hence

$$S_n = \frac{1 - r^n}{1 - r}. \tag{18.5}$$

Note that

$$S = \lim_{n \to \infty} S_n = 1 + r + r^2 + \cdots + r^n + \cdots.$$

Therefore

$$S = \lim_{n \to \infty} S_n = \lim_{n \to \infty} \frac{1 - r^n}{1 - r} = \frac{1}{1 - r}$$

since $r \in [0, 1)$. We leave the proof of the last inequality to the reader. □

18.3 Hyers' Theorem

Hyers (1941) obtained the first important result on stability theory that stemmed from Ulam's problem. We present the original theorem of Hyers in the following theorem.

Theorem 18.3. *If $f : \mathbb{R} \to \mathbb{R}$ is a real function satisfying*

$$|f(x+y) - f(x) - f(y)| \leq \delta$$

for some $\delta > 0$ and for all $x, y \in \mathbb{R}$, then there exists a unique additive function $A : \mathbb{R} \to \mathbb{R}$ such that

$$|f(x) - A(x)| \leq \delta$$

for all $x \in \mathbb{R}$.

Proof. Let $f : \mathbb{R} \to \mathbb{R}$ be a real function such that

$$|f(x+y) - f(x) - f(y)| \leq \delta \qquad (18.6)$$

for all $x, y \in \mathbb{R}$ and for some $\delta > 0$. To establish this theorem we have to show that

(i) $\left\{ \frac{f(2^n x)}{2^n} \right\}_{n=1}^{\infty}$ is a Cauchy sequence for every fixed $x \in \mathbb{R}$;

(ii) if

$$A(x) = \lim_{n \to \infty} \frac{f(2^n x)}{2^n},$$

then A is additive on $\mathbb{R}$;

(iii) further A satisfies

$$|f(x) - A(x)| \leq \delta$$

for $x \in \mathbb{R}$;

(iv) A is unique.

Letting $y = x$ in (18.6), we have

$$|f(2x) - 2f(x)| \leq \delta \qquad (18.7)$$

for all $x \in \mathbb{R}$. Replacing x by $2^{k-1}x$ (where k is a positive integer greater than or equal to 1), we obtain

$$|f(2^k x) - 2f(2^{k-1}x)| \leq \delta$$

for all $x \in \mathbb{R}$ and $k = 1, 2, \ldots, n$, where $n \in \mathbb{N}$. Multiplying both sides of the above inequality by $\frac{1}{2^k}$ and adding the resulting n inequalities, we have

$$\sum_{k=1}^{n} \frac{1}{2^k} |f(2^k x) - 2f(2^{k-1}x)| \leq \sum_{k=1}^{n} \frac{1}{2^k} \delta \qquad (18.8)$$

which yields

$$\sum_{k=1}^{n} \frac{1}{2^k} |f(2^k x) - 2f(2^{k-1}x)| \le \delta \left(1 - \frac{1}{2^n}\right).$$
(18.9)

Using the triangle inequality

$$|a + b| \le |a| + |b|$$

in (18.9), we obtain

$$\left| \frac{1}{2^n} f(2^n x) - f(x) \right| \le \delta \left(1 - \frac{1}{2^n}\right)$$
(18.10)

for all $x \in \mathbb{R}$ and $n \in \mathbb{N}$. Using induction it can be easily shown that (18.10) holds for all positive integers $n \in \mathbb{N}$.

Now if $n > m > 0$, then $n - m$ is a natural number, and n can be replaced by $n - m$ in (18.10) to obtain

$$\left| \frac{f(2^{n-m}x)}{2^{n-m}} - f(x) \right| \le \delta \left(1 - \frac{1}{2^{n-m}}\right).$$

Multiplying both sides by $\frac{1}{2^m}$ and simplifying, we get

$$\left| \frac{f(2^{n-m}x)}{2^n} - \frac{f(x)}{2^m} \right| \le \delta \left(\frac{1}{2^m} - \frac{1}{2^n}\right)$$

for all $x \in \mathbb{R}$. Now we replace x by $2^m x$ to have

$$\left| \frac{f(2^n x)}{2^n} - \frac{f(2^m x)}{2^m} \right| \le \delta \left(\frac{1}{2^m} - \frac{1}{2^n}\right).$$

If $m \to \infty$, then

$$\left(\frac{1}{2^m} - \frac{1}{2^n} \right) \to 0$$

and therefore

$$\lim_{m \to \infty} \left| \frac{f(2^n x)}{2^n} - \frac{f(2^m x)}{2^m} \right| = 0.$$

Hence

$$\left\{ \frac{f(2^n x)}{2^n} \right\}_{n=1}^{\infty}$$

is a Cauchy sequence in $\mathbb{R}$. Hence the limit of this sequence exists. Define $A : \mathbb{R} \to \mathbb{R}$ by

$$A(x) := \lim_{n \to \infty} \frac{f(2^n x)}{2^n}.$$
(18.11)

Now we show that $A : \mathbb{R} \to \mathbb{R}$ defined by (18.11) is additive.

Consider

$$|A(x+y) - A(x) - A(y)| = \left| \lim_{n \to \infty} \left\{ \frac{f(2^n(x+y))}{2^n} - \frac{f(2^n x)}{2^n} - \frac{f(2^n y)}{2^n} \right\} \right|$$

$$= \left| \lim_{n \to \infty} \frac{1}{2^n} \{ f(2^n(x+y)) - f(2^n x) - f(2^n y) \} \right|$$

$$= \lim_{n \to \infty} \frac{1}{2^n} |f(2^n x + 2^n y) - f(2^n x) - f(2^n y)|$$

$$\leq \lim_{n \to \infty} \frac{\delta}{2^n} \qquad \text{(by (18.2))}$$

$$= 0.$$

Therefore

$$A(x+y) = A(x) + A(y)$$

for all $x, y \in \mathbb{R}$.

Our next goal is to show that

$$|A(x) - f(x)| \leq \delta.$$

Thus consider

$$|A(x) - f(x)| = \left| \lim_{n \to \infty} \frac{f(2^n x)}{2^n} - f(x) \right|$$

$$= \lim_{n \to \infty} \left| \frac{f(2^n x)}{2^n} - f(x) \right|$$

$$\leq \lim_{n \to \infty} \delta \left(1 - \frac{1}{2^n} \right) \qquad \text{(by (18.10))}$$

$$= \delta.$$

Hence we obtain

$$|A(x) - f(x)| \leq \delta$$

for all $x \in \mathbb{R}$.

Finally we prove that A is unique. Suppose A is not unique, then there exists another additive function $B : \mathbb{R} \to \mathbb{R}$ such that

$$|B(x) - f(x)| \leq \delta \qquad (18.12)$$

for all $x \in \mathbb{R}$. Note that

$$|B(x) - A(x)| = |B(x) - f(x) + f(x) - A(x)|$$

$$\leq |B(x) - f(x)| + |f(x) - A(x)|$$
$$= \delta + \delta.$$

Therefore

$$|B(x) - A(x)| \leq 2\delta. \tag{18.13}$$

Further, since A and B are additive, we have

$$|A(x) - B(x)| = \left| \frac{nA(x)}{n} - \frac{nB(x)}{n} \right|$$
$$= \left| \frac{A(nx)}{n} - \frac{B(nx)}{n} \right|$$
$$= \frac{1}{n}|A(nx) - B(nx)|$$
$$\leq \frac{2\delta}{n},$$

where $n \in \mathbb{N}$.

Hence

$$|A(x) - B(x)| \leq \frac{2\delta}{n}.$$

Taking the limit on both sides, we get

$$\lim_{n \to \infty} |A(x) - B(x)| \leq \lim_{n \to \infty} \frac{2\delta}{n}$$

which is

$$|A(x) - B(x)| \leq 0.$$

Hence

$$A(x) = B(x) \qquad \forall\, x \in \mathbb{R}.$$

Therefore the additive map A is unique and the proof of the theorem is now complete. $\qquad\square$

Remark 18.1. *In general the proof of Theorem 18.3 works for functions* $f : E_1 \to E_2$ *where* E_1 *and* E_2 *are Banach spaces.*

This pioneer result of D. H. Hyers can be expressed in the following way: Cauchy functional equation $f(x+y) = f(x) + f(y)$ is stable for any pair of Banach spaces. The function $(x, y) \mapsto f(x+y) - f(x) - f(y)$ is called the Cauchy difference of the function f. Functions with a bounded Cauchy difference are called approximately additive (or ϵ-additive if the Cauchy difference is bounded by the constant ϵ). The sequence $\left\{ \frac{f(2^n x)}{2^n} \right\}$ is called the Hyers-Ulam sequence.

Remark 18.2. *Any result similar to Theorem 18.3 is known as the Hyers-Ulam stability of the corresponding functional equation.*

18.4 Generalizations of Hyers' Theorem

It is possible to prove a stability result similar to Hyers (that is, Theorem 18.3) for functions that do not have bounded Cauchy difference. Aoki (1950) first proved such a result for additive function. Then Rassias (1978) proved such a result for linear mappings on Banach spaces. In the following theorem, we present Rassias' result that generated a lot of activities in the stability theory of functional equations.

Theorem 18.4. *If* $f : \mathbb{R} \to \mathbb{R}$ *is a real map satisfying*

$$|f(x + y) - f(x) - f(y)| \leq \delta(|x|^p + |y|^p)$$

for some $\delta > 0$, $p \in [0, 1)$ *and for all* $x, y \in \mathbb{R}$, *then there exists a unique additive function* $A : \mathbb{R} \to \mathbb{R}$ *such that*

$$|f(x) - A(x)| \leq \frac{2\delta}{2 - 2^p} |x|^p$$

for all $x \in \mathbb{R}$.

Proof. Let $f : \mathbb{R} \to \mathbb{R}$ be a real function satisfying

$$|f(x + y) - f(x) - f(y)| \leq \delta(|x|^p + |y|^p) \tag{18.14}$$

for all $x, y \in \mathbb{R}$ and for some $\delta > 0$ and $p \in [0, 1)$. Letting $y = x$ in (18.14), we get

$$|f(2x) - 2f(x)| \leq 2\delta|x|^p \tag{18.15}$$

for all $x \in \mathbb{R}$. We replace x by $2^{k-1}x$ (for $k \in \mathbb{N}$ and $k \geq 1$), and we obtain

$$|f(2^k x) - 2f(2^{k-1}x)| \leq 2^{kp-p+1}\delta|x|^p.$$

Multiplying both sides of the above inequality by $\frac{1}{2^k}$ and then adding the resulting n inequalities, we get

$$\sum_{k=1}^{n} \frac{1}{2^k} |f(2^k x) - 2f(2^{k-1}x)| \leq \delta|x|^p \sum_{k=1}^{n} \frac{2^{kp-p+1}}{2^k}.$$

Using the triangle inequality

$$|a + b| \leq |a| + |b|$$

and simplifying the left side of the inequality, we get

$$\left| \frac{1}{2^n} f(2^n x) - f(x) \right| \leq \delta|x|^p \sum_{k=1}^{n} 2^{k(p-1)} \cdot 2^{1-p}. \tag{18.16}$$

Since

$$\sum_{k=1}^{n} 2^{k(p-1)} \le \sum_{k=1}^{\infty} 2^{k(p-1)},$$

the inequality (18.16) yields

$$\left| \frac{1}{2^n} f(2^n x) - f(x) \right| \le \delta |x|^p 2^{1-p} \sum_{k=1}^{\infty} 2^{k(p-1)}$$

which is

$$\left| \frac{1}{2^n} f(2^n x) - f(x) \right| \le \frac{2\delta}{2 - 2^p} |x|^p \tag{18.17}$$

for all $x \in \mathbb{R}$. By induction it can be shown that (18.17) is valid for all natural numbers. If $m > n > 0$, then $m - n$ is a natural number and replacing n by $m - n$ in (18.17), we get

$$\left| \frac{1}{2^{m-n}} f(2^{m-n} x) - f(x) \right| \le \frac{2\delta}{2 - 2^p} |x|^p \tag{18.18}$$

which is

$$\left| \frac{1}{2^m} f(2^{m-n} x) - \frac{1}{2^n} f(x) \right| \le \frac{1}{2^n} \left(\frac{2\delta}{2 - 2^p} \right) |x|^p, \tag{18.19}$$

for all $x \in \mathbb{R}$. Replacing x by $2^n x$ in (18.19), we obtain

$$\left| \frac{1}{2^m} f(2^m x) - \frac{1}{2^n} f(2^n x) \right| \le \left(\frac{2\delta}{2 - 2^p} \right) \frac{2^{np}}{2^n} |x|^p. \tag{18.20}$$

Since $0 \le p < 1$,

$$\lim_{n \to \infty} 2^{n(p-1)} = 0,$$

and hence from (18.20), we obtain

$$\lim_{n \to \infty} \left| \frac{1}{2^m} f(2^m x) - \frac{1}{2^n} f(2^n x) \right| = 0.$$

Therefore

$$\left\{ \frac{f(2^n x)}{2^n} \right\}_{n=1}^{\infty}$$

is a Cauchy sequence. This Cauchy sequence has a limit in $\mathbb{R}$. We define

$$A(x) = \lim_{n \to \infty} \frac{f(2^n x)}{2^n} \tag{18.21}$$

for $x \in \mathbb{R}$. First we show that $A : \mathbb{R} \to \mathbb{R}$ is additive.

Consider

$$|A(x+y) - A(x) - A(y)| = \lim_{n\to\infty} \frac{1}{2^n} |f(2^n x + 2^n y) - f(2^n x) - f(2^n y)|$$

$$\leq \lim_{n\to\infty} \frac{\delta(|x|^p + |y|^p) 2^{np}}{2^n} \quad \text{(by (18.20))}$$

$$= 0$$

since $p \in [0, 1)$.

Hence

$$A(x+y) = A(x) + A(y)$$

for all $x, y \in \mathbb{R}$. Next, we consider

$$|A(x) - f(x)| = \left| \lim_{n\to\infty} \frac{f(2^n x)}{2^n} - f(x) \right|$$

$$= \lim_{n\to\infty} \left| \frac{f(2^n x)}{2^n} - f(x) \right|$$

$$\leq \lim_{n\to\infty} \frac{2\delta}{2 - 2^p} |x|^p.$$

Hence, we get

$$|A(x) - f(x)| \leq \frac{2\delta}{2 - 2^p} |x|^p$$

for all $x \in \mathbb{R}$.

Next, we show A is unique. Suppose A is not unique. Then there exists another additive function $B : \mathbb{R} \to \mathbb{R}$ such that

$$|B(x) - f(x)| \leq \frac{2\delta}{2 - 2^p} |x|^p.$$

Hence

$$|B(x) - A(x)| \leq |B(x) - f(x)| + |A(x) - f(x)|$$

$$\leq \frac{2\delta}{2 - 2^p} |x|^p + \frac{2\delta}{2 - 2^p} |x|^p$$

$$= \frac{4\delta}{2 - 2^p} |x|^p.$$

Further, since A and B are additive, we have

$$|A(x) - B(x)| = \frac{1}{n} |A(nx) - B(nx)|$$

$$\leq \frac{1}{n} \frac{4\delta}{2 - 2^p} |x|^p. \quad (18.22)$$

Hence taking the limit as $n \to \infty$, we get from (18.22)

$$\lim_{n\to\infty} |A(x) - B(x)| \leq \lim_{n\to\infty} \frac{1}{n} \frac{4\delta}{2 - 2^p} |x|^p.$$

Hence

$$|A(x) - B(x)| \leq 0.$$

Therefore $A(x) = B(x)$ for all $x \in \mathbb{R}$. Hence A is unique. This completes the proof of the theorem. $\qquad\square$

Remark 18.3. *If $p = 0$, then Theorem 18.4 implies Theorem 18.3.*

Remark 18.4. *Theorem 18.4 holds for all $p \in \mathbb{R}\backslash\{1\}$. Gajda (1991) gave an example to show that Theorem 18.4 fails if $p = 1$. Gajda succeeded in constructing an example of a bounded continuous function $g : \mathbb{R} \to \mathbb{R}$ satisfying*

$$|g(x + y) - g(x) - g(y)| \leq |x| + |y|$$

for any $x, y \in \mathbb{R}$, with

$$\lim_{x\to 0} \frac{g(x)}{x} = \infty.$$

Gajda's function g behaves badly near 0.

The function g which Gajda (1991) constructed is the following. For a fixed $\theta > 0$, let $g : \mathbb{R} \to \mathbb{R}$ be defined by

$$g(x) = \sum_{n=0}^{\infty} 2^{-n} \phi(2^n x), \qquad x \in \mathbb{R},$$

where the function $\phi : \mathbb{R} \to \mathbb{R}$ is given by

$$\phi(x) = \begin{cases} \frac{1}{6}\theta & \text{if } 1 \leq x < \infty \\ \frac{1}{6}\theta x & \text{if } -1 < x < 1 \\ -\frac{1}{6}\theta & \text{if } -\infty < x \leq -1. \end{cases} \qquad (18.23)$$

This construction shows that Theorem 18.4 is false for $p = 1$, as we see in the following theorem.

Theorem 18.5. *The function g defined above satisfies the inequality*

$$|g(x + y) - g(x) - g(y)| \leq \theta\,(|x| + |y|)$$

for all $x, y \in \mathbb{R}$. But there is no constant $\delta \in [0, \infty)$ and no additive function $A : \mathbb{R} \to \mathbb{R}$ satisfying the inequality

$$|f(x) - A(x)| \leq \delta\,|x|$$

for all $x \in \mathbb{R}$.

Rassias and Šemrl (1992) has also constructed a continuous function f satisfying

$$|f(x + y) - f(x) - f(y)| \leq |x| + |y|$$

for any $x, y \in \mathbb{R}$, with

$$\lim_{x \to \infty} \frac{f(x)}{x} = \infty.$$

It follows that the set

$$\left\{ \frac{|f(x) - A(x)|}{|x|} : x \neq 0 \right\}$$

is unbounded for any linear mapping $A : \mathbb{R} \to \mathbb{R}$ defined by $A(x) = ax$ for a given $a \in \mathbb{R}$. In other words, an analogue of Theorem 18.4 cannot be obtained.

The following theorem is due to Rassias and Šemrl (1992).

Theorem 18.6. *There exists a continuous function $f : \mathbb{R} \to \mathbb{R}$ satisfying*

$$|f(x + y) - f(x) - f(y)| \leq |x| + |y| \tag{18.24}$$

for any $x, y \in \mathbb{R}$, with

$$\lim_{x \to \infty} \frac{f(x)}{x} = \infty.$$

Proof. The function f defined by

$$f(x) = \begin{cases} x \log_2(x + 1) & \text{if } x \geq 0 \\ x \log_2 |x - 1| & \text{if } x < 0 \end{cases} \tag{18.25}$$

satisfies the condition

$$\lim_{x \to \infty} \frac{f(x)}{x} = \infty.$$

Further, it can be verified that f is continuous, odd and convex on $\mathbb{R}$.

Let x, y be any two positive real numbers. Since f is convex, it follows that

$$|f(x + y) - f(x) - f(y)| \leq f(x + y) - 2f\left(\frac{x + y}{2}\right). \tag{18.26}$$

From (18.26) and (18.25), we get

$$|f(x + y) - f(x) - f(y)| \leq (x + y) \log_2 \left(\frac{x + y + 1}{\frac{x+y}{2} + 1} \right). \tag{18.27}$$

Since

$$0 < \log_2 \left(\frac{x+y+1}{\frac{x+y}{2}+1} \right) \leq 1$$

from the inequality (18.27), we see that

$$|f(x+y) - f(x) - f(y)| \leq x + y \qquad (18.28)$$

which is the inequality (18.24). Since f is an odd function, the inequality (18.24) holds for $x < 0$ and $y < 0$ as well. It remains to consider the case when $x > 0$ and $y < 0$. There is no loss of generality in assuming that $|x| > |y|$ holds. Because f is odd and convex on the set $\mathbb{R}_+$, we have

$$|f(x+y)-f(x)-f(y)| = -f(x+y)+f(x)+f(y) = f(x)-f(x+y)-f(-y).$$

But $x+y$ and $-y$ are positive real numbers. Thus, the previous argument completes the proof. □

Remark 18.5. *Any result similar to Theorem 18.4 is known as the Hyers-Ulam-Rassias stability of the corresponding functional equation.*

18.5   Concluding Remarks

In many areas of mathematics one can ask the following question: When is it true that a mathematical object satisfying a certain property approximately must be close to an object satisfying the property exactly. In functional equations, one can ask when the solutions of a functional equation differing slightly from a given one must be close to the solution of a given equation. The stability theory for functional equation was initiated by a problem of S. M. Ulam in 1940 (see Ulam (1960)).

Hyers (1941) proved that if X and Y are Banach spaces and f is a function mapping X into Y satisfying $\|f(x+y) - f(x) - f(y)\| \leq \varepsilon$ for some $\varepsilon > 0$ and for all $x, y \in X$, then there exists a unique additive map $A : X \to Y$ such that $\|f(x) - A(x)\| \leq \varepsilon$ for all $x \in X$. Moreover, if $f(tx)$ is continuous in t for each fixed $x \in X$, then A is linear. To prove this result, Hyers constructed the additive map A explicitly from the given function f. His method is called a *direct method* and has been widely used for studying stability of many functional equations.

The inequality $\|f(x+y) - f(x) - f(y)\| \leq \varepsilon$ can be stated differently as the Cauchy difference of the function f is bounded. Aoki (1950) and also independently Rassias (1978) proved a generalized version of Hyers' theorem where the Cauchy difference was allowed to be unbounded. Both the proofs exploited the direct method of Hyers. Rassias (1978) proved the following theorem.

Theorem 18.7. *If a function* $f : X \to Y$ *between two Banach spaces satisfies the inequality*

$$||f(x+y) - f(x) - f(y)|| \leq \varepsilon (||x||^p + ||y||^p) \qquad (18.29)$$

for some $\varepsilon \geq 0$, $0 \leq p < 1$, *and for all* $x, y \in X$, *then there exists a unique additive function* $A : X \to Y$ *such that*

$$||f(x) - A(x)|| \leq \frac{2\varepsilon}{2 - 2^p} ||x||^p$$

for any $x \in X$. *Moreover, if* $f(tx)$ *is continuous in* t *for each fixed* $x \in X$, *then* A *is linear.*

We would like to point out that Aoki (1950) did not prove the last part of the above theorem; that is, if $f(tx)$ is continuous in t for each fixed $x \in X$ then A is linear, but he did prove the first part of the theorem.

Works of Aoki (1950) and Rassias (1978) have been generalized by many mathematicians by weakening the condition (18.29) for the Cauchy difference. A generalization of Theorems 18.3 and 18.4 can be initiated by considering the following inequality

$$|f(x+y) - f(x) - f(y)| \leq \phi(x, y)$$

where $\phi : X^2 \to \mathbb{R}_+$, where X is a Banach space.

It seems that D.G. Bourgin in 1951 was the first to consider this inequality and stated without proof that if ϕ depends on $|x|$ and $|y|$; is monotonic, nondecreasing and symmetric in $|x|$ and $|y|$; and moreover the series

$$\frac{1}{2} \sum_{k=0}^{\infty} \left(\frac{1}{2}\right)^k \phi(2^k x, 2^k y)$$

converges for each $x \in \mathbb{R}$, then

$$|f(x) - A(x)| \leq \psi(x),$$

where

$$\psi(x) = \frac{1}{2} \sum_{k=0}^{\infty} \left(\frac{1}{2}\right)^k \phi(2^k x, 2^k x).$$

Forti (1980) and independently also Z. Kominek obtained the following result.

Theorem 18.8. *Suppose $(X, +)$ is a commutative semigroup and E is a Banach space. If the functions $f : X \to E$ and $\phi : X^2 \to \mathbb{R}_+$ satisfy the inequality*

$$|f(x + y) - f(x) - f(y)| \leq \phi(x, y)$$

and for all $x, y \in X$

$$\sum_{k=0}^{\infty} \left(\frac{1}{2}\right)^k \phi(2^k x, 2^k x) < \infty$$

and

$$\lim_{n \to \infty} \left(\frac{1}{2}\right)^n \phi(2^n x, 2^n y) = 0,$$

then there exists a unique additive function $A : X \to E$ such that

$$|f(x) - A(x)| \leq \frac{1}{2} \sum_{k=0}^{\infty} \left(\frac{1}{2}\right)^k \phi(2^k x, 2^k x)$$

for every $x \in X$.

The following theorem due to Găvruta (1994) generalizes Theorem 18.8 due to Forti and Kominek. We state the theorem without a proof.

Theorem 18.9. *Let G and E be an abelian group and a Banach space, respectively. Let $\phi : G \times G \to [0, \infty)$ be a function satisfying*

$$\Phi(x, y) = \sum_{k=1}^{\infty} 2^{-(k+1)} \phi(2^k x, 2^k y) < \infty$$

for all $x, y \in G$. If a function $f : G \to E$ satisfies the inequality

$$\|f(x + y) - f(x) - f(y)\| \leq \phi(x, y)$$

for any $x, y \in G$, then there exists a unique additive function $A : G \to E$ such that

$$\|f(x) - A(x)\| \leq \Phi(x, x)$$

for all $x \in G$. If moreover $f(tx)$ is continuous in t for fixed $x \in G$, then A is linear.

The stability of additive Cauchy functional equation with special control function φ occurring in the inequality

$$\|f(x + y) - f(x) - f(y)\| \leq \phi(x, y)$$

has been studied by Ger (1992). Among other results he proved the following interesting result.

Theorem 18.10. *Let X be a real Banach space and let Y be a real normed linear space. Suppose that $\varphi : X \to \mathbb{R}$ is a nonnegative subadditive functional on X and $f : X \to Y$ is a mapping such that*

$$\|f(x + y) - f(x) - f(y)\| \leq \varphi(x) + \varphi(y) - \varphi(x + y)$$

holds true for all $x, y \in X$. If, moreover, f and φ have a common continuity point or if the function $X \ni x \to \|f(x)\| + \varphi(x)$ is bounded above on a second category Baire set, then there exists a nonnegative constant c such that

$$\|f(x)\| \leq c\,\|x\| + \varphi(x), \quad x \in X.$$

The following result of this kind was proved by Šemrl (1994).

Theorem 18.11. *Let $\delta > 0$ and assume that a continuous mapping $f : \mathbb{R} \to \mathbb{R}$ satisfies*

$$\left| f\left(\sum_{n=1}^{m} x_n \right) - \sum_{n=1}^{m} f(x_n) \right| \leq \delta \sum_{n=1}^{m} |x_n|, \qquad x_1, x_2, ..., x_m \in \mathbb{R}$$

for all $m \in \mathbb{N}$. Then there exists an additive mapping $A : \mathbb{R} \to \mathbb{R}$ such that

$$|f(x) - A(x)| \leq \delta\,|x|, \qquad x \in \mathbb{R}.$$

Recently, Arriola and Beyer (2005) studied the Hyers-Ulam stability of the Cauchy functional equation $f(x + y) = f(x) + f(y)$ on a non-Archimedean field, namely, the p-adic field numbers $\mathbb{Q}_p$. The Hyers-Ulam-Rassias type stability of Cauchy functional equation was studied by Moslehian and Rassias (2007) in a complete non-Archimedean normed space.

The domain of the function f in Theorem 18.3 (that is, Hyers' theorem) or its generalizations we have discussed so far is a normed space, a Banach space or, in general, an abelian group. Let G be an arbitrary group and E be a Banach space. Let $B(G, E)$ denote the space of all bounded functions $f : G \to E$. The Hyers-Ulam stability of the Cauchy functional equation can be restated as follows. The Cauchy functional equation is said to be stable for the pair (G, E) if for any $f : G \to E$ satisfying

$$\|f(xy) - f(x) - f(y)\| \leq \delta \qquad \forall\, x, y \in G$$

(for some $\delta \geq 0$), there is a solution $a : G \to E$ of the Cauchy functional equation $a(xy) = a(x) + a(y)$ such that $f - a \in B(G, E)$.

Let G be a group and E and B be any two arbitrary Banach spaces over reals. Forti (1987) proved that the Cauchy functional equation is stable for the pair (G, E) if and only if it is stable for the pair (G, B). In view of this result it is not important which Banach space is used on the range. Thus one may consider the stability of the Cauchy functional equation on the pair $(G, \mathbb{R})$ or $(G, \mathbb{C})$. In 1985, Székelyhidi (see Forti (1987) and Székelyhidi (1988)) replaced the original proof given by Hyers with a new one based on the use of invariant means. He proved the following theorem.

Theorem 18.12. *Let G be a left amenable semigroup and M be a left invariant mean on the space of all bounded complex valued functions defined on G. Let $f : G \to \mathbb{C}$ be a mapping satisfying*

$$\|f(xy) - f(x) - f(y)\| \leq \epsilon$$

for all $x, y \in G$. Then the map $\phi : G \to \mathbb{C}$ defined by

$$\phi(x) = M_y\big(f(xy) - f(y)\big)$$

for any x in the semigroup G is a unique homomorphism for which

$$\|f(x) - \phi(x)\| \leq \epsilon$$

holds for all $x \in G$.

This is the first result of its kind on non-abelian groups (or semi-groups). Faiziev (1993) proved that if $E = B(T)$, a Banach space of all bounded complex functions on the nonempty set T, then the Cauchy functional equation is stable of the pair (G, E) for every solvable group G. Faiziev, Rassias and Sahoo (2002) studied the (ψ, γ)-stability of the Cauchy functional equation on non-abelian groups such as metabelian groups. They also proved that any group can be embedded into a group G where the Cauchy functional equation is $\psi, \gamma)$-stable, and the Cauchy functional equation is $\psi, \gamma)$-stable on metabelian groups.

For more on the stability of additive Cauchy functional equation the interested reader is referred to the survey papers of Forti (1995) and Rassias (2000), and the books of Hyers, Isac and Rassias (1998) and Jung (2001).

18.6 Exercises

1. Suppose for some $\delta > 0$ the continuous function $f : \mathbb{R} \to \mathbb{R}$ satisfies the inequality $|f(x + y) - f(x) - f(y)| \leq \delta$ for any $x, y \in \mathbb{R}$. Prove that f can be represented as the sum of a linear function and a function not exceeding δ in absolute value.

2. Show that if $f : \mathbb{R} \to \mathbb{R}$ satisfies

$$|f(x+y) + f(z) - f(x) - f(y+z)| \leq \delta$$

for some positive δ and for all $x, y, z \in \mathbb{R}$, then there exists a unique solution $a : \mathbb{R} \to \mathbb{R}$ of the functional equation $a(x+y) = a(x) + a(y)$ for all $x, y \in \mathbb{R}$ such that

$$|f(x) - a(x) - f(0)| \leq \delta$$

for all $x \in \mathbb{R}$.

3. Show that if $f : \mathbb{R} \to \mathbb{R}$ satisfies

$$|f(x+y) + f(x-y) - 2f(x)| \leq \delta$$

for some positive δ and for all $x, y \in \mathbb{R}$, then there exists a unique solution $a : \mathbb{R} \to \mathbb{R}$ of the functional equation $a(x+y) = a(x) + a(y)$ for all $x, y \in \mathbb{R}$ such that

$$|f(x) - a(x) - f(0)| \leq 2\delta$$

for all $x \in \mathbb{R}$.

4. Show that if $f : \mathbb{R} \to \mathbb{R}$ satisfies

$$|f(x+y) - f(x-y) - 2f(y)| \leq \delta$$

for some positive δ and for all $x, y \in \mathbb{R}$, then there exists a unique solution $a : \mathbb{R} \to \mathbb{R}$ of the functional equation $a(x+y) = a(x) + a(y)$ for all $x, y \in \mathbb{R}$ such that

$$|f(x) - a(x)| \leq \frac{3}{2}\delta$$

for all $x \in \mathbb{R}$.

5. Show that if $f, g : \mathbb{R} \to \mathbb{R}$ satisfies

$$|f(x+y) - f(x-y) - 2g(y)| \leq \delta$$

for some positive δ and for all $x, y \in \mathbb{R}$, then there exists a unique solution $a : \mathbb{R} \to \mathbb{R}$ of the functional equation $a(x+y) = a(x) + a(y)$ for all $x, y \in \mathbb{R}$ such that

$$|f(x) - a(x) - f(0)| \leq 6\delta,$$
$$|g(x) - a(x) - g(0)| \leq 2\delta$$

for all $x \in \mathbb{R}$.

6. Show that if $f, g, h : \mathbb{R} \to \mathbb{R}$ satisfies

$$|f(x + y) + g(x - y) - h(x)| \leq \delta$$

for some positive δ and for all $x, y \in \mathbb{R}$, then there exists a unique solution $a : \mathbb{R} \to \mathbb{R}$ of the functional equation $a(x + y) = a(x) + a(y)$ for all $x, y \in \mathbb{R}$ and constants α and β such that

$$\begin{aligned}
|f(x) - a(x) - \alpha| &\leq 5\delta, \\
|g(x) - a(x) - \beta| &\leq 5\delta, \\
|h(x) - a(x) - \alpha - \beta| &\leq 11\delta
\end{aligned}$$

for all $x \in \mathbb{R}$.

7. Show that if $f, g, h : \mathbb{R} \to \mathbb{R}$ satisfies

$$|f(x + y) - g(x) - h(y)| \leq \delta$$

for some positive δ and for all $x, y \in \mathbb{R}$, then there exists a unique solution $a : \mathbb{R} \to \mathbb{R}$ of the functional equation $a(x + y) = a(x) + a(y)$ for all $x, y \in \mathbb{R}$ such that

$$\begin{aligned}
|f(x) - a(x) - g(0) - h(0)| &\leq 3\delta, \\
|g(x) - a(x) - g(0)| &\leq 4\delta, \\
|h(x) - a(x) - h(0)| &\leq 4\delta
\end{aligned}$$

for all $x \in \mathbb{R}$.

8. Show that if $f, g, h : \mathbb{R} \to \mathbb{R}$ satisfies

$$|f(x + y) - g(x) - h(y)| \leq \theta \left(|x|^p + |y|^p\right)$$

for some positive δ, $p \in [0, 1)$ and for all $x, y \in \mathbb{R}$, then there exists a unique solution $a : \mathbb{R} \to \mathbb{R}$ of the functional equation $a(x + y) = a(x) + a(y)$ for all $x, y \in \mathbb{R}$ such that

$$|f(x) - a(x) - g(0) - h(0)| \leq \frac{4}{2 - 2^p} \theta |x|^p,$$

$$|g(x) - a(x) - g(0)| \leq \frac{6 - 2^p}{2 - 2^p} \theta |x|^p,$$

$$|h(x) - a(x) - h(0)| \leq \frac{6 - 2^p}{2 - 2^p} \theta |x|^p$$

for all $x \in \mathbb{R}$.

9. Show that if $f : \mathbb{R} \to \mathbb{R}$ satisfies

$$|f(x + y) - f(x) - f(y) - xy| \leq \theta \left(|x|^p + |y|^p\right)$$

for some positive δ, $p \in [0, 1)$ and for all $x, y \in \mathbb{R}$, then there exists a unique solution $a : \mathbb{R} \to \mathbb{R}$ of the functional equation $a(x + y) = a(x) + a(y)$ for all $x, y \in \mathbb{R}$ such that

$$\left| f(x) - a(x) - \frac{1}{2}x^2 \right| \le \frac{2}{2 - 2^p} \theta |x|^p$$

for all $x \in \mathbb{R}$.

10. If the functions $f : \mathbb{R} \to \mathbb{R}$ and $\phi : \mathbb{R}^2 \to \mathbb{R}_+$ satisfy the inequality

$$|f(x + y) - f(x) - f(y)| \le \phi(x, y)$$

for all $x, y \in \mathbb{R}$ with

$$\sum_{k=0}^{\infty} \left(\frac{1}{2} \right)^k \phi(2^k x, 2^k x) < \infty$$

and

$$\lim_{n \to \infty} \left(\frac{1}{2} \right)^n \phi(2^n x, 2^n y) = 0,$$

then show that there exists a unique additive function $A : \mathbb{R} \to \mathbb{R}$ such that

$$|f(x) - A(x)| \le \frac{1}{2} \sum_{k=0}^{\infty} \left(\frac{1}{2} \right)^k \phi(2^k x, 2^k x)$$

for every $x \in \mathbb{R}$.

11. Let $\phi : \mathbb{R}^2 \to \mathbb{R}_+$ be a function satisfying

$$\Phi(x, y) := \sum_{k=1}^{\infty} 2^{-(k+1)} \phi(2^k x, 2^k y) < \infty$$

for all $x, y \in \mathbb{R}$. If a function $f : \mathbb{R} \to \mathbb{R}$ satisfies the inequality

$$|f(x + y) - f(x) - f(y)| \le \phi(x, y)$$

for any $x, y \in \mathbb{R}$, then show that there exists a unique additive function $A : \mathbb{R} \to \mathbb{R}$ such that

$$|f(x) - A(x)| \le \Phi(x, x)$$

for all $x \in \mathbb{R}$.

Chapter 19

Stability of Exponential Cauchy Equations

19.1   Introduction

Professor Eugene Lukacs (1906–1987) posed the following problem: Does the exponential equation

$$f(x + y) = f(x)f(y) \quad \forall\, x, y \in \mathbb{R} \qquad (19.1)$$

have an analogous stability theorem, whereby f is approximated by an exponential function? Baker, Lawrence and Zorzitto (1979) examined the stability of the exponential equation to provide an answer to the problem asked by Lukacs.

The goal of this chapter is to introduce the notion of superstability of functional equations and to examine the superstability of the exponential Cauchy functional equation, the multiplicative Cauchy functional equation, and a functional equation connected with the Reynolds operator. In this chapter, the Ger type stability of the exponential equation is also examined.

19.2   Stability of Exponential Equation

In this section, first we examine the stability of the exponential Cauchy functional equation (19.1). The original proof of Baker, Lawrence and Zorzitto (1979) was greatly simplified by Baker (1980). Here we present the proof given by Baker to the following theorem regarding the stability of the exponential Cauchy functional equation (19.1).

Theorem 19.1. *Let* $f : \mathbb{R} \to \mathbb{R}$ *be a real function satisfying*

$$|f(x + y) - f(x)f(y)| \leq \delta \qquad (19.2)$$

for all $x, y \in \mathbb{R}$ and some $\delta > 0$. Then either

$$|f(x)| \leq \frac{1 + \sqrt{1 + 4\delta}}{2} \qquad (19.3)$$

or

$$f(x) = E(x), \qquad (19.4)$$

where $E : \mathbb{R} \to \mathbb{R}$ is an exponential function.

Proof. Letting

$$\epsilon = \frac{1 + \sqrt{1 + 4\delta}}{2}, \qquad (19.5)$$

we note that $2\epsilon = 1 + \sqrt{1 + 4\delta}$ which can be written as

$$(2\epsilon - 1)^2 = 1 + 4\delta.$$

Therefore, from the last equation after solving for δ, we obtain

$$\epsilon^2 - \epsilon = \delta. \qquad (19.6)$$

Since $\delta > 0$, by (19.6), we have

$$\epsilon(\epsilon - 1) > 0.$$

Hence

$$\epsilon > 0 \quad \text{and} \quad \epsilon > 1,$$

or

$$\epsilon < 0 \quad \text{and} \quad \epsilon < 1.$$

Since by (19.5) $\epsilon > 0$, therefore we have $\epsilon > 1$.

Suppose f does not satisfy the inequality

$$|f(x)| \leq \frac{1 + \sqrt{1 + 4\delta}}{2} = \epsilon.$$

Then there exists $a \in \mathbb{R}$ such that $|f(a)| > \epsilon$, say,

$$|f(a)| = \epsilon + p$$

for some $p > 0$. Note that

$$
\begin{aligned}
|f(2a)| &= |f(a)^2 - \{f(a)^2 - f(2a)\}| \\
&\geq |f(a)^2| - |f(a)^2 - f(2a)| \\
&\geq |f(a)^2| - \delta \qquad \text{(by (19.2))} \\
&= (\epsilon + p)^2 - \delta
\end{aligned}
$$

$$
\begin{aligned}
&= (\epsilon + p)^2 - \epsilon^2 + \epsilon && \text{(by (19.6))}\\
&= 2p\epsilon + p^2 + \epsilon\\
&= (\epsilon + 2p) + 2p(\epsilon - 1) + p^2\\
&> \epsilon + 2p,
\end{aligned}
$$

since $\epsilon > 1$ and $p > 0$. Hence, we get

$$
|f(2a)| > \epsilon + 2p. \tag{19.7}
$$

Next consider

$$
\begin{aligned}
|f(2^2a)| &= |f(2a)^2 - [f(2a)^2 - f(2^2a)]|\\
&\geq |f(2a)|^2 - |f(2a)^2 - f(2^2a)|\\
&\geq |f(2a)|^2 - \delta\\
&> (\epsilon + 2p)^2 - (\epsilon^2 - \epsilon)\\
&= \epsilon^2 + 4p\epsilon + 4p^2 - \epsilon^2 + \epsilon\\
&= \epsilon + 4p\epsilon + 4p^2\\
&= \epsilon + 3p + 4p^2 + 4p\left(\epsilon - \frac{3}{4}\right)\\
&> \epsilon + (2+1)p,
\end{aligned}
$$

since $\epsilon > 1$ and $p > 0$. Similarly

$$
\begin{aligned}
|f(2^3a)| &= |f(2^2a)^2 - [f(2^2a) - f(2^3a)]|\\
&\geq |f(2^2a)^2| - |f(2^2a) - f(2^3a)|\\
&\geq |f(2^2a)|^2 - \delta\\
&> (\epsilon + 3p)^2 - (\epsilon^2 - \epsilon)\\
&= \epsilon + (3+1)p + 6p\left(\epsilon - \frac{2}{3}\right) + 9p^2\\
&> \epsilon + (3+1)p.
\end{aligned}
$$

Hence, by induction, one can obtain

$$
|f(2^n a)| > \epsilon + (n+1)p \tag{19.8}
$$

for all positive integers $n \in \mathbb{N}$.

Now for every $x, y, z \in \mathbb{R}$, we have

$$
|f(x + y + z) - f(x + y)f(z)| \leq \delta \tag{19.9}
$$

and

$$|f(x + y + z) - f(x)f(y + z)| \leq \delta. \qquad (19.10)$$

Hence (19.9) and (19.10) yield

$$|f(x + y)f(z) - f(x)f(y + z)| \leq 2\delta$$

by triangle inequality. Therefore

$$|f(x + y)f(z) - f(x)f(y)f(z)|$$
$$\leq |f(x + y)f(z) - f(x)f(y + z)| + |f(x)f(y + z) - f(x)f(y)f(z)|$$
$$\leq 2\delta + |f(x)| \, |f(y + z) - f(y)f(z)|$$
$$\leq 2\delta + |f(x)|\delta$$

or

$$|f(x + y) - f(x)f(y)| \cdot |f(z)| \leq 2\delta + |f(x)|\delta \qquad (19.11)$$

for all $x, y, z \in \mathbb{R}$. Letting $z = 2^n a$, we obtain

$$|f(x + y) - f(x)f(y)| \leq \frac{2\delta + |f(x)|\delta}{f(2^n a)} \qquad (19.12)$$

that is (by (19.8))

$$|f(x + y) - f(x)f(y)| \leq \frac{\delta[2 + |f(x)|]}{\epsilon + (n + 1)p}. \qquad (19.13)$$

Letting $n \to \infty$ in (19.13), we obtain

$$|f(x + y) - f(x)f(y)| \leq 0$$

for all $x, y \in \mathbb{R}$. Hence

$$f(x + y) = f(x)f(y)$$

for all $x, y \in \mathbb{R}$. Thus
$$f(x) = E(x),$$

where $E : \mathbb{R} \to \mathbb{R}$ is an exponential function. This completes the proof of the theorem.   □

From the celebrated theorem of Hyers (see Theorem 18.3), we have seen that if $f : \mathbb{R} \to \mathbb{R}$ satisfies

$$|f(x + y) - f(x) - f(y)| \leq \epsilon$$

for all $x, y \in \mathbb{R}$ and some $\epsilon > 0$, then there exists an additive function $A : \mathbb{R} \to \mathbb{R}$ such that

$$|f(x) - A(x)| \leq \epsilon.$$

This says f is stable on the pair $(\mathbb{R}, \mathbb{R})$. However, Theorem 19.1 says that if $f : \mathbb{R} \to \mathbb{R}$ satisfies

$$|f(x + y) - f(x)f(y)| \leq \epsilon$$

for all $x, y \in \mathbb{R}$ and some $\epsilon > 0$, then either f is bounded or f is exponential. This means f is *superstable* on the pair $(\mathbb{R}, \mathbb{R})$. The mapping f satisfying the last inequality is called approximately exponential (or ϵ-exponential). Hence every approximately exponential map $f : \mathbb{R} \to \mathbb{R}$ is either bounded or exponential.

Remark 19.1. *In general, the above proof of Theorem 19.1 works for functions $f : G \to E$, where G is a semigroup and E is a normed algebra in which the norm is multiplicative, that is, $\|xy\| = \|x\| \|y\|$ for any $x, y \in E$. Examples of such normed algebras are the quaternions and the Cayley numbers.*

Remark 19.2. *If the domain of the function f is a semigroup, we may identify the multiplicative Cauchy functional equation*

$$f(xy) = f(x)f(y)$$

with the exponential Cauchy functional equation (19.1). Hence, for the results of the stability of the multiplicative Cauchy functional equation, we can refer to the theorem presented above.

From Remark 19.2 and Theorem 19.1, we have the following superstability theorem for the multiplicative Cauchy functional equation.

Theorem 19.2. *Let $f : \mathbb{R} \to \mathbb{R}$ be a real function satisfying*

$$|f(xy) - f(x)f(y)| \leq \delta$$

for all $x, y \in \mathbb{R}$ and some $\delta > 0$. Then either f is bounded or f is a multiplicative function.

In the remainder of this section, we consider the stability of a functional equation connected with the Reynolds operator, namely, the functional equation

$$f(x\,g(y)) = f(x)\,f(y) \quad \text{for all } x, y \in \mathbb{R}. \tag{19.14}$$

Here $f, g : \mathbb{R} \to \mathbb{R}$ are unknown functions. In the case g is an identity function, that is, $g(y) = y$, then the functional equation (19.14) reduces to multiplicative Cauchy functional equation $f(xy) = f(x)f(y)$. If $g = f$, then the functional equation (19.14) yields the following composite functional equation

$$f(x\, f(y)) = f(x)\, f(y) \quad \text{for all } x, y \in \mathbb{R}. \tag{19.15}$$

The origin of the equation (19.15) is in the averaging theory applied to turbulent fluid motion.

Following Najdecki (2007), next we prove the superstability of the functional equation (19.14).

Theorem 19.3. *Let $f, g : \mathbb{R} \to \mathbb{R}$ be real functions satisfying*

$$|f(x\, g(y)) - f(x)\, f(y)| \le \delta \tag{19.16}$$

for all $x, y \in \mathbb{R}$ and some $\delta > 0$. Then either f is bounded or the functional equation (19.14) holds.

Proof. Suppose that f is unbounded. Then we can choose a sequence $\{x_n : n \in \mathbb{N}\}$ of elements of $\mathbb{R}$ such that $0 \ne |f(x_n)| \to \infty$ as $n \to \infty$. Letting $y = x_n$ in (19.16), we obtain

$$\left| \frac{f(x\, g(x_n))}{f(x_n)} - f(x) \right| \le \frac{\delta}{|f(x_n)|}. \tag{19.17}$$

Since $|f(x_n)| \to \infty$ as $n \to \infty$, from (19.17), we obtain

$$f(x) = \lim_{n \to \infty} \frac{f(x\, g(x_n))}{f(x_n)} \tag{19.18}$$

for all $x \in \mathbb{R}$. Replacing in (19.16) x by $x\, g(x_n)$, we have

$$|f(x\, g(x_n)\, g(y)) - f(x\, g(x_n))\, f(y)| \le \delta \tag{19.19}$$

for all $x, y \in \mathbb{R}$. From (19.19), it is easy to see that

$$\lim_{n \to \infty} \frac{f(x\, g(y)\, g(x_n)) - f(x\, g(x_n))\, f(y)}{f(x_n)} = 0. \tag{19.20}$$

Thus from (19.18) and (19.20), for every $x, y \in \mathbb{R}$, we obtain

$$
\begin{aligned}
f(x\, g(y)) &= \lim_{n \to \infty} \frac{f(x\, g(y)\, g(x_n))}{f(x_n)} \\
&= \lim_{n \to \infty} \frac{f(x\, g(y)\, g(x_n)) - f(x\, g(x_n))\, f(y)}{f(x_n)}
\end{aligned}
$$

$$+ \lim_{n \to \infty} \frac{f(x\,g(x_n))}{f(x_n)} f(y)$$

$$= f(x)\,f(y).$$

This completes the proof of the theorem. $\qquad\square$

Theorem (19.2) can be obtained as a corollary of the last theorem. The next theorem generalizes Theorem (19.2).

Theorem 19.4. *Suppose* $\phi : \mathbb{R} \to \mathbb{R}$ *to be any real function. Let* $f : \mathbb{R} \to \mathbb{R}$ *be a real function satisfying*

$$|f(xy) - f(x)f(y)| \le \phi(x) \tag{19.21}$$

for all $x, y \in \mathbb{R}$. *Then either* f *is bounded or* f *is a multiplicative function.*

Proof. Suppose that f is unbounded. Then we can choose a sequence $\{x_n : n \in \mathbb{N}\}$ of elements of $\mathbb{R}$ such that $0 \ne |f(x_n)| \to \infty$ as $n \to \infty$. Letting $y = x_n$ in (19.21), we obtain

$$\left| \frac{f(x\,x_n)}{f(x_n)} - f(x) \right| \le \frac{\phi(x)}{|f(x_n)|}. \tag{19.22}$$

Since $|f(x_n)| \to \infty$ as $n \to \infty$, from (19.22), we obtain

$$f(x) = \lim_{n \to \infty} \frac{f(x\,x_n)}{f(x_n)} \tag{19.23}$$

for all $x \in \mathbb{R}$. Replacing in (19.21) y by $x_n\,y$, we have

$$|f(x\,x_n\,y) - f(x)\,f(y\,x_n)| \le \phi(x) \tag{19.24}$$

for all $x, y \in \mathbb{R}$. From (19.24), it is easy to see that

$$\lim_{n \to \infty} \frac{f(x\,x_n\,y) - f(x)\,f(x_n\,y)}{f(x_n)} = 0. \tag{19.25}$$

Thus from (19.23) and (19.25), for every $x, y \in \mathbb{R}$, we obtain

$$f(x\,y) = \lim_{n \to \infty} \frac{f(x\,y\,x_n)}{f(x_n)}$$

$$= \lim_{n \to \infty} \frac{f(x\,y\,x_n) - f(x)\,f(x_n\,y)}{f(x_n)} + f(x) \lim_{n \to \infty} \frac{f(x_n\,y)}{f(x_n)}$$

$$= f(x)\,f(y).$$

This completes the proof of the theorem. $\qquad\square$

Remark 19.3. *The above theorem is true if we replace the right side of the inequality* (19.21) *by* $\phi(y)$.

19.3   Ger Type Stability of Exponential Equation

Theorem 19.1 says that the exponential Cauchy functional equation (19.1), that is, $f(x+y) = f(x) f(y)$, is superstable. This superstability of the exponential functional equation is caused by the fact that the operations on both sides of (19.1) are addition and multiplication whereas the distance between the two sides of (19.1) is measured by their difference. In the next theorem, we measure the distance between the two sides of (19.1) by using a quotient and show that (19.1) is not superstable.

The following result is due to Ger and Šemrl (1996) and is stated here without a proof.

Corollary 19.1. *Let $\varepsilon \in \left(0, \frac{1}{4}\right)$ and $g : \mathbb{R} \to \mathbb{R}$ satisfy the congruence*

$$g(x + y) - g(x) - g(y) \in \mathbb{Z} + (-\varepsilon, \varepsilon)$$

for all $x, y \in \mathbb{R}$. Then there exists a function $a : \mathbb{R} \to \mathbb{R}$ such that

$$a(x + y) - a(x) - a(y) \in \mathbb{Z}$$

for all $x, y \in \mathbb{R}$, and

$$|g(x) - a(x)| \leq \varepsilon$$

for all $x \in \mathbb{R}$.

Next, following Ger and Šemrl (1996), we present the Ger type stability of the exponential Cauchy functional equation.

Theorem 19.5. *Let $f : \mathbb{R} \to \mathbb{C}\backslash\{0\}$ satisfy the inequality*

$$\left| \frac{f(x + y)}{f(x)f(y)} - 1 \right| \leq \varepsilon \tag{19.26}$$

for some $\varepsilon \in [0, 1)$ and for all $x, y \in \mathbb{R}$. Then, there exists a unique exponential function $E : \mathbb{R} \to \mathbb{C}\backslash\{0\}$ such that

$$\max \left\{ \left| \frac{E(x)}{f(x)} - 1 \right|, \left| \frac{f(x)}{E(x)} - 1 \right| \right\} \leq \sqrt{1 + \frac{1}{(1 - \varepsilon)^2} - 2\sqrt{\frac{1 + \varepsilon}{1 - \varepsilon}}} \tag{19.27}$$

for any $x \in \mathbb{R}$.

Proof. Every nonzero complex number λ can be uniquely expressed as

$$\lambda = |\lambda| \exp(i \arg \lambda)$$

with $-\pi < \arg \lambda \leq \pi$. Then (19.26) implies

$$\left| \frac{|f(x+y)|}{|f(x)|\,|f(y)|} \exp[i(\arg f(x+y) - \arg f(x) - \arg f(y))] - 1 \right| \leq \varepsilon$$

for all $x, y \in \mathbb{R}$. From this inequality we can obtain the following two relations:

$$1 - \varepsilon \leq \frac{|f(x+y)|}{|f(x)|\,|f(y)|} \leq 1 + \varepsilon \tag{19.28}$$

and

$$\arg f(x+y) - \arg f(x) - \arg f(y) \in 2\pi\mathbb{Z} + [-\sin^{-1}\varepsilon, \sin^{-1}\varepsilon]$$

for $x, y \in \mathbb{R}$. Since we assumed $\varepsilon < 1$, it is clear that $\sin^{-1}\varepsilon < \pi/2$. According to Corollary 19.1, there exists a function $a : \mathbb{R} \to \mathbb{R}$ such that

$$a(xy) - a(x) - a(y) \in 2\pi\mathbb{Z} \tag{19.29}$$

for $x, y \in \mathbb{R}$ and

$$|a(x) - \arg f(x)| \leq \sin^{-1}\varepsilon \tag{19.30}$$

for $x \in \mathbb{R}$.

It follows from (19.28) that

$$\big|\ln|f(x+y)| - \ln|f(x)| - \ln|f(y)|\big| \leq -\ln(1-\varepsilon)$$

for any $x, y \in \mathbb{R}$. Hence there exists an additive function $h : \mathbb{R} \to \mathbb{R}$ such that

$$\big|h(x) - \ln|f(x)|\big| \leq -\ln(1-\varepsilon) \tag{19.31}$$

for $x \in \mathbb{R}$.

Let us define a function $E : \mathbb{R} \to \mathbb{C} \setminus \{0\}$ by

$$E(x) := \exp(h(x) + i\,a(x)).$$

Since h is a homomorphism, it follows from (19.29) that

$$\begin{aligned}
E(x+y) &= \exp(h(x+y) + i\,a(x+y)) \\
&= \exp(h(x) + h(y) + i\,a(x) + i\,a(y) + i\,2\pi k) \\
&= E(x)\,E(y)
\end{aligned}$$

for any $x, y \in \mathbb{R}$, where $k \in \mathbb{Z}$ is an appropriate constant.

Moreover, we have

$$\left| \frac{f(x)}{E(x)} - 1 \right| = \big|\exp[\ln|f(x)| - h(x)]\exp[i(\arg f(x) - a(x))] - 1\big|$$

for $x \in \mathbb{R}$. By (19.30) and (19.31) we see that the complex number $f(x)/E(x)$ belongs to the set

$$\Lambda = \left\{ \lambda \in \mathbb{C} : 1 - \varepsilon \leq |\lambda| \leq (1 - \varepsilon)^{-1}; \ -\sin^{-1}\varepsilon \leq \arg \lambda \leq \sin^{-1}\varepsilon \right\}.$$

It is not difficult to see

$$\sup\{|\lambda - 1| \ : \ \lambda \in \Lambda\} = \left|(1 - \varepsilon)^{-1}\exp(i\sin^{-1}\varepsilon) - 1\right|$$

$$= \sqrt{1 + \frac{1}{(1 - \varepsilon)^2} - 2\sqrt{\frac{1 + \varepsilon}{1 - \varepsilon}}}$$

$$=: \eta.$$

Analogously, we may obtain the same upper bound η for $|E(x)/f(x) - 1|$, that is,

$$\left|\frac{f(x)}{E(x)} - 1\right| \leq \eta \quad \text{and} \quad \left|\frac{E(x)}{f(x)} - 1\right| \leq \eta \tag{19.32}$$

for $x \in \mathbb{R}$. These establish the inequality (19.27).

Now, let $E_1 : \mathbb{R} \to \mathbb{C} \setminus \{0\}$ be another exponential function satisfying the inequality (19.27) in place of E. Since E and E_1 are exponential, we have

$$E(x) = E(nx)^{\frac{1}{n}} \quad \text{and} \quad E_1(x) = E_1(nx)^{\frac{1}{n}}$$

for any $x \in \mathbb{R}$ and $n \in \mathbb{N}$. Hence, these and (19.32) imply

$$\frac{E(x)}{E_1(x)} = \left(\frac{E(nx)}{f(nx)}\right)^{\frac{1}{n}}\left(\frac{f(nx)}{E_1(nx)}\right)^{\frac{1}{n}}$$

$$\leq (1 + \eta)^{\frac{1}{n}}(1 + \eta)^{\frac{1}{n}}$$

$$\to 1 \quad \text{as} \ \ n \to \infty,$$

which implies the uniqueness of E and the proof of the theorem is now complete. $\qquad\qquad\qquad\qquad\qquad\qquad\qquad\qquad\qquad\qquad\qquad\square$

19.4   Concluding Remarks

For the first time, the stability of the exponential Cauchy functional equation was investigated by Baker, Lawrence and Zorzitto (1979) to answer a question of Professor Eugene Lukacs. They proved the following result.

Theorem 19.6. *Let V be a vector space over the rationals $\mathbb{Q}$ and let $f : V \to \mathbb{R}$ be a real-valued function such that*

$$|f(x + y) - f(x) f(y)| \le \delta \quad \text{for some } \delta > 0 \text{ and all } x, y \in V.$$

Then either $|f(x)| \le \max(4, 4\delta)$ or there is a $\mathbb{Q}$-linear map $\ell : V \to \mathbb{R}$ such that $f(x) = exp(\ell(x))$ for all $x \in V$.

Baker (1980) simplified the proof of Baker, Lawrence and Zorzitto (1979) and established the following theorem.

Theorem 19.7. *Let $(S, +)$ be an arbitrary semigroup, and $f : S \to \mathbb{C}$ be a complex valued function defined on S such that*

$$|f(x + y) - f(x) f(y)| \le \delta \quad \text{for some } \delta > 0 \text{ and all } x, y \in S.$$

Then either

$$|f(x)| \le \frac{1 + \sqrt{1 + 4\delta}}{2}$$

or $f(x)$ is exponential (that is, $f(x + y) = f(x) f(y)$) on S.

Theorem 19.7 is also true for functions f with values in a normed algebra $\mathcal{A}$ with the property that the norm is multiplicative, that $||xy|| = ||x|| \, ||y||$ for all $x, y \in \mathcal{A}$. Baker (1980) gave an example that Theorem 19.7 is false if the algebra does not have the multiplicative norm. Let $\delta > 0$, choose $\epsilon > 0$ so that $|\epsilon - \epsilon^2| = \delta$, and let $f : \mathbb{C} \to \mathbb{C} \oplus \mathbb{C}$ be defined as

$$f(\lambda) = (e^\lambda, \epsilon), \quad \lambda \in \mathbb{C}.$$

Then with the multiplicative norm given by $||(\lambda, \mu)|| = \max\{|\lambda|, |\mu|\}$, we have $||f(\lambda + \mu) - f(\lambda)f(\mu)|| = \delta$, for all complex numbers λ, μ; f is unbounded, but it is not true that $f(\lambda + \mu) = f(\lambda)f(\mu)$ for all complex numbers λ and μ. In this counterexample the algebra $\mathcal{A} = \mathbb{C} \oplus \mathbb{C}$ can be decomposed as a direct sum of two ideals $\mathcal{A} = I_1 \oplus I_2$, where $I_1 = \{(\lambda, 0) \mid \lambda \in \mathbb{C}\}$ and $I_1 = \{(0, \lambda) \mid \lambda \in \mathbb{C}\}$. If one denotes by P_1 and P_2 the projections corresponding to this direct sum decomposition, then the mapping $P_1 f$ is exponential, while $P_2 f$ is bounded. Ger and Šemrl (1996) have shown that such behavior is typical for approximately exponential mappings with values in an arbitrary semisimple complex commutative Banach algebra.

A function $f : S \to \mathcal{A}$ is called approximately exponential if f satisfies the inequality $||f(x + y) - f(x) f(y)|| \le \delta$ for some $\delta > 0$ and all $x, y \in S$. We say $f : S \to \mathcal{A}$ is approximated by an exponential function $E : S \to \mathcal{A}$ if E satisfies $E(x + y) = E(x) E(y)$ for all $x, y \in S$ and $||f(x) - E(x)|| \le \epsilon$ for some $\epsilon \ge 0$ and all $x \in S$. We have seen from the above counterexample that the theorem of Baker (1980) is not true for all Banach algebras. Lawrence (1985) studied approximately exponential mapping defined on a semigroup or group S and taking values in a Banach algebra. To state some of his important results we introduce the following notations:

$$U = \{f(x) \mid x \in S\},$$
$$V = \{f(x + y) - f(x) f(y) \mid x, y \in S\},$$
$$B = \text{subalgebra of } \mathcal{A} \text{ generated by } U,$$
$$L = \text{the two-sided ideal of } B \text{ generated by } V.$$

Some important results of Lawrence are the following.

Theorem 19.8. *If the algebra B is simple and $f : S \to \mathcal{A}$ is approximately exponential, then f is either bounded or exponential.*

Theorem 19.9. *If $f : S \to M_n(\mathbb{C})$ is approximately exponential and $B = M_n(\mathbb{C})$, then f is either bounded or exponential.*

The following result gives a partial answer to the problem of stability of exponential mappings from a semigroup S to $M_n(\mathbb{C})$.

Theorem 19.10. *If $f : S \to M_n(\mathbb{C})$ is approximately exponential, then there exists a function $h : S \to M_n(\mathbb{C})$ such that $f - h$ is bounded on S and $[h(x + y) - h(x) h(y)]^2 = 0$ for all $x, y \in S$.*

Theorem 19.11. *If S is a commutative group and $f : S \to M_2(\mathbb{C})$ is approximately exponential, then f is approximated by a multiplicative function from S to $M_2(\mathbb{C})$.*

Jung (1997) has extended the theorem of Baker, Lawrence and Zorzitto (1979) to cases in which the Cauchy difference $f(x + y) - f(x) f(y)$ is not bounded. He considers the complex valued functions f defined on a complex normed space X. Let $H : \mathbb{R}_+ \times \mathbb{R}_+ \to \mathbb{R}_+$ be a mapping which is monotonically increasing in each variable, and suppose further that for given $u, v \in \mathbb{R}_+$ there exist $\alpha = \alpha(u, v)$ and $w_o = w_o(u, v)$ such that

$$H(u, v + w) \le \alpha H(w, w) \qquad \text{for all } w \ge w_o. \tag{19.33}$$

The function f and the mapping H are to be subjected to the following conditions. There exist $z \ne 0$ in X and a real number $\beta \in (0, 1)$ such that

$$\sum_{j=1}^{\infty} H(j\|z\|, \|z\|) \, |f(z)|^{j-1} < \beta \tag{19.34}$$

and

$$H(n||z||, n||z||) = o(|f(z)|^n) \quad \text{as } n \to \infty. \tag{19.35}$$

Moreover, the Cauchy difference is assumed to satisfy the following inequality for all $x, y \in X$:

$$|f(x + y) - f(x) f(y)| \leq H(||x||, ||y||). \tag{19.36}$$

Theorem 19.12. *Let f and H satisfy the conditions (19.33), (19.34), (19.35) and (19.36). Then f is exponential in X.*

Let $(G, \circ)$ be an abelian group, $g : G \to G$, and let $\mathbb{K}$ be either $\mathbb{R}$ or $\mathbb{C}$. A generalization of the exponential functional equation is the following:

$$f(x \circ g(y)) = f(x) f(y) \quad \text{for } x, y \in G. \tag{19.37}$$

Najdecki (2007) proved the following stability theorem.

Theorem 19.13. *Let $f : G \to \mathbb{K}$ be a function satisfying*

$$|f(x \circ g(y)) - f(x) f(y)| \leq \epsilon$$

for all $x, y \in G$. Then either f is bounded or (19.37) holds.

The notion of superstability arose from Theorem 19.1. The group operation in the range of exponential functions is "multiplication." Ger (1993) noticed that the superstability of the functional inequality (19.2) is caused by the fact that the natural group structure in the range is disregarded. Hence the stability for the exponential equation can be posed more naturally as

$$\left\| \frac{f(x + y)}{f(x) f(y)} - 1 \right\| \leq \delta. \tag{19.38}$$

Ger and Šemrl (1996) investigated the stability problem given by the inequality (19.38) and proved the following theorem

Theorem 19.14. *Let $(G, +)$ be a cancellative abelian semigroup. If a function $f : G \to \mathbb{C} \setminus \{0\}$ satisfies the inequality (19.38) for a given $\delta \in [0, 1)$ and for all $x, y \in G$, then there exists a unique exponential function $m : G \to \mathbb{C} \setminus \{0\}$ such that*

$$\max\left\{ \left\| \frac{f(x)}{m(x)} - 1 \right\|, \left\| \frac{m(x)}{f(x)} - 1 \right\| \right\} \leq \left[1 + \frac{1}{(1 - \delta)^2} - 2\left(\frac{1 + \delta}{1 - \delta} \right)^{\frac{1}{2}} \right]^{\frac{1}{2}}$$

for all $x \in G$.

19.5 Exercises

1. If $f : \mathbb{R} \to \mathbb{R}$ is a function satisfying

$$|f(xy) - f(x)f(y)| \le \delta$$

for some positive δ and for all $x, y \in \mathbb{R}$, then show that either f is bounded or f is a solution of the multiplicative Cauchy functional equation $f(xy) = f(x)f(y)$ for all $x, y \in \mathbb{R}$.

2. If $f : \mathbb{R} \to \mathbb{R}$ is a function satisfying

$$|f(x + y + xy) - f(x) + f(y) + f(x)f(y)| \le \delta$$

for some positive δ and for all $x, y \in \mathbb{R}$, then show that either f is bounded or f is a solution of the equation

$$f(x + y + xy) = f(x) + f(y) + f(x)f(y)$$

for all $x, y \in \mathbb{R}$.

3. If $f : (0, 1)^2 \to \mathbb{R}$ is a function satisfying

$$|f(pr, qs) + f(ps, qr) - f(p, q)\, f(r, s)| \le \delta$$

for all $p, q, r, s \in I$ and for some positive δ, then show that either f is bounded or f satisfies

$$f(pr, qs) + f(ps, qr) = f(p, q)\, f(r, s)$$

for all $p, q, r, s \in I$.

4. If $f : (0, 1)^2 \to \mathbb{R}$ is a function satisfying

$$|f(pr, qs) + f(ps, qr) - f(p, q)\, f(r, s)| \le \phi(r, s)$$

for all $p, q, r, s \in I$ and for some $\phi : I^2 \to \mathbb{R}_+$, then show that either f is bounded or f satisfies

$$f(pr, qs) + f(ps, qr) = f(p, q)\, f(r, s)$$

for all $p, q, r, s \in I$.

5. If $f : \mathbb{R} \to \mathbb{R}$ is a function satisfying

$$|f(x + y) - a^{xy}\, f(x)\, f(y)| \le \delta$$

for all $x, y \in \mathbb{R}$ and for some positive δ, where a is a positive real constant, then show that either the function $f(x) \, a^{-\frac{1}{2}x^2}$ is bounded or f is a solution of $f(x + y) = a^{xy} \, f(x) \, f(y)$ for all $x, y \in \mathbb{R}$.

6. If $f, g, h : \mathbb{R} \times (0, \infty) \to \mathbb{C}$ are functions satisfying

$$|f(x + y, u + v) - g(x, u) \, h(y, v)| \leq \delta$$

for all $x, y \in \mathbb{R}$, $u, v \in (0, \infty)$ and for some positive δ, then show that either f, g, h are all bounded or

$$g(x, u) = a \, E_1(u) \, E_2(x)$$
$$h(x, u) = b \, E_1(u) \, E_2(x)$$
$$|f(x, u) - a \, b E_1(u) \, E_2(x)| \leq \delta,$$

where $a, b \in \mathbb{C}$ and $E_1, E_2 : \mathbb{R} \to \mathbb{C}$ are exponential functions.

7. If $f : \mathbb{R} \to \mathbb{R}$ is a function satisfying

$$\left| f\left(\frac{x + y}{2}\right)^2 - f(x) \, f(y) \right| \leq \delta$$

for some $\delta > 0$ and for all $x, y \in \mathbb{R}$, then show that either f is bounded or f is a solution of the functional equation

$$f\left(\frac{x + y}{2}\right)^2 = f(x) \, f(y)$$

for all $x, y \in \mathbb{R}$.

Chapter 20

Stability of d'Alembert and Sine Equations

20.1 Introduction

The goal of this chapter is to examine the stability of the d'Alembert functional equation (also known as the cosine functional equation)

$$f(x + y) + f(x - y) = 2 f(x) f(y) \quad \text{for all } x, y \in \mathbb{R} \qquad (20.1)$$

and the sine functional equation

$$f(x + y) f(x - y) = f(x)^2 - f(y)^2 \quad \text{for all } x, y \in \mathbb{R}. \qquad (20.2)$$

In this chapter, we will show that if the function

$$(x, y) \mapsto f(x + y) + f(x - y) - 2f(x)f(y)$$

is bounded on $\mathbb{R}^2$, then either f is bounded on $\mathbb{R}$ or f is a solution of the d'Alembert functional equation (20.1). This result was established for the first time by Baker (1980). In this chapter, we will also prove that if the function

$$(x, y) \mapsto f(x + y)f(x - y) - f(x)^2 + f(y)^2$$

is bounded on $\mathbb{R}^2$, then either f is bounded on $\mathbb{R}$ or f is a solution of the sine functional equation (20.2). Cholewa (1983) was the first mathematician to investigate the stability of sine functional equation.

20.2 Stability of d'Alembert Equation

In this section, we present a result concerning the stability of the d'Alembert equation. The proof of the following theorem is based on the proof presented by Baker (1980).

Theorem 20.1. *If the function $f : \mathbb{R} \to \mathbb{C}$ satisfies the inequality*

$$|f(x+y) + f(x-y) - 2f(x)f(y)| \leq \delta \tag{20.3}$$

for all $x, y \in \mathbb{R}$ and some $\delta > 0$, then either

$$|f(x)| \leq \frac{1 + \sqrt{1+2\delta}}{2}$$

or there exists a function $m : \mathbb{R} \to \mathbb{C}$ such that

$$f(x) = \frac{m(x) + m(-x)}{2}$$

and

$$|m(x+y) - m(x)m(y)| \leq \frac{\delta}{2}$$

for all $x, y \in \mathbb{R}$.

Proof. Letting $x = 0 = y$ in (20.3), we obtain

$$|2f(0) - 2f(0)^2| \leq \delta.$$

Letting $z = f(0)$, we obtain

$$|z - z^2| \leq \frac{\delta}{2}. \tag{20.4}$$

Since

$$\left| |z| - |z|^2 \right| \leq |z - z^2|, \tag{20.5}$$

we obtain from (20.4)

$$\left| |z| - |z|^2 \right| \leq \frac{\delta}{2} \tag{20.6}$$

which is

$$-\frac{\delta}{2} \leq |z|^2 - |z| \leq \frac{\delta}{2}. \tag{20.7}$$

From the inequality

$$|z|^2 - |z| - \frac{\delta}{2} \leq 0,$$

we have

$$2|z|^2 - 2|z| - \delta \leq 0.$$

Therefore

$$|z| < \frac{1 + \sqrt{1+2\delta}}{2}. \tag{20.8}$$

Hence

$$|f(0)| < \epsilon, \tag{20.9}$$

where

$$\epsilon = \frac{1 + \sqrt{1 + 2\delta}}{2}. \tag{20.10}$$

Next we consider

$$\begin{aligned}
|f(2x)| &= |2f(x)^2 - f(0) - \{2f(x)^2 - f(0) - f(2x)\}| \\
&\geq |2f(x)^2 - f(0)| - |f(2x) + f(0) - 2f(x)^2| \\
&> |2f(x)^2 - f(0)| - \delta \\
&> |2f(x)^2| - |f(0)| - \delta \\
&> 2|f(x)|^2 - \epsilon - \delta. \tag{20.11}
\end{aligned}$$

Note that here we have used the fact that $|a - b| \geq |a| - |b|$.

Now we show that if $|f(x)| > \epsilon$ for some $x \in \mathbb{R}$, then

$$|f(2^n x)| \to \infty \text{ as } n \to \infty.$$

Let $x \in \mathbb{R}$ and

$$y = |f(x)| = \epsilon + p$$

for some $p > 0$. Then we have

$$\begin{aligned}
2y^2 - y - \delta - \epsilon &= 2(\epsilon + p^2) - \epsilon - p - \delta - \epsilon \\
&= 2\epsilon^2 + 4p\epsilon + 2p^2 - \epsilon - p - \epsilon - \delta \\
&= 2(\epsilon^2 - \epsilon) + (4\epsilon - 1)p + 2p^2 - \delta \\
&= 2\delta + (4\epsilon - 1)p + 2p^2 - \delta \\
&= \delta + (4\epsilon - 1)p + 2p^2 \\
&= \delta + (4\epsilon - 1)p + 2p^2 - 3p + 3p \\
&= \delta + 4(\epsilon - 1)p + 2p^2 + 3p \\
&> 3p,
\end{aligned}$$

since $\delta > 0$, $p > 0$ and $\epsilon > 1$. Thus we have shown that

$$2y^2 - y - \mu > 3p$$

which is

$$2y^2 - \mu > y + 3p, \tag{20.12}$$

where $\mu = \delta + \epsilon$. Hence

$$2|f(x)|^2 - \mu > |f(x)| + 3p. \tag{20.13}$$

Further, we have seen that

$$|f(2x)| > 2|f(x)|^2 - \epsilon - \delta.$$

Hence by (20.13), we get

$$
\begin{aligned}
|f(2x)| &> 2|f(x)|^2 - \epsilon - \delta \\
&> |f(x)| + 3p \\
&= \epsilon + p + 3p \\
&> \epsilon + 2p.
\end{aligned}
$$

Therefore

$$|f(2x)| > \epsilon + 2p. \tag{20.14}$$

By induction it can be shown that

$$|f(2^n x)| > \epsilon + 2^n p \tag{20.15}$$

for $n = 1, 2, 3, \ldots$.

Hence $f(x)$ is unbounded,

$$\lim_{n \to \infty} |f(2^n x)| \to \infty.$$

Now letting $y = 0$ in (20.3), we obtain

$$|f(x) + f(x) - 2f(x)f(0)| \le \delta,$$

and thus we have

$$|1 - f(0)| \le \frac{\delta}{2|f(x)|}. \tag{20.16}$$

Since f is unbounded, (20.16) yields

$$|1 - f(0)| \le 0.$$

Therefore

$$f(0) = 1. \tag{20.17}$$

For any $x, y \in \mathbb{R}$,

$$
\begin{aligned}
2|f(x)|\,|f(y) - f(-y)| &= |2f(x)f(y) - 2f(x)f(-y)| \\
&= |2f(x)f(y) - f(x+y) - f(x-y) \\
&\quad + f(x+y) + f(x-y) - 2f(x)f(-y)| \\
&\le |2f(x)f(y) - f(x+y) - f(x-y)| \\
&\quad + |f(x+y) + f(x-y) - 2f(x)f(-y)| \\
&\le \delta + \delta \\
&= 2\delta.
\end{aligned}
$$

Thus we have

$$|f(y) - f(-y)| \le \frac{\delta}{|f(x)|}.$$

Since f is unbounded, we obtain

$$|f(y) - f(-y)| = 0$$

and

$$f(y) = f(-y) \tag{20.18}$$

for all $y \in \mathbb{R}$.

Since f is unbounded and

$$|f(2x) + f(0) - 2f(x)^2| \le \delta$$

for all $x \in \mathbb{R}$, we may choose $a \in \mathbb{R}$ and α such that

$$2\alpha^2 [f(2a) - 1] = 1. \tag{20.19}$$

Define

$$g(x) = f(x + a) - f(x - a) \tag{20.20}$$

and

$$m(x) = f(x) + \alpha\, g(x). \tag{20.21}$$

Since f is even,

$$\begin{aligned}
g(-x) &= f(-x + a) - f(-x - a) \\
&= f(x - a) - f(x + a) \\
&= -g(x).
\end{aligned}$$

That is, g is odd. Further

$$\begin{aligned}
m(-x) &= f(-x) + \alpha\, g(-x) \\
&= f(x) - \alpha\, g(x). \tag{20.22}
\end{aligned}$$

Adding (20.21) and (20.22), we get

$$f(x) = \frac{m(x) + m(-x)}{2}$$

as asserted in the theorem.

Let

$$2f(x)f(y) = f(x + y) + f(x - y) + E(x, y)$$

so that

$$|E(x, y)| \le \delta$$

for all $x, y \in \mathbb{R}$. Next, we compute

$$
\begin{aligned}
2f(x)g(y) &+ 2f(y)g(x) \\
&= 2f(x)f(y+a) - 2f(x)f(y-a) \\
&\quad + 2f(y)f(x+a) - 2f(y)f(x-a) \\
&= f(x+y+a) + f(x-y-a) + E(x, y+a) \\
&\quad - f(x+y-a) - f(x-y+a) - E(x, y-a) \\
&\quad + f(x+y+a) + f(y-x-a) + E(x+a, y) \\
&\quad - f(y+x-a) - f(y-x+a) - E(x-a, y) \\
&= 2f(x+y+a) - 2f(x+y-a) + f(x-y-a) \\
&\quad - f(y-x+a) + f(y-x-a) - f(y-x+a) \\
&\quad + E(x, y+a) - E(x, y-a) \\
&\quad + E(x+a, y) - E(x-a, y) \\
&= 2g(x+y) + E(x, y+a) - E(x, y-a) \\
&\quad + E(x+a, y) - E(x-a, y),
\end{aligned}
$$

since f is even. Hence, we have

$$
\begin{aligned}
2g(x+y) &- 2f(x)g(y) - 2f(y)g(x) \\
&= E(x, y-a) - E(x, y+a) + E(x-a, y) - E(x+a, y).
\end{aligned}
$$

Since $|E(x, y)| \leq \delta$, we have from the above equation

$$
|g(x+y) - f(x)g(y) - f(y)g(x)| \leq 2\delta \qquad (20.23)
$$

for all $x, y \in \mathbb{R}$. Similarly

$$
\begin{aligned}
2g(x)g(y) &= 2[f(x+a) - f(x-a)][f(y+a) - f(y-a)] \\
&= 2f(x+a)f(y+a) - 2f(x+a)f(y-a) \\
&\quad - 2f(x-a)f(y+a) + 2f(x-a)f(y-a) \\
&= f(x+y+2a) + f(x-y) + E(x+a, y+a) \\
&\quad - f(x+y) - f(x-y+2a) + E(x+a, y-a) \\
&\quad - f(x+y) - f(x-y-2a) - E(x-a, y+a) \\
&\quad + f(x+y-2a) + f(x-y) + E(x-a, y-a) \\
&= 2f(x-y) - 2f(x+y) + f(x+y+2a) \\
&\quad + f(x+y-2a) - f(x-y+2a) - f(x-y-2a) \\
&\quad + E(x+a, y+a) - E(x+a, y-a) \\
&\quad + E(x-a, y-a) - E(x-a, y+a) \\
&= 2f(x-y) - 2f(x+y) + 2f(x+y)f(2a)
\end{aligned}
$$

$$- E(x + y + 2a, x + y - 2a) - 2f(x - y)f(2a)$$
$$+ E(x - y + 2a, x - y - 2a) + E(x + a, y + a)$$
$$- E(x + a, y - a) + E(x - a, y - a) - E(x - a, y + a).$$

Hence

$$2g(x)g(y) - 2(f(2a) - 1)[f(x + y) - f(x - y)]$$
$$= E(x - y + 2a, x - y - 2a) - E(x + y + 2a, x + y - 2a)$$
$$+ E(x + a, y + a) - E(x + a, y - a)$$
$$+ E(x - a, y - a) - E(x - a, y + a).$$

Since $|E(x, y)| \leq \delta$, we get

$$|g(x)g(y) - (f(2a) - 1)(f(x + y) - f(x - y))| \leq 3\delta. \qquad (20.24)$$

Further, we get

$$\left| f(x)f(y) - \frac{f(x + y) + f(x - y)}{2} \right| \leq \frac{\delta}{2}. \qquad (20.25)$$

Now using (20.23), (20.24) and (20.25), we compute

$$|m(x + y) - m(x)m(y)| = |f(x + y) + \alpha g(x + y)$$
$$- f(x)f(y) - \alpha^2 g(x)g(y)$$
$$- \alpha f(x)g(y) - \alpha f(y)g(x)|$$
$$\leq |\alpha| \cdot |g(x + y) - f(x)g(y) - f(y)g(x)|$$
$$+ |f(x + y) - f(x)f(y) - \alpha^2 g(x)g(y)|$$
$$\leq 3|\alpha|\delta + \left| f(x + y) - \alpha^2 g(x)g(y) \right.$$
$$\left. - \frac{f(x + y) + f(x - y)}{2} \right| + \frac{\delta}{2}$$
$$\leq 3|\alpha|\delta + \frac{\delta}{2} + \left| \frac{f(x + y) - f(x - y)}{2} \right.$$
$$\left. - \alpha^2(f(a) - 1)[f(x + y) - f(x - y)] \right|$$
$$+ 2|\alpha|^2\delta$$
$$= \delta \left(3|\alpha| + |\alpha|^2 + \frac{1}{2} \right)$$

for all $x, y \in \mathbb{R}$. Since f is unbounded, a can be chosen so that

$$|f(a)| \quad \text{and} \quad |f(2a)|$$

can be as large as desired. Thus α can be chosen so that $|\alpha|$ is as small as we wish. Thus we must have

$$|m(x+y) - m(x)m(y)| \leq \frac{\delta}{2}$$

for all $x, y \in \mathbb{R}$. This completes the proof of the theorem. $\qquad\square$

Now by Theorem 19.1 and Theorem 20.1, we have the following theorem.

Theorem 20.2. *If $f : \mathbb{R} \to \mathbb{C}$ satisfies*

$$|f(x+y) + f(x-y) - 2f(x)f(y)| \leq \delta$$

for all $x, y \in \mathbb{R}$ and some $\delta > 0$, then either

$$|f(x)| \leq \frac{1 + \sqrt{1+2\delta}}{2}$$

or f satisfies

$$f(x+y) + f(x-y) = 2f(x)f(y)$$

for all $x, y \in \mathbb{R}$.

20.3   Stability of Sine Equation

In this section, we study the stability of the sine functional equation (20.2). This section is adapted from Cholewa (1983).

Let $f : \mathbb{R} \to \mathbb{C}$ be a complex valued function defined on the set of reals. Suppose $f : \mathbb{R} \to \mathbb{C}$ satisfies the inequality

$$|f(x+y)\,f(x-y) - f(x)^2 + f(y)^2| \leq \delta \qquad (20.26)$$

for all $x, y \in \mathbb{R}$ and for some real $\delta > 0$. Moreover, we assume that f is an unbounded function.

We need the following three lemmas to establish the stability of the sine functional equation (20.26).

Lemma 20.1. *Let $f : \mathbb{R} \to \mathbb{C}$ be an unbounded function satisfying the inequality (20.26) for all $x, y \in \mathbb{R}$. Then $f(0) = 0$.*

Proof. Letting $y = x$ in (20.26) and then letting $u = 2x$ in the resulting inequality, we obtain $|f(u) f(0)| \leq \delta$. Hence

$$|f(0)| \leq \frac{\delta}{|f(u)|}.$$

Since f is unbounded, $|f(u)|$ can be made as large as possible, and therefore we have $|f(0)| \leq 0$. Thus the assertion of the lemma follows. $\square$

Lemma 20.2. *Let* $f : \mathbb{R} \to \mathbb{C}$ *be an unbounded function satisfying the inequality* (20.26) *for all* $x, y \in \mathbb{R}$. *Then* f *satisfies the inequality*

$$|f(x + y) + f(x - y) - 2 f(x) g(y)| \leq \delta \qquad (20.27)$$

for all $x, y \in \mathbb{R}$ *and for some* $g : \mathbb{R} \to \mathbb{C}$.

Proof. Letting $x = \frac{u+v}{2}$ and $y = \frac{u-v}{2}$ in (20.26), we have

$$\left| f(u) f(v) - f\left(\frac{u + v}{2}\right)^2 + f\left(\frac{u - v}{2}\right)^2 \right| \leq \delta \qquad (20.28)$$

for all $u, v \in \mathbb{R}$. Since f is unbounded, there exists $a \in \mathbb{R}$ such that $|f(a)| \geq 4$. Let $g : \mathbb{R} \to \mathbb{C}$ be defined as

$$g(x) = \frac{f(x + a) - f(x - a)}{2 f(a)} \qquad (20.29)$$

for all $x \in \mathbb{R}$. Using (20.28) and (20.29), we compute

$$|f(x + y) + f(x - y) - 2 f(x) g(y)|$$

$$= \frac{1}{|f(a)|} |f(x + y) f(a) + f(x - y) f(a) - 2 f(a) f(x) g(y)|$$

$$\leq \frac{1}{|f(a)|} \left| f(x + y) f(a) - f\left(\frac{x + y + a}{2}\right)^2 + f\left(\frac{x + y - a}{2}\right)^2 \right|$$

$$+ \frac{1}{|f(a)|} \left| f(x - y) f(a) - f\left(\frac{x - y + a}{2}\right)^2 + f\left(\frac{x - y - a}{2}\right)^2 \right|$$

$$+ \frac{1}{|f(a)|} \left| f\left(\frac{x + y + a}{2}\right)^2 - f\left(\frac{x - y - a}{2}\right)^2 - f(x) f(y + a) \right|$$

$$+ \frac{1}{|f(a)|} \left| f(x) f(y - a) - f\left(\frac{x + y - a}{2}\right)^2 + f\left(\frac{x - y + a}{2}\right)^2 \right|$$

$$+ \left| 2 f(x) \frac{f(y + a) - f(y - a)}{2 f(a)} - 2 f(x) g(y) \right|$$

$$\leq \frac{4\,\delta}{|f(a)|}$$

$$\leq \delta \qquad (\text{since } |f(a)| \geq 4).$$

Hence

$$|f(x+y) + f(x-y) - 2\,f(x)\,g(y)| \leq \delta,$$

and the proof of the lemma is complete. $\qquad\qquad\qquad\qquad\square$

Lemma 20.3. *Let* $f : \mathbb{R} \to \mathbb{C}$ *be an unbounded function satisfying the inequality* (20.26) *for all* $x, y \in \mathbb{R}$. *Then* f *satisfies the inequality*

$$f(x+y) + f(x-y) = 2\,f(x)\,g(y) \qquad\qquad (20.30)$$

for all $x, y \in \mathbb{R}$ *and for some* $g : \mathbb{R} \to \mathbb{C}$.

Proof. Let x, y be two arbitrarily fixed points in $\mathbb{R}$. Then by Lemma 20.2, we see that f satisfies the inequality (20.27) and g is given by (20.29). Using (20.27) and (20.29), for all $z \in \mathbb{R}$, we have

$$|f(z)|\,|f(x+y) + f(x-y) - 2f(x)\,g(y)|$$
$$= |f(z)\,f(x+y) + f(z)\,f(x-y) - 2\,f(x)\,f(z)\,g(y)|$$
$$\leq \left| f(z)\,f(x+y) - f\left(\frac{z+x+y}{2}\right)^2 + f\left(\frac{z-x-y}{2}\right)^2 \right|$$
$$+ \left| f(z)\,f(x-y) - f\left(\frac{z+x-y}{2}\right)^2 + f\left(\frac{z-x+y}{2}\right)^2 \right|$$
$$+ \left| f\left(\frac{z+x+y}{2}\right)^2 - f\left(\frac{z-x+y}{2}\right)^2 - f(z+y)\,f(x) \right|$$
$$+ \left| f\left(\frac{z+x-y}{2}\right)^2 - f\left(\frac{z-x-y}{2}\right)^2 - f(z-y)\,f(x) \right|$$
$$+ |\{f(z+y) + f(z-y)\}f(x) - 2\,f(z)\,g(y)\,f(x)|$$
$$\leq 4\,\delta + \delta\,|f(x)|$$
$$= \delta\,(4 + |f(x)|).$$

Therefore

$$|f(x+y) + f(x-y) - 2\,f(x)\,g(y)| \leq \frac{(4 + |f(x)|)\delta}{|f(z)|},$$

and since f is unbounded and x is a fixed element, the right-hand side of the above inequality can be made as small as possible. Hence, we must have

$$|f(x+y) + f(x-y) - 2\,f(x)\,g(y)| = 0.$$

Therefore $f(x+y)+f(x-y) = 2\,f(x)\,g(y)$. Since x and y were arbitrarily fixed elements of $\mathbb{R}$, the last equation holds for all $x,y \in \mathbb{R}$ and the proof of the lemma is now complete. $\qquad\qquad\qquad\qquad\qquad\qquad\qquad\square$

Now we prove the stability of the sine functional equation.

Theorem 20.3. *Let the unbounded function* $f : \mathbb{R} \to \mathbb{C}$ *satisfy inequality*

$$|f(x + y)\,f(x - y) - f(x)^2 + f(y)^2| \le \delta$$

for all $x, y \in \mathbb{R}$ *and for some real* $\delta > 0$. *Then* $f : \mathbb{R} \to \mathbb{C}$ *is a solution of*

$$f(x + y)\,f(x - y) = f(x)^2 - f(y)^2$$

for all $x, y \in \mathbb{R}$.

Proof. Let f be an unbounded solution of inequality (20.26). By Lemma 20.3, f satisfies

$$f(x + y) + f(x - y) = 2\,f(x)\,g(y) \qquad\qquad\text{(FE)}$$

for all $x, y \in \mathbb{R}$. Letting $x = 0$ in the last equation and using Lemma 20.1, we get

$$f(y) + f(-y) = 0$$

for all $x, y \in \mathbb{R}$. Hence f is an odd function. Now, put $x = \frac{u+v}{2}$ and $y = \frac{u-v}{2}$ in (FE). Hence we have

$$f(u) + f(v) = 2f\left(\frac{u + v}{2}\right) g\left(\frac{u - v}{2}\right) \qquad\qquad (20.31)$$

for all $u, v \in \mathbb{R}$. From (20.31) and Lemma 20.1, we infer that

$$f(x + y) = f(x + y) + f(0) = 2f\left(\frac{x + y}{2}\right) g\left(\frac{x + y}{2}\right) \qquad\qquad (20.32)$$

and

$$f(x - y) = f(x - y) + f(0) = 2f\left(\frac{x - y}{2}\right) g\left(\frac{x - y}{2}\right) \qquad\qquad (20.33)$$

for all $x, y \in \mathbb{R}$. Using (20.31) and the fact that f is odd, we have

$$f(x) - f(y) = f(x) + f(-y) = 2f\left(\frac{x - y}{2}\right) g\left(\frac{x + y}{2}\right) \qquad\qquad (20.34)$$

for all $x, y \in \mathbb{R}$. Now, using (20.31)–(20.34), we obtain

$f(x+y)f(x-y)$
$$= \left[2f\left(\frac{x+y}{2}\right)g\left(\frac{x+y}{2}\right)\right]\left[2f\left(\frac{x-y}{2}\right)g\left(\frac{x-y}{2}\right)\right]$$
$$= \left[2f\left(\frac{x+y}{2}\right)g\left(\frac{x-y}{2}\right)\right]\left[2f\left(\frac{x-y}{2}\right)g\left(\frac{x+y}{2}\right)\right]$$
$$= [f(x)+f(y)][f(x)-f(y)]$$
$$= f(x)^2 - f(y)^2$$

for all $x, y \in \mathbb{R}$. This completes the proof of the theorem. $\square$

In Theorem 20.1, we have seen that if a function $f : \mathbb{R} \to \mathbb{R}$ satisfies $|f(x+y) + f(x-y) - 2f(x)f(y)| \le \delta$ for all $x, y \in \mathbb{R}$ and some $\delta > 0$, then either f is bounded by a constant depending on δ only or it is a solution of the cosine functional equation $f(x+y)+f(x-y) = 2f(x)f(y)$. However, it is not the case for the sine functional equation; indeed, the bounded functions

$$f_n(x) = n\,\sin(x) + \frac{1}{n}$$

satisfy the inequality (20.26) with $\delta = 3$,

$$|f_n(x+y)f_n(x-y) - f_n(x)^2 + f_n(y)^2| \le 3$$

for all real numbers x, y and all positive integers n. Nevertheless, for each positive real number M, the inequality $|f_n(x)| \le M$ fails to hold for certain x and n.

20.4 Concluding Remarks

Baker (1980) proved the following theorem concerning the stability of the d'Alembert functional equation.

Theorem 20.4. *Let $\delta > 0$, let G be an abelian group and let f be a complex valued function defined on G such that*

$$|f(x+y) + f(x-y) - 2f(x)f(y)| \le \delta \qquad \text{for all } x, y \in G. \quad (20.35)$$

Then either $|f(x)| \le (1 + \sqrt{1+2\delta})/2$ for all $x \in G$ or there exists a complex valued function E on G such that

$$f(x) = \frac{E(x) + E(-x)}{2} \qquad \text{for all } x \in G$$

and

$$|E(x + y) - E(x) E(y)| \leq \frac{\delta}{2}$$

for all $x, y \in G$.

Badora (1998), instead of the commutativity of the group G, assumed the Kannappan condition $f(x + y + z) = f(x + z + y)$ for all $x, y, z \in G$ and gave the following generalization of Baker's result

Theorem 20.5. *Let $\delta > 0$, $\epsilon > 0$, and let $(G, +)$ be a group. Suppose f is a complex valued function defined on G such that*

$$|f(x + y) + f(x - y) - 2f(x)f(y)| \leq \delta \qquad \text{for all } x, y \in G$$

and

$$|f(x + y + z) - f(x + z + y)| \leq \epsilon \qquad \text{for all } x, y, z \in G.$$

Then either f is bounded or f satisfies the d'Alembert functional equation.

For vector-valued mappings, Badora (1998) proved the following interesting theorem.

Theorem 20.6. *Let $(G, +)$ be an abelian group and let $\mathcal{A}$ be a complex normed algebra. For some $\epsilon, \delta \geq 0$, suppose $f : G \to \mathcal{A}$ satisfies*

$$\|f(x + y) + f(x - y) - 2f(x)f(y)\| \leq \delta \qquad \text{for all } x, y \in G$$

and

$$\|f(x) - f(-x)\| \leq \epsilon \qquad \text{for all } x, \in G.$$

If there exists a $z_o \in G$ such that the map $G \ni x \mapsto \|f(x) f(z_o)\| \in \mathbb{R}$ is bounded, then there exist a function $E : G \to \mathcal{A}$ and constants $c, d \in \mathbb{R}$ such that

$$\left\| f(x) - \frac{E(x) + E(-x)}{2} \right\| \leq c \qquad \text{for all } x \in G$$

and

$$\|E(x + y) - E(x) E(y)\| \leq d \qquad \text{for all } x, y \in G.$$

Tyrala (2005) proved the following stability result for the matrix-valued function f defined on an abelian group G.

Theorem 20.7. *Let G be an abelian group written additively and $M_n(\mathbb{C})$ be the normed algebra of $n \times n$ matrices with complex entries. Suppose $f : G \to M_n(\mathbb{C})$ satisfies the inequality*

$$\|f(x + y) + f(x - y) - 2f(x)f(y)\| \le \delta$$

for some $\delta > 0$ and for all $x, y \in G$. Then there is a map $h : G \to M_n(\mathbb{C})$ such that

(a) $\|f(x) - h(x)\| \le \epsilon$ for some $\epsilon \ge 0$ and all $x \in G$,
(b) $[h(x + y) + h(x - y) - 2\,h(x)\,h(y)]^2 = 0$ for all $x, y \in G$.

Badora and Ger (2002) replaced the constant δ from the right side of the inequality (20.35) by a control function $\phi(x)$ and studied the stability of the cosine functional equation. They proved the following theorem.

Theorem 20.8. *Let $(G, +)$ be an abelian group written additively. For some $\phi : G \to \mathbb{R}$, let the function $f :\to \mathbb{C}^*$ (the set of nonzero complex numbers) satisfy the inequality*

$$|f(x + y) + f(x - y) - 2\,f(x)\,f(y)| \le \phi(x)$$

for all $x, y \in G$. Then either f is bounded or f satisfies $f(x+y) + f(x - y) = 2\,f(x)\,f(y)$ for all $x, y \in G$.

Two generalizations of the cosine functional equations are the following:

$$f(x + y) + f(x - y) = 2\,f(x)\,g(y) \tag{20.36}$$

and

$$f(x + y) + f(x - y) = 2\,g(x)\,f(y). \tag{20.37}$$

The superstability of these two functional equations were studied by Kannappan and Kim (2001). Kim and Dragomir (2006) replaced the constant δ by a control function ϕ and studied the superstability of the equations (20.36) and (20.36). They proved the following theorems.

Theorem 20.9. *Let $(G, +)$ be an abelian group. For some $\phi : G \to \mathbb{R}$, let the functions $f, g :\to \mathbb{C}^*$ satisfy the inequality*

$$|f(x + y) + f(x - y) - 2\,f(x)\,g(y)| \le \phi(y)$$

for all $x, y \in G$. Then either f is bounded or g is a solution of $g(x + y) + g(x - y) = 2g(x)g(y)$ for all $x, y \in G$.

Theorem 20.10. *Let $(G, +)$ be an abelian group. For some $\phi : G \to \mathbb{R}$, let the functions $f, g :\to \mathbb{C}^*$ satisfy the inequality*

$$|f(x + y) + f(x - y) - 2\,f(x)\,g(y)| \le \phi(x)$$

for all $x, y \in G$. Further, assume that f is even in G. Then either g (or f) is bounded or g is a solution of $g(x + y) + g(x - y) = 2g(x)g(y)$ along with f and g satisfying (20.36) and (20.36) for all $x, y \in G$.

Theorem 20.11. *Let $(G, +)$ be an abelian group. For some $\phi : G \to \mathbb{R}$, let the functions $f, g :\to \mathbb{C}^*$ satisfy the inequality*

$$|f(x + y) + f(x - y) - 2\,g(x)\,f(y)| \le \phi(x)$$

for all $x, y \in G$. Then either f is bounded or g is a solution of $g(x + y) + g(x - y) = 2g(x)g(y)$ for all $x, y \in G$.

Theorem 20.12. *Let $(G, +)$ be an abelian group. For some $\phi : G \to \mathbb{R}$, let the functions $f, g :\to \mathbb{C}^*$ satisfy the inequality*

$$|f(x + y) + f(x - y) - 2\,g(x)\,f(y)| \le \phi(y)$$

for all $x, y \in G$. Then either g (or f) is bounded or g is a solution of $g(x + y) + g(x - y) = 2g(x)g(y)$ along with f and g satisfying (20.36) and (20.36) for all $x, y \in G$.

So far we have commented on the various stability of cosine functional equations and their generalization. Now we make some comments regarding sine functional equation. The stability of the sine functional equation was studied by Cholewa (1983). He proved the following theorem.

Theorem 20.13. *Let $(G, +)$ be an abelian group in which division by 2 is uniquely performable. Every unbounded function $f : G \to \mathbb{C}$ satisfying the inequality*

$$|f(x + y) f(x - y) - f(x)^2 + f(y)^2| \le \delta \quad \text{for all } x, y \in G$$

for some $\delta > 0$ has to be a solution of the equation

$$f(x + y) f(x - y) = f(x)^2 - f(y)^2$$

for all $x, y \in G$.

Notice that the equation $f(x + y) f(x - y) = f(x)^2 - f(y)^2$ can be rewritten as

$$f(x) f(y) = f\left(\frac{x + y}{2}\right)^2 - f\left(\frac{x - y}{2}\right)^2. \tag{20.38}$$

Badora and Ger (2002) proved the following two theorems concerning the stability of the last functional equation.

Theorem 20.14. *Let* $(G, +)$ *be a uniquely 2-divisible abelian group and let* $f : G \to \mathbb{C}$, $\phi : G \to \mathbb{R}$ *satisfy the inequality*

$$\left| f(x) f(y) - f\left(\frac{x+y}{2}\right)^2 + f\left(\frac{x-y}{2}\right)^2 \right| \leq \phi(x) \quad \text{for all } x, y \in G.$$

Then either f *is bounded or* f *satisfies* (20.38) *for all* $x, y \in G$.

Theorem 20.15. *Let* $(G, +)$ *be a uniquely 2-divisible abelian group and let* $f : G \to \mathbb{C}$, $\phi : G \to \mathbb{R}$ *satisfy the inequality*

$$\left| f(x) f(y) - f\left(\frac{x+y}{2}\right)^2 + f\left(\frac{x-y}{2}\right)^2 \right| \leq \phi(y) \quad \text{for all } x, y \in G.$$

Then either f *is bounded or* f *satisfies* (20.38) *for all* $x, y \in G$.

Kim (2006) investigated the stability of the following generalizations of the above sine functional equation:

$$g(x) f(y) = f\left(\frac{x+y}{2}\right)^2 - f\left(\frac{x+\sigma y}{2}\right)^2,$$

$$f(x) g(y) = f\left(\frac{x+y}{2}\right)^2 - f\left(\frac{x+\sigma y}{2}\right)^2,$$

$$g(x) g(y) = f\left(\frac{x+y}{2}\right)^2 - f\left(\frac{x+\sigma y}{2}\right)^2.$$

Here σ is an endomorphism of order two of the uniquely 2-divisible abelian group G.

Nakmahachalasint (2007) investigated the stability of a cosine functional equation and proved the following theorem.

Theorem 20.16. *Let* δ *be a positive real constant and let* $(G, +)$ *be an abelian group. If* $f : G \to \mathbb{C}$ *satisfies*

$$|f(x+y+z) + f(x+y-z) + f(y+z-x) + f(z+x-y) - 4f(x)f(y)f(z)| \leq \delta$$

for all $x, y, z \in G$, *then either* f *is bounded or* f *is a solution of the functional equation*

$$f(x+y+z) + f(x+y-z) + f(y+z-x) + f(z+x-y) = 4f(x)f(y)f(z)$$

for all $x, y, z \in G$.

In Chapter 11, we dealt with the functional equation

$$f(x + y) = f(x)f(y) - d\sin(x)\sin(y)$$

for all $x, y \in \mathbb{R}$, where $d < -1$ is an a priopri chosen real number. The stability of this functional equation was studied by several mathematicians, namely Jung (2005, 2006), Jung and Cheng (2006) and Takahasi, Miura and Takagi (2007).

20.5 Exercises

1. If $f, g : \mathbb{R} \to \mathbb{C}$ satisfy the functional inequality

$$|f(x + y) - f(x - y) - 2g(x)f(y)| \le \delta$$

for some positive δ and for all $x, y \in \mathbb{R}$, then show that either f is bounded or g is a solution of the equation

$$g(x + y) + g(x - y) = 2g(x)g(y)$$

for all $x, y \in \mathbb{R}$.

2. If $f, g : \mathbb{R} \to \mathbb{C}$ satisfy the functional inequality

$$|f(x + y) - f(x - y) - 2g(x)g(y)| \le \delta$$

for some positive δ and for all $x, y \in \mathbb{R}$, then show that either g is bounded or g is a solution of the equation

$$g\left(\frac{x + y}{2}\right)^2 - g\left(\frac{x + y}{2}\right)^2 = g(x)g(y)$$

for all $x, y \in \mathbb{R}$.

3. If $f : \mathbb{R} \to \mathbb{C}$ satisfies the functional inequality

$$|f(x + y) - f(x - y) - 2f(x)f(y)| \le \phi(x)$$

for some $\phi : \mathbb{R} \to (0, \infty)$ and for all $x, y \in \mathbb{R}$, then show that either f is bounded or f is a solution of the equation

$$f(x + y) + f(x - y) = 2f(x)f(y)$$

for all $x, y \in \mathbb{R}$.

4. If $f, g : \mathbb{R} \to \mathbb{C}$ satisfy the functional inequality

$$|f(x + y) - f(x - y) - 2g(x)f(y)| \le \phi(x)$$

for some $\phi : \mathbb{R} \to (0, \infty)$ and for all $x, y \in \mathbb{R}$, then show that either f is bounded or g is a solution of the equation $g(x+y)+g(x-y) = 2g(x)g(y)$ for all $x, y \in \mathbb{R}$.

5. If $f, g : \mathbb{R} \to \mathbb{C}$ satisfy the functional inequality

$$|f(x + y) + f(x - y) - 2f(x)g(y)| \le \delta$$

for some positive δ and for all $x, y \in \mathbb{R}$, then show that either f is bounded or g is a solution of the equation $g(x+y)+g(x-y) = 2g(x)g(y)$ for all $x, y \in \mathbb{R}$.

6. If $f : \mathbb{R} \to \mathbb{C}$ satisfies the functional inequality

$$|f(x+y+z)+f(x+y-z)+f(y+z-x)+f(z+x-y)-4f(x)f(y)f(z)| \le \delta$$

for some positive δ and for all $x, y, z \in \mathbb{R}$, then show that either f is bounded or f is a solution of the equation

$$f(x+y+z) + f(x+y-z) + f(y+z-x) + f(z+x-y) = 4f(x)f(y)f(z)$$

for all $x, y, z \in \mathbb{R}$.

7. Let λ be a nonnegative real constant. If $f, g : \mathbb{R} \to \mathbb{C}$ satisfy the functional inequality

$$|f(x + y) + f(x - y) - \lambda \, f(x)g(y)| \le \phi(x)$$

for some function $\phi : \mathbb{R} \to \mathbb{R}$ and for all $x, y \in \mathbb{R}$, then show that either g with $f(0) = 0$ is bounded or g is a solution of the equation

$$g \left(\frac{x+y}{2} \right)^2 - g \left(\frac{x+y}{2} \right)^2 = g(x)g(y)$$

for all $x, y \in \mathbb{R}$.

8. Let λ be a nonnegative real constant. If $f, g : \mathbb{R} \to \mathbb{C}$ satisfy the functional inequality

$$|f(x + y) + g(x - y) - \lambda \, g(x)g(y)| \le \phi(x)$$

for some function $\phi : \mathbb{R} \to \mathbb{R}$ and for all $x, y \in \mathbb{R}$, then show that either g with $g(0) = 0$ is bounded or g is a solution of the equation

$$g \left(\frac{x+y}{2} \right)^2 - g \left(\frac{x+y}{2} \right)^2 = g(x)g(y)$$

for all $x, y \in \mathbb{R}$.

9. If $f : \mathbb{R} \to \mathbb{C}$ satisfies the functional inequality

$$\left| f(x)f(y) - f\left(\frac{x+y}{2}\right)^2 + f\left(\frac{x+y}{2}\right)^2 \right| \leq \phi(x)$$

for some function $\phi : \mathbb{R} \to \mathbb{R}$ and for all $x, y \in \mathbb{R}$, then show that either f is bounded or f is a solution of the equation

$$f\left(\frac{x+y}{2}\right)^2 - f\left(\frac{x+y}{2}\right)^2 = f(x)f(y)$$

for all $x, y \in \mathbb{R}$.

10. If $f : \mathbb{R} \to \mathbb{C}$ satisfies the functional inequality

$$\left| f(x)f(y) - f\left(\frac{x+y}{2}\right)^2 + f\left(\frac{x+y}{2}\right)^2 \right| \leq \phi(y)$$

for some function $\phi : \mathbb{R} \to \mathbb{R}$ and for all $x, y \in \mathbb{R}$, then show that either f is bounded or f is a solution of the equation

$$f\left(\frac{x+y}{2}\right)^2 - f\left(\frac{x+y}{2}\right)^2 = f(x)f(y)$$

for all $x, y \in \mathbb{R}$.

Chapter 21

Stability of Quadratic Functional Equations

21.1   Introduction

The quadratic functional equation

$$f(x + y) + f(x - y) = 2f(x) + 2f(y) \tag{21.1}$$

is an important equation in the theory of functional equations and it plays an important role in the characterization of inner product spaces. Every solution of the quadratic equation (21.1) is called a quadratic function. In this chapter, we present some important results concerning the stability of the quadratic functional equation, the generalized quadratic functional equations and some other related functional equations. We will treat the Hyers-Ulam type stability with the bounded Cauchy difference as well as with the unbounded Cauchy difference. This chapter is primarily based on the works found in Skof (1983), Cholewa (1984), Ger (1992), Czerwik (1992, 1994), Borelli and Forti (1995) and Jung and Sahoo (2001, 2002b) related to the stability of quadratic functional equation.

21.2   Stability of the Quadratic Equation

Now we examine the Ulam-Hyers type stability for the quadratic functional equation. The following theorem is due to Skof (1983) and also independently to Cholewa (1984).

Theorem 21.1. *If the function $f : \mathbb{R} \to \mathbb{R}$ satisfies the inequality*

$$|f(x + y) + f(x - y) - 2f(x) - 2f(y)| \leq \delta$$

for all $x, y \in \mathbb{R}$ and some $\delta > 0$, then there exists a unique quadratic function $q : \mathbb{R} \to \mathbb{R}$ such that

$$|f(x) - q(x)| \leq \frac{\delta}{2}$$

for all $x \in \mathbb{R}$.

Proof. Let $f : \mathbb{R} \to \mathbb{R}$ satisfy

$$|f(x + y) + f(x - y) - 2f(x) - 2f(y)| \leq \delta \tag{21.2}$$

for all $x, y \in \mathbb{R}$ and some $\delta > 0$. Letting $x = 0 = y$ in (21.2), we see that

$$|f(0)| \leq \frac{\delta}{2}. \tag{21.3}$$

Further, letting $y = x$ in (21.2), we see that

$$|f(2x) + f(0) - 4f(x)| \leq \delta,$$

that is,

$$|f(2x) - 4f(x)| - |f(0)| \leq \delta$$

which is

$$|f(2x) - 4f(x)| \leq \frac{3}{2}\delta \tag{21.4}$$

for all $x \in \mathbb{R}$. We replace x by $2^{k-1}x$ in (21.4) to get

$$\left| f(2^k x) - 2^2 f(2^{k-1}x) \right| \leq \frac{3}{2}\delta.$$

Multiplying both sides of the last inequality by $\frac{1}{2^{2k}}$ and then summing both sides of the resulting inequality as k goes from 1 to n, we get

$$\sum_{k=1}^{n} \frac{1}{2^{2k}} \left| f(2^k x) - 2^2 f(2^{k-1}x) \right| \leq \sum_{k=1}^{n} \frac{3}{2} \frac{1}{2^{2k}} \delta.$$

Using the inequality

$$|x| - |y| \leq |x - y|,$$

we see that

$$\left| \frac{1}{2^{2n}} f(2^n x) - f(x) \right| \leq \frac{\delta}{2}$$

for all positive integers n.

If $m > n > 0$, then $m - n$ is a natural number and n can be replaced by $m - n$ in the above inequality. Thus, we have

$$\left| \frac{1}{2^{2(m-n)}} f(2^{m-n}x) - f(x) \right| \leq \frac{\delta}{2}$$

or

$$\left| \frac{1}{2^{2m}} f(2^{m-n}x) - \frac{1}{2^{2n}} f(x) \right| \le \frac{\delta}{2 \cdot 2^{2n}}.$$

Replacing x by $x2^n$, we get

$$\left| \frac{1}{2^{2m}} f(2^m x) - \frac{1}{2^{2n}} f(2^n x) \right| \le \frac{\delta}{2^{(2n+1)}}.$$

Letting $n \to \infty$, we see that

$$\lim_{n \to \infty} \frac{\delta}{2^{2n+1}} = 0,$$

and hence

$$\left| \frac{1}{2^{2m}} f(2^m x) - \frac{1}{2^{2n}} f(2^n x) \right| \to 0 \qquad \text{as } n \to \infty.$$

Thus

$$\left\{ \frac{f(2^n x)}{2^{2n}} \right\}_{n=1}^{\infty}$$

is a Cauchy sequence. Hence this sequence has a limit in $\mathbb{R}$. We define a function $q : \mathbb{R} \to \mathbb{R}$ using this limit by

$$q(x) = \lim_{n \to \infty} \frac{f(2^n x)}{2^{2n}} \tag{21.5}$$

for $x \in \mathbb{R}$. Next, we show that $q(x)$ is quadratic. Since

$$|q(x+y) + q(x-y) - 2q(x) - 2q(y)|$$

$$= \lim_{n \to \infty} \frac{1}{2^{2n}} |f(2^n x + 2^n y) + f(2^n x - 2^n y) - 2f(2^n x) - 2f(2^n y)|$$

$$= \lim_{n \to \infty} \frac{\delta}{2^{2n}} = 0 \qquad \text{(by (21.2))},$$

therefore, we have

$$q(x+y) + q(x-y) = 2q(x) + 2q(y)$$

for all $x, y \in \mathbb{R}$. Hence q is a quadratic function.

Next, we consider

$$|q(x) - f(x)| = \left| \lim_{n \to \infty} \frac{f(2^n x)}{2^{2n}} - f(x) \right|$$

$$= \lim_{n \to \infty} \left| \frac{f(2^n x)}{2^{2n}} - f(x) \right|$$

$$\leq \lim_{n \to \infty} \frac{\delta}{2}$$
$$= \frac{\delta}{2}.$$

Hence

$$|q(x) - f(x)| \leq \frac{\delta}{2}$$

for all $x \in \mathbb{R}$.

Finally, we prove that q is unique. Suppose $q : \mathbb{R} \to \mathbb{R}$ is not unique. Then there exists another quadratic function $s : \mathbb{R} \to \mathbb{R}$ such that

$$|s(x) - f(x)| \leq \frac{\delta}{2}$$

for all $x \in \mathbb{R}$. Note that

$$|s(x) - q(x)| \leq |s(x) - f(x)| + |f(x) - q(x)|$$
$$\leq \frac{\delta}{2} + \frac{\delta}{2}$$
$$= \delta.$$

Therefore

$$|s(x) - q(x)| \leq \delta \tag{21.6}$$

for all $x \in \mathbb{R}$.

Since a quadratic function is rationally homogeneous of degree two, we have

$$|s(x) - q(x)| = \left| \frac{n^2 s(x)}{n^2} - \frac{n^2 q(x)}{n^2} \right|$$
$$= \left| \frac{s(nx)}{n^2} - \frac{q(nx)}{n^2} \right|$$
$$= \frac{1}{n^2} |s(nx) - q(nx)|$$
$$\leq \frac{\delta}{n^2}.$$

Taking the limit $n \to \infty$, we get

$$\lim_{n \to \infty} |s(x) - q(x)| \leq \lim_{n \to \infty} \frac{\delta}{n^2}.$$

Hence

$$|s(x) - q(x)| \leq 0.$$

Therefore

$$s(x) = q(x)$$

for all $x \in \mathbb{R}$. Therefore q is unique. This completes the proof. $\qquad \square$

Remark 21.1. *The proof of the above theorem goes over without any changes if one replaces the real function $f : \mathbb{R} \to \mathbb{R}$ by a function from a normed space into a Banach space.*

Czerwik (1992) proved the Hyers-Ulam-Rassias stability of the quadratic functional equation which includes the following theorem as a special case:

Theorem 21.2. *If a function $f : \mathbb{R} \to \mathbb{R}$ satisfies the inequality*

$$|f(x+y) + f(x-y) - 2f(x) - 2f(y)| \leq \varepsilon$$

for some $\varepsilon \geq 0$ and for all $x, y \in \mathbb{R}$, then there exists a unique quadratic function $Q : \mathbb{R} \to \mathbb{R}$ such that

$$|f(x) - Q(x)| \leq \frac{1}{3}(\varepsilon + |f(0)|)$$

for all $x \in \mathbb{R}$. Moreover, if $f(tx)$ is continuous in t for each fixed $x \in \mathbb{R}$, then $Q(tx) = t^2 Q(x)$ for all $t \in \mathbb{R}$ and $x \in \mathbb{R}$.

21.3   Stability of Generalized Quadratic Equation

In this section, we will prove the Hyers-Ulam type stability of the quadratic functional equation of Pexider type, $f_1(x+y) + f_2(x-y) = f_3(x) + f_4(y)$. The following theorem is due to Jung and Sahoo (2001b).

Theorem 21.3. *If functions $f_1, f_2, f_3, f_4 : \mathbb{R} \to \mathbb{R}$ satisfy the inequality*

$$|f_1(x+y) + f_2(x-y) - f_3(x) - f_4(y)| \leq \varepsilon \qquad (21.7)$$

for some $\varepsilon \geq 0$ and for all $x, y \in \mathbb{R}$, then there exists a unique quadratic function $Q : \mathbb{R} \to \mathbb{R}$ and exactly two additive functions $A_1, A_2 : \mathbb{R} \to \mathbb{R}$ such that

$$\left.\begin{array}{l} |f_1(x) - Q(x) - A_1(x) - A_2(x) - f_1(0)| \leq \frac{137}{3}\varepsilon, \\[2mm] |f_2(x) - Q(x) - A_1(x) + A_2(x) - f_2(0)| \leq \frac{125}{3}\varepsilon, \\[2mm] |f_3(x) - 2Q(x) - 2A_1(x) - f_3(0)| \leq \frac{136}{3}\varepsilon, \\[2mm] |f_4(x) - 2Q(x) - 2A_2(x) - f_4(0)| \leq \frac{124}{3}\varepsilon \end{array}\right\} \qquad (21.8)$$

for all $x \in \mathbb{R}$. Moreover, if $f_3(tx)$ and $f_4(tx)$ are continuous in $t \in \mathbb{R}$ for each $x \in \mathbb{R}$, then the Q satisfies $Q(tx) = t^2 Q(x)$ for all $x \in \mathbb{R}$ and A_1, A_2 are linear.

Proof. Let us define $F_i(x) = f_i(x) - f_i(0)$, and by F_i^e and F_i^o denote the even part and the odd part of F_i for $i = 1, 2, 3, 4$. Then, we get $F_i(0) = F_i^e(0) = F_i^o(0) = 0$ for $i = 1, 2, 3, 4$.

By putting $x = y = 0$ in (21.7) and using the resulting inequality and (21.7), we have

$$|F_1(x+y) + F_2(x-y) - F_3(x) - F_4(y)| \leq 2\varepsilon \qquad (21.9)$$

for all $x, y \in \mathbb{R}$. First we replace x and y in (21.9) by $-x$ and $-y$, respectively, to get

$$|F_1(-x-y) + F_2(-x+y) - F_3(-x) - F_4(-y)| \leq 2\varepsilon. \qquad (21.10)$$

Next we add (subtract) the argument of the norm of the inequality (21.10) to (from) that of the inequality (21.9), and then taking the norm and manipulating the resulting expression, we obtain

$$|F_1^e(x+y) + F_2^e(x-y) - F_3^e(x) - F_4^e(y)| \leq 2\varepsilon, \qquad (21.11)$$

$$|F_1^o(x+y) + F_2^o(x-y) - F_3^o(x) - F_4^o(y)| \leq 2\varepsilon \qquad (21.12)$$

for all $x, y \in \mathbb{R}$.

If we put $y = 0$, $x = 0$ (and replace y by x), $y = x$, and $y = -x$ in (21.11), respectively, then we get

$$|F_1^e(x) + F_2^e(x) - F_3^e(x)| \leq 2\varepsilon, \qquad (21.13)$$

$$|F_1^e(x) + F_2^e(x) - F_4^e(x)| \leq 2\varepsilon, \qquad (21.14)$$

$$|F_1^e(2x) - F_3^e(x) - F_4^e(x)| \leq 2\varepsilon, \qquad (21.15)$$

$$|F_2^e(2x) - F_3^e(x) - F_4^e(x)| \leq 2\varepsilon \qquad (21.16)$$

for all $x \in \mathbb{R}$, respectively.

In view of (21.13) and (21.14), we see that

$$|F_3^e(x) - F_4^e(x)| \leq 4\varepsilon, \qquad (21.17)$$

and it follows from (21.15) and (21.16) that

$$|F_1^e(x) - F_2^e(x)| \leq 4\varepsilon \qquad (21.18)$$

for any x in $\mathbb{R}$. By using (21.11), (21.17) and (21.18), we have

$$|F_2^e(x+y) + F_2^e(x-y) - F_4^e(x) - F_4^e(y)|$$
$$\leq |F_1^e(x+y) + F_2^e(x-y) - F_3^e(x) - F_4^e(y)|$$
$$+ |F_2^e(x+y) - F_1^e(x+y)| + |F_3^e(x) - F_4^e(x)|$$
$$\leq 10\varepsilon. \tag{21.19}$$

By putting $y = 0$ in (21.19), we get

$$|2F_2^e(x) - F_4^e(x)| \leq 10\varepsilon. \tag{21.20}$$

Hence, (21.19) and (21.20) imply

$$|F_4^e(x+y) + F_4^e(x-y) - 2F_4^e(x) - 2F_4^e(y)|$$
$$\leq 2|F_2^e(x+y) + F_2^e(x-y) - F_4^e(x) - F_4^e(y)|$$
$$+ |F_4^e(x+y) - 2F_2^e(x+y)| + |F_4^e(x-y) - 2F_2^e(x-y)|$$
$$\leq 40\varepsilon$$

for all $x, y \in \mathbb{R}$.

By Theorem 21.2, there exists a unique quadratic function $Q : \mathbb{R} \to \mathbb{R}$ such that

$$|F_4^e(x) - 2Q(x)| \leq \frac{40}{3}\varepsilon \tag{21.21}$$

for all $x \in \mathbb{R}$. Furthermore, if $f_4(tx)$ is continuous in $t \in \mathbb{R}$ for each $x \in \mathbb{R}$, then the quadratic function Q satisfies $Q(tx) = t^2 Q(x)$ for all $x \in \mathbb{R}$.

Using (21.17), (21.18), (21.20) and (21.21), we obtain

$$|F_1^e(x) - Q(x)|$$
$$\leq |F_1^e(x) - F_2^e(x)| + \left|F_2^e(x) - \frac{1}{2}F_4^e(x)\right| + \left|\frac{1}{2}F_4^e(x) - Q(x)\right|$$
$$\leq \frac{47}{3}\varepsilon, \tag{21.22}$$

$$|F_2^e(x) - Q(x)|$$
$$\leq \left|F_2^e(x) - \frac{1}{2}F_4^e(x)\right| + \left|\frac{1}{2}F_4^e(x) - Q(x)\right|$$
$$\leq \frac{35}{3}\varepsilon, \tag{21.23}$$

$$|F_3^e(x) - 2Q(x)|$$

$$\leq |F_3^e(x) - F_4^e(x)| + |F_4^e(x) - 2Q(x)|$$
$$\leq \frac{52}{3}\varepsilon \tag{21.24}$$

for any $x \in \mathbb{R}$.

As before, if we put $y = 0$, $x = 0$ (and replace y by x), $y = x$, and $y = -x$ in (21.12), respectively, then we obtain

$$|F_1^o(x) + F_2^o(x) - F_3^o(x)| \leq 2\varepsilon, \tag{21.25}$$

$$|F_1^o(x) - F_2^o(x) - F_4^o(x)| \leq 2\varepsilon, \tag{21.26}$$

$$|F_1^o(2x) - F_3^o(x) - F_4^o(x)| \leq 2\varepsilon, \tag{21.27}$$

$$|F_2^o(2x) - F_3^o(x) + F_4^o(x)| \leq 2\varepsilon \tag{21.28}$$

for all $x \in \mathbb{R}$, respectively. Due to (21.25) and (21.26), we have

$$|2F_1^o(x) - F_3^o(x) - F_4^o(x)|$$
$$\leq |F_1^o(x) + F_2^o(x) - F_3^o(x)| + |F_1^o(x) - F_2^o(x) - F_4^o(x)|$$
$$\leq 4\varepsilon \tag{21.29}$$

and

$$|2F_2^o(x) - F_3^o(x) + F_4^o(x)|$$
$$\leq |F_1^o(x) + F_2^o(x) - F_3^o(x)| + |F_2^o(x) + F_4^o(x) - F_1^o(x)|$$
$$\leq 4\varepsilon \tag{21.30}$$

for each $x \in \mathbb{R}$. Combining (21.27) with (21.29) yields

$$|F_3^o(2x) + F_4^o(2x) - 2F_3^o(x) - 2F_4^o(x)|$$
$$\leq |F_3^o(2x) + F_4^o(2x) - 2F_1^o(2x)| + |2F_1^o(2x) - 2F_3^o(x) - 2F_4^o(x)|$$
$$\leq 8\varepsilon. \tag{21.31}$$

Analogously, by (21.28) and (21.30), we get

$$|F_3^o(2x) - F_4^o(2x) - 2F_3^o(x) + 2F_4^o(x)|$$
$$\leq |F_3^o(2x) - F_4^o(2x) - 2F_2^o(2x)| + |2F_2^o(2x) - 2F_3^o(x) + 2F_4^o(x)|$$
$$\leq 8\varepsilon \tag{21.32}$$

for any $x \in \mathbb{R}$. Now it follows from (21.31) and (21.32) that

$$|F_3^o(2x) - 2F_3^o(x)|$$

$$\leq \left| \frac{1}{2} F_3^o(2x) + \frac{1}{2} F_4^o(2x) - F_3^o(x) - F_4^o(x) \right|$$

$$+ \left| \frac{1}{2} F_3^o(2x) - \frac{1}{2} F_4^o(2x) - F_3^o(x) + F_4^o(x) \right|$$

$$\leq 8\varepsilon \tag{21.33}$$

and analogously

$$|F_4^o(2x) - 2F_4^o(x)| \leq 8\varepsilon \tag{21.34}$$

for all $x \in \mathbb{R}$. In view of (21.12), (21.29), (21.30), (21.33) and (21.34), we have

$$|F_3^o(x+y) + F_4^o(x+y) + F_3^o(x-y) - F_4^o(x-y) - F_3^o(2x) - F_4^o(2y)|$$
$$\leq |2F_1^o(x+y) + 2F_2^o(x-y) - 2F_3^o(x) - 2F_4^o(y)|$$
$$+ |F_3^o(x+y) + F_4^o(x+y) - 2F_1^o(x+y)|$$
$$+ |F_3^o(x-y) - F_4^o(x-y) - 2F_2^o(x-y)|$$
$$+ |2F_3^o(x) - F_3^o(2x)| + |2F_4^o(y) - F_4^o(2y)|$$
$$\leq 28\varepsilon \tag{21.35}$$

for all $x, y \in \mathbb{R}$. If we replace y in (21.35) by $-y$ and then use the fact that F_4^o is an odd function, we get

$$\left| F_3^o(x-y) + F_4^o(x-y) + F_3^o(x+y) \right.$$
$$\left. - F_4^o(x+y) - F_3^o(2x) + F_4^o(2y) \right| \leq 28\varepsilon. \tag{21.36}$$

From (21.35) and (21.36), we get

$$|F_3^o(x+y) + F_3^o(x-y) - F_3^o(2x)|$$
$$= \frac{1}{2} |F_3^o(x+y) + F_4^o(x+y) + F_3^o(x-y)$$
$$- F_4^o(x-y) - F_3^o(2x) - F_4^o(2y)$$
$$+ F_3^o(x-y) + F_4^o(x-y) + F_3^o(x+y)$$
$$- F_4^o(x+y) - F_3^o(2x) + F_4^o(2y)|$$
$$\leq \frac{1}{2} |F_3^o(x-y) + F_4^o(x-y) + F_3^o(x+y)$$
$$- F_4^o(x+y) - F_3^o(2x) + F_4^o(2y)|$$
$$+ \frac{1}{2} |F_3^o(x-y) + F_4^o(x-y) + F_3^o(x+y)$$
$$- F_4^o(x+y) - F_3^o(2x) + F_4^o(2y)|$$
$$\leq 28\varepsilon. \tag{21.37}$$

Similarly, from (21.35) and (21.36), we get

$$|F_4^o(x+y) - F_4^o(x-y) - F_4^o(2y)|$$
$$= \frac{1}{2}|F_3^o(x+y) + F_4^o(x+y) + F_3^o(x-y)$$
$$\quad - F_4^o(x-y) - F_3^o(2x) - F_4^o(2y)$$
$$\quad - \{F_3^o(x-y) + F_4^o(x-y) + F_3^o(x+y)$$
$$\quad - F_4^o(x+y) - F_3^o(2x) + F_4^o(2y)\}|$$
$$\leq \frac{1}{2}|F_3^o(x-y) + F_4^o(x-y) + F_3^o(x+y)$$
$$\quad - F_4^o(x+y) - F_3^o(2x) + F_4^o(2y)|$$
$$\quad + \frac{1}{2}|F_3^o(x-y) + F_4^o(x-y) + F_3^o(x+y)$$
$$\quad - F_4^o(x+y) - F_3^o(2x) + F_4^o(2y)|$$
$$\leq 28\varepsilon. \tag{21.38}$$

By letting $u = x + y$ and $v = x - y$ in (21.37), we obtain

$$|F_3^o(u) + F_3^o(v) - F_3^o(u+v)| \leq 28\varepsilon$$

for all $u, v \in \mathbb{R}$. According to Theorem 18.3, there exists a unique additive function $A_1 : \mathbb{R} \to \mathbb{R}$ such that

$$|F_3^o(x) - 2A_1(x)| \leq 28\varepsilon \tag{21.39}$$

for all x in $\mathbb{R}$. If, moreover, $f_3(tx)$ is continuous in $t \in \mathbb{R}$ for every fixed $x \in \mathbb{R}$, then A_1 is a linear function.

By putting $u = x - y$ and $v = 2y$ in (21.38), we get

$$|F_4^o(u+v) - F_4^o(u) - F_4^o(v)| \leq 28\varepsilon$$

for all $u, v \in \mathbb{R}$. By Theorem 18.3 again, there exists a unique additive function $A_2 : \mathbb{R} \to \mathbb{R}$ such that

$$|F_4^o(x) - 2A_2(x)| \leq 28\varepsilon \tag{21.40}$$

for any x in $\mathbb{R}$. Furthermore, if $f_4(tx)$ is continuous in $t \in \mathbb{R}$ for all $x \in \mathbb{R}$, then A_2 is also linear.

From (21.29), (21.30), (21.39) and (21.40) it follows that

$$|F_1^o(x) - A_1(x) - A_2(x)|$$
$$\leq \left|F_1^o(x) - \frac{1}{2}F_3^o(x) - \frac{1}{2}F_4^o(x)\right| + \left|\frac{1}{2}F_3^o(x) - A_1(x)\right|$$

$$+ \left| \frac{1}{2} F_4^o(x) - A_2(x) \right| \leq 30\varepsilon \tag{21.41}$$

and

$$|F_2^o(x) - A_1(x) + A_2(x)|$$

$$\leq \left| F_2^o(x) - \frac{1}{2} F_3^o(x) + \frac{1}{2} F_4^o(x) \right| + \left| \frac{1}{2} F_3^o(x) - A_1(x) \right|$$

$$+ \left| A_2(x) - \frac{1}{2} F_4^o(x) \right| \leq 30\varepsilon \tag{21.42}$$

for each x in $\mathbb{R}$.

The inequalities in (21.8) are direct consequences of the inequalities (21.21), (21.22), (21.23), (21.24), (21.39), (21.40), (21.41) and (21.42).

Now, let Q', A_1', $A_2' : \mathbb{R} \to \mathbb{R}$ be another quadratic function and additive functions, respectively, satisfying the inequalities in (21.8) instead of Q, A_1 and A_2. Then, we have

$$|Q(x) + A_2(x) - Q'(x) - A_2'(x)|$$

$$\leq \left| -\frac{1}{2} f_4(x) + Q(x) + A_2(x) + \frac{1}{2} f_4(0) \right|$$

$$+ \left| \frac{1}{2} f_4(x) - Q'(x) - A_2'(x) - \frac{1}{2} f_4(0) \right|$$

$$\leq \frac{124}{3} \varepsilon \tag{21.43}$$

for each $x \in \mathbb{R}$. Replacing x by $-x$ in (21.43) and using the fact that quadratic functions are even and additive functions are odd, we get

$$|Q(x) - A_2(x) - Q'(x) + A_2'(x)| \leq \frac{124}{3} \varepsilon. \tag{21.44}$$

From (21.43) and (21.44), we see that

$$|Q(x) - Q'(x)| = \frac{1}{2} |Q(x) + A_2(x) - Q'(x) - A_2'(x)$$

$$+ Q(x) - A_2(x) - Q'(x) + A_2'(x)|$$

$$\leq \frac{1}{2} |Q(x) + A_2(x) - Q'(x) - A_2'(x)|$$

$$+ \frac{1}{2} |Q(x) - A_2(x) - Q'(x) + A_2'(x)|$$

$$\leq \frac{124}{3} \varepsilon. \tag{21.45}$$

Similarly, again from (21.43) and (21.44), we get

$$|A_2(x) - A_2'(x)|$$
$$= \frac{1}{2}|Q(x) + A_2(x) - Q'(x) - A_2'(x)$$
$$\qquad - \{Q(x) - A_2(x) - Q'(x) + A_2'(x)\}|$$
$$\leq \frac{1}{2}|Q(x) + A_2(x) - Q'(x) - A_2'(x)|$$
$$\qquad + \frac{1}{2}|Q(x) - A_2(x) - Q'(x) + A_2'(x)|$$
$$\leq \frac{124}{3}\varepsilon. \tag{21.46}$$

Hence (21.45) and (21.46) imply that

$$Q(x) = Q'(x) \quad \text{and} \quad A_2(x) = A_2'(x)$$

for every $x \in \mathbb{R}$. Similarly, we can show that $A_1(x) = A_1'(x)$ for any $x \in \mathbb{R}$. Now the proof of the theorem is complete. $\qquad\square$

Remark 21.2. *We can easily verify the following statements:*

(i)   *If $f_1 \equiv f_2$ in Theorem 21.3, then $A_2 \equiv 0$.*

(ii) *If $f_3 \equiv f_4$ in Theorem 21.3, then $A_1 \equiv A_2$.*

(iii) *If $2f_1 \equiv f_3$ in Theorem 21.3, then $A_2 \equiv 0$.*

(iv) *If $2f_2 \equiv f_4$ in Theorem 21.3, then $A_1 \equiv 2A_2$.*

(v)   *If $2f_1 \equiv f_4$ in Theorem 21.3, then $A_1 \equiv 0$.*

(vi) *If $2f_2 \equiv f_3$ in Theorem 21.3, then $A_2 \equiv 0$.*

Remark 21.3. *The proof of the above theorem goes over without any changes if one replaces the real functions $f_1, f_2, f_3, f_4 : \mathbb{R} \to \mathbb{R}$ by functions $f_1, f_2, f_3, f_4 : E_1 \to E_2$ where E_1 is a real normed space and E_2 is a Banach space.*

21.4   Stability of a Functional Equation of Drygas

To obtain a Jordan and von Neumann type characterization theorem for the quasi-inner-product spaces, Drygas (1987) considered the functional equation

$$f(x+y) + f(x-y) = 2f(x) + f(y) + f(-y) \tag{21.47}$$

for all $x, y \in \mathbb{R}$. However, the general solution of this functional equation was given by Ebanks, Kannappan and Sahoo (1992a) as

$$f(x) = Q(x) + A(x),$$

where $A : \mathbb{R} \to \mathbb{R}$ is an additive function and $Q : \mathbb{R} \to \mathbb{R}$ is a quadratic function.

In this section, we will prove the Hyers-Ulam stability of the functional equation

$$f(x+y) + f(x-y) = 2f(x) + g(2y) \tag{21.48}$$

which includes the functional equation of the Drygas as a special case.

First, we establish some results which will be instrumental in proving the stability of the Drygas equation.

Lemma 21.1. *If functions $f, g : \mathbb{R} \to \mathbb{R}$ satisfy the inequality*

$$|f(x+y) + f(x-y) - 2f(x) - g(2y)| \leq \varepsilon \tag{21.49}$$

for all $x, y \in \mathbb{R}$ and some $\varepsilon > 0$, then there exists a unique quadratic function $Q : \mathbb{R} \to \mathbb{R}$ such that

$$|f(x) + f(-x) - 2f(0) - 4Q(x)| \leq 3\varepsilon + \frac{1}{3}|g(0)| \tag{21.50}$$

and

$$|g(x) - Q(x)| \leq 2\varepsilon + \frac{1}{3}|g(0)| \tag{21.51}$$

for all $x \in \mathbb{R}$.

Proof. Substituting z in place of x and $\frac{x+y}{2}$ in place of y in (21.49), we get

$$\left| f\left(z + \frac{x+y}{2}\right) + f\left(z - \frac{x+y}{2}\right) - 2f(z) - g(x+y) \right| \leq \varepsilon. \tag{21.52}$$

Similarly, replacing x by z and y by $\frac{x-y}{2}$ in (21.49), we have

$$\left| f\left(z + \frac{x-y}{2}\right) + f\left(z - \frac{x-y}{2}\right) - 2f(z) - g(x-y) \right| \leq \varepsilon. \tag{21.53}$$

Again we replace x by $z + \frac{x}{2}$ and y by $\frac{y}{2}$ in (21.49) to obtain

$$\left| f\left(z + \frac{x+y}{2}\right) + f\left(z + \frac{x-y}{2}\right) - 2f\left(z + \frac{x}{2}\right) - g(y) \right| \le \varepsilon. \quad (21.54)$$

Finally, we let $z - \frac{x}{2}$ for x and $\frac{y}{2}$ for y in (21.49) to have

$$\left| f\left(z - \frac{x-y}{2}\right) + f\left(z - \frac{x+y}{2}\right) - 2f\left(z - \frac{x}{2}\right) - g(y) \right| \le \varepsilon. \quad (21.55)$$

Using (21.52) and (21.53), we see that

$$\left| f\left(z + \frac{x+y}{2}\right) + f\left(z - \frac{x+y}{2}\right) + f\left(z + \frac{x-y}{2}\right) \right.$$
$$\left. + f\left(z - \frac{x-y}{2}\right) - 4f(z) - g(x+y) - g(x-y) \right|$$
$$\le \left| f\left(z + \frac{x+y}{2}\right) + f\left(z - \frac{x+y}{2}\right) - 2f(z) - g(x+y) \right|$$
$$+ \left| f\left(z + \frac{x-y}{2}\right) + f\left(z - \frac{x-y}{2}\right) - 2f(z) - g(x-y) \right|$$
$$\le 2\varepsilon.$$

Similarly, using (21.54) and (21.55), we have

$$\left| f\left(z + \frac{x+y}{2}\right) + f\left(z - \frac{x+y}{2}\right) + f\left(z + \frac{x-y}{2}\right) \right.$$
$$\left. + f\left(z - \frac{x-y}{2}\right) - 2f\left(z + \frac{x}{2}\right) - 2f\left(z - \frac{x}{2}\right) - 2g(y) \right|$$
$$\le 2\varepsilon.$$

From (21.49) and the last two inequalities, we see that

$$|g(x+y) + g(x-y) - 2g(y) - 2g(x)|$$
$$\le \left\| g(x+y) + g(x-y) + 4f(z) - f\left(z + \frac{x+y}{2}\right) \right.$$
$$\left. - f\left(z - \frac{x+y}{2}\right) - f\left(z + \frac{x-y}{2}\right) - f\left(z - \frac{x-y}{2}\right) \right\|$$
$$+ \left| f\left(z + \frac{x+y}{2}\right) + f\left(z - \frac{x+y}{2}\right) + f\left(z + \frac{x-y}{2}\right) \right.$$
$$\left. + f\left(z - \frac{x-y}{2}\right) - 2f\left(z + \frac{x}{2}\right) - 2f\left(z - \frac{x}{2}\right) - 2g(y) \right|$$
$$+ \left| 2f\left(z + \frac{x}{2}\right) + 2f\left(z - \frac{x}{2}\right) - 2g(x) - 4f(z) \right|$$
$$\le 6\varepsilon.$$

Hence by Theorem 21.2 there exists a unique quadratic map $Q : \mathbb{R} \to \mathbb{R}$ such that

$$|g(x) - Q(x)| \leq 2\varepsilon + \frac{1}{3}|g(0)| \tag{21.56}$$

for all $x \in \mathbb{R}$. Next, from (21.49) and (21.56), we obtain

$$
\begin{aligned}
|f(x) &+ f(-x) - 2f(0) - Q(2x)| \\
&= |f(x) + f(-x) - 2f(0) - g(2x) + g(2x) - Q(2x)| \\
&\leq |f(x) + f(-x) - 2f(0) - g(2x)| + |g(2x) - Q(2x)| \\
&\leq 3\varepsilon + \frac{1}{3}|g(0)|,
\end{aligned}
$$

and the fact that $Q(2x) = 4Q(x)$ completes the proof. $\qquad\square$

In the following two lemmas, we determine the Hyers-Ulam stability of the functional equations $f(x + y) - f(x - y) = 2f(y)$ and $f(x + y) - f(x - y) = 2g(y)$, where $f, g : \mathbb{R} \to \mathbb{R}$ are unknown functions.

Lemma 21.2. *If the function $f : \mathbb{R} \to \mathbb{R}$ satisfies the inequality*

$$|f(x + y) - f(x - y) - 2f(y)| \leq \varepsilon \tag{21.57}$$

for all $x, y \in \mathbb{R}$ and some $\varepsilon > 0$, then there exists a unique additive function $A : \mathbb{R} \to \mathbb{R}$ such that

$$|f(x) - A(x)| \leq \frac{3}{2}\varepsilon \tag{21.58}$$

for all $x \in \mathbb{R}$.

Proof. Interchanging x with y in (21.57), we have

$$|f(x + y) - f(y - x) - 2f(x)| \leq \varepsilon \tag{21.59}$$

for all $x, y \in \mathbb{R}$. Next, letting $x = 0$ in (21.57), we see that

$$|f(y) + f(-y)| \leq \varepsilon. \tag{21.60}$$

Now using (21.57), (21.59) and (21.60), we have

$$
\begin{aligned}
|2f(x + y) &- 2f(x) - 2f(y)| \\
&\leq |f(x + y) - f(x - y) - 2f(y)| + |f(x + y) - f(y - x) - 2f(x)| \\
&\quad + |f(x - y) + f(y - x)| \\
&\leq 3\varepsilon.
\end{aligned}
$$

Using the well-known stability result of the additive Cauchy equation due to Hyers (1941) (see Theorem 18.3), we have the asserted inequality (21.58) and the proof of the lemma is now complete. $\qquad\square$

Next, we consider the Hyers-Ulam stability of a pexiderized version of the functional equation in the following lemma.

Lemma 21.3. *If functions $f, g : \mathbb{R} \to \mathbb{R}$ satisfy the inequality*

$$|f(x + y) - f(x - y) - 2g(y)| \leq \varepsilon \qquad (21.61)$$

for all $x, y \in \mathbb{R}$ and some $\varepsilon > 0$, then there exists a unique additive function $A : \mathbb{R} \to \mathbb{R}$ such that

$$|f(x) - A(x) - f(0)| \leq 6\varepsilon \qquad (21.62)$$

and

$$|g(x) - A(x) - g(0)| \leq 2\varepsilon \qquad (21.63)$$

for all $x \in \mathbb{R}$.

Proof. Letting $y = 0$ in (21.61), we obtain

$$2\,|g(0)| \leq \varepsilon. \qquad (21.64)$$

Now substituting $y = x$ in (21.61), we have

$$|f(2x) - 2g(x) - f(0)| \leq \varepsilon \qquad (21.65)$$

for all $x \in \mathbb{R}$.

Using (21.61), we get

$$
\begin{aligned}
2\,|2g(x) &- g(x + y) - g(x - y)| \\
&\leq |f(z + (x + y)) - f(z - (x + y)) - 2g(x + y)| \\
&\quad + |f(z + (x - y)) - f(z - (x - y)) - 2g(x - y)| \\
&\quad + |2g(x) - f((z + y) + x) + f((z + y) - x)| \\
&\quad + |2g(x) - f((z - y) + x) + f((z - y) - x)| \\
&\leq 4\varepsilon
\end{aligned}
$$

for any $z \in \mathbb{R}$. Thus, we have

$$|g(x + y) + g(x - y) - 2g(x)| \leq 2\varepsilon \qquad (21.66)$$

for all $x, y \in \mathbb{R}$. Letting $s = x + y$ and $t = x - y$, we see that

$$\left| g(s) + g(t) - 2g\left(\frac{s + t}{2} \right) \right| \leq 2\varepsilon$$

for all $s, t \in \mathbb{R}$. Using a stability result of Jung (1996), we have

$$|g(x) - A(x) - g(0)| \leq 2\varepsilon \qquad (21.67)$$

for some unique additive function $A : \mathbb{R} \to \mathbb{R}$. Using (21.64), (21.65) and (21.67), we have

$$|f(2x) - 2A(x) - f(0)|$$
$$\leq |f(2x) - 2g(x) - f(0)| + 2\,|g(x) - A(x) - g(0)| + 2\,|g(0)|$$
$$\leq 6\varepsilon$$

for all $x \in \mathbb{R}$, and hence

$$|f(x) - A(x) - f(0)| \leq 6\varepsilon.$$

Now the proof of the theorem is complete.   □

In the following theorem, we will prove the Hyers-Ulam stability of the functional equation (21.48) that includes the functional equation (21.47) as a special case.

Theorem 21.4. *If functions f, $g : \mathbb{R} \to \mathbb{R}$ satisfy the inequality (21.49) for some $\varepsilon > 0$ and all $x, y \in \mathbb{R}$, then there exists a unique quadratic function $Q : \mathbb{R} \to \mathbb{R}$ and a unique additive function $A : \mathbb{R} \to \mathbb{R}$ such that*

$$|f(x) - 2Q(x) - A(x) - f(0)| \leq \frac{37}{6}\,\varepsilon \qquad (21.68)$$

and

$$|g(x) - Q(x)| \leq \frac{13}{3}\,\varepsilon \qquad (21.69)$$

for all $x \in \mathbb{R}$.

Proof. By $f^e(x)$, $f^o(x)$, $g^e(x)$ and $g^o(x)$, we denote the even part of $f(x)$, odd part of $f(x)$, even part of $g(x)$ and odd part of $g(x)$, respectively.

In (21.49), replace x by $-x$ and y by $-y$, respectively, to get

$$|f(-x - y) + f(-x + y) - 2f(-x) - g(-2y)| \leq \varepsilon. \qquad (21.70)$$

Next, we add (subtract) the argument of the norm of the inequality (21.70) to (from) that of the inequality (21.49), and then taking the norm and manipulating the resulting expression, we obtain

$$|f^e(x + y) + f^e(x - y) - 2f^e(x) - g^e(2y)| \leq \varepsilon \qquad (21.71)$$

and

$$|f^o(x + y) + f^o(x - y) - 2f^o(x) - g^o(2y)| \leq \varepsilon \qquad (21.72)$$

for all $x, y \in \mathbb{R}$.

Since f^e is even, $f^e(0) = f(0)$ and $g^e(0) = g(0)$, Lemma 21.1 and

(21.71) imply that there exists a unique quadratic function $Q : \mathbb{R} \to \mathbb{R}$ such that

$$|f^e(x) - f(0) - 2Q(x)| \leq \frac{3}{2}\varepsilon + \frac{1}{6}|g(0)| \tag{21.73}$$

and

$$|g^e(x) - Q(x)| \leq 2\varepsilon + \frac{1}{3}|g(0)| \tag{21.74}$$

for each x in $\mathbb{R}$.

Similarly, since f^o is odd and $f^o(0) = g^o(0) = 0$, Lemma 21.1 and (21.72) yield that there exists a unique quadratic function $q : \mathbb{R} \to \mathbb{R}$ with

$$|q(x)| \leq \frac{3}{4}\varepsilon \tag{21.75}$$

and

$$|g^o(x) - q(x)| \leq 2\varepsilon. \tag{21.76}$$

The inequality (21.75) implies that $q(x) = 0$ for any $x \in \mathbb{R}$. This fact, together with (21.76), means that

$$|g^o(x)| \leq 2\varepsilon. \tag{21.77}$$

From (21.72) and (21.77), it follows that

$$|f^o(x+y) + f^o(x-y) - 2f^o(x)| \leq 3\varepsilon.$$

Interchanging x with y in the last inequality, we get

$$|f^o(x+y) - f^o(x-y) - 2f^o(y)| \leq 3\varepsilon \tag{21.78}$$

for all $x, y \in \mathbb{R}$.

Hence, according to Lemma 21.2, the inequality (21.78) implies that there exists a unique additive function $A : \mathbb{R} \to \mathbb{R}$ such that

$$|f^o(x) - A(x)| \leq \frac{9}{2}\varepsilon \tag{21.79}$$

for all $x \in \mathbb{R}$. Letting $x = y = 0$ in (21.49), we get

$$|g(0)| \leq \varepsilon. \tag{21.80}$$

By (21.73), (21.79) and (21.80), we have

$$\begin{aligned} |f(x) - 2Q(x) &- A(x) - f(0)| \\ &\leq |f^e(x) - f(0) - 2Q(x)| + |f^o(x) - A(x)| \\ &\leq \frac{37}{6}\varepsilon \end{aligned}$$

which proves the inequality (21.68). Analogously, (21.74), (21.77) and (21.80) yield

$$|g(x) - Q(x)| \leq |g^e(x) - Q(x)| + |g^o(x)| \leq \frac{13}{3}\varepsilon$$

which proves the inequality (21.69).

It now remains to prove the uniqueness of the functions Q and A. Let $A' : \mathbb{R} \to \mathbb{R}$ and $Q' : \mathbb{R} \to \mathbb{R}$ be another additive function and another quadratic function, respectively, which satisfy the inequalities (21.68) and (21.69) in place of A and Q. If there were some $x_0 \in \mathbb{R}$ with $Q(x_0) \neq Q'(x_0)$, then we would have

$$|Q(2^n x_0) - Q'(2^n x_0)| = 4^n |Q(x_0) - Q'(x_0)|$$
$$\to \infty \quad \text{as } n \to \infty.$$

On the other hand, it follows from (21.69) that

$$|Q(2^n x_0) - Q'(2^n x_0)| \leq |Q(2^n x_0) - g(2^n x_0)| + |g(2^n x_0) - Q'(2^n x_0)|$$
$$\leq \frac{26}{3}\varepsilon$$

which is a contradiction. Hence, the Q is uniquely determined.

Assume analogously that $A(x_0) \neq A'(x_0)$ for some $x_0 \in \mathbb{R}$. Then,

$$|A(2^n x_0) - A'(2^n x_0)| = 2^n |A(x_0) - A'(x_0)|$$
$$\to \infty \quad \text{as } n \to \infty.$$

However, from (21.68), we see that

$$|A(2^n x_0) - A'(2^n x_0)| \leq |-f(2^n x_0) + 2Q(2^n x_0) + A(2^n x_0) + f(0)|$$
$$+ |f(2^n x_0) - 2Q(2^n x_0) - A'(2^n x_0) - f(0)|$$
$$\leq \frac{37}{3}\varepsilon$$

which is also a contradiction. This means that A is also uniquely determined. $\square$

Remark 21.4. *The proof of the above theorem goes over without any changes if one replaces the real functions $f, g : \mathbb{R} \to \mathbb{R}$ by functions $f, g : E_1 \to E_2$ where E_1 is a normed space and E_2 is a Banach space.*

The Hyers-Ulam stability of the functional equation (21.47) of Drygas is a consequence of the above theorem as we will see in the following corollary.

Corollary 21.1. *If the function* $f : \mathbb{R} \to \mathbb{R}$ *satisfies the inequality*

$$|f(x+y) + f(x-y) - 2f(x) - f(y) - f(-y)| \leq \varepsilon \qquad (21.81)$$

for some $\varepsilon > 0$ *and for all* $x, y \in \mathbb{R}$, *then there exist a unique additive function* $A : \mathbb{R} \to \mathbb{R}$ *and a unique quadratic function* $Q : \mathbb{R} \to \mathbb{R}$ *such that*

$$|f(x) - Q(x) - A(x)| \leq \frac{25}{3}\varepsilon \qquad (21.82)$$

for all $x \in \mathbb{R}$.

Proof. Define a function $g : \mathbb{R} \to \mathbb{R}$ by $g(x) = f(\frac{x}{2}) + f(-\frac{x}{2})$. Then, the inequality (21.81) is transformed into

$$|f(x+y) + f(x-y) - 2f(x) - g(2y)| \leq \varepsilon.$$

By Theorem 21.4, there exist a unique additive function $A : \mathbb{R} \to \mathbb{R}$ and a unique quadratic function $Q_1 : \mathbb{R} \to \mathbb{R}$ such that

$$|f(x) - 2Q_1(x) - A(x) - f(0)| \leq \frac{37}{6}\varepsilon \qquad (21.83)$$

and

$$|g(x) - Q_1(x)| = \left| f\left(\frac{x}{2}\right) + f\left(-\frac{x}{2}\right) - Q_1(x) \right| \leq \frac{13}{3}\varepsilon \qquad (21.84)$$

for all x in $\mathbb{R}$.

It follows from (21.84) that

$$|f^e(x) - 2Q_1(x)| \leq \frac{13}{6}\varepsilon, \qquad (21.85)$$

where $f^e(x)$ denotes the even part of $f(x)$. Replacing x by $-x$ in (21.83), we have

$$|f(-x) - 2Q_1(x) + A(x) - f(0)| \leq \frac{37}{6}\varepsilon,$$

since Q_1 is even and A is odd. From (21.83) and the last inequality, we obtain

$$|f^o(x) - A(x)| \leq \frac{37}{6}\varepsilon \qquad (21.86)$$

for any $x \in \mathbb{R}$, where f^o is the odd part of f.

Using (21.85) and (21.86), we get

$$|f(x) - 2Q_1(x) - A(x)| \leq |f^e(x) - 2Q_1(x)| + |f^o(x) - A(x)| \leq \frac{25}{3}\varepsilon$$

which proves the validity of (21.82) by defining $Q(x) = 2Q_1(x)$.

Finally, it remains to prove the uniqueness of the Q and A. Let $A' : \mathbb{R} \to \mathbb{R}$ and $Q' : \mathbb{R} \to \mathbb{R}$ be another additive function and another quadratic function satisfying (21.82) instead of A and Q. Then, we have

$$|Q(x) - Q'(x) + A(x) - A'(x)| \leq \frac{50}{3}\,\varepsilon$$

for all $x \in \mathbb{R}$. Defining $q(x) = Q(x) - Q'(x)$ and $a(x) = A(x) - A'(x)$, we see

$$|q(x) + a(x)| \leq \frac{50}{3}\,\varepsilon$$

for every $x \in \mathbb{R}$. Since q is a quadratic function and a is an additive function, we can necessarily conclude that $q(x) = a(x) = 0$ for all $x \in \mathbb{R}$ and this fact implies the uniqueness of the Q and A. $\qquad\square$

21.5 Concluding Remarks

The Hyers-Ulam type stability of the quadratic functional equation (21.1) was first proved by Skof (1983) for functions from a normed space into a Banach space. Cholewa (1984) demonstrated that Skof's theorem is also valid if the relevant domain is replaced by an abelian group.

Let G be an abelian group and let E be a Banach space. A function $f : G \to E$ is called an ε-quadratic (or approximately quadratic) if for given $\varepsilon > 0$ it satisfies the inequalities

$$\|f(x+y) + f(x-y) - 2f(x) - 2f(y)\| \leq \varepsilon.$$

The following result regarding the stability of the quadratic functional equation is due to Skof (1983) and also independently to Cholewa (1984).

Theorem 21.5. *If $f : G \to E$ is an ε-quadratic for all x and y in G, then there exists a unique quadratic function $Q : G \to E$ such that*

$$\|f(x) - Q(x)\| \leq \frac{\varepsilon}{2}$$

for all $x \in G$. Moreover the function Q is given by

$$Q(x) = \lim_{n\to\infty} 4^{-n} f(2^n x).$$

Fenyö (1987) improved the bound obtained by Skof (1983) and Cholewa (1984) from $\frac{\varepsilon}{2}$ to $\frac{\varepsilon + \|f(0)\|}{3}$ (cf. Czerwik (1992)).

So far the Hyers-Ulam type stability of the quadratic functional equation has been investigated when the domain of the function is a normed space, or Banach space or in general an abelian group. Let G be an arbitrary group and E be a Banach space. Let $B(G, E)$ denote the space of all bounded functions $f : G \to E$. The Hyers-Ulam stability of the quadratic functional equation can be restated as follows. The quadratic functional equation is said to be stable for the pair (G, E) if for any $f : G \to E$ satisfying

$$\|f(xy) + f(xy^{-1}) - 2f(x) - 2f(y)\| \leq \delta \qquad \forall\, x, y \in G$$

(for some $\delta \geq 0$), there is a solution $q : G \to E$ of the quadratic functional equation $q(xy) + q(xy^{-1}) = 2q(x) + 2q(y)$ such that $f - q \in B(G, E)$.

Let G be a group and E and B be any two arbitrary Banach spaces over reals. Faiziev and Sahoo (2007b) proved that the quadratic functional equation is stable for the pair (G, E) if and only if it is stable for the pair (G, B). In view of this result it is not important which Banach space is used on the range. Thus one may consider the stability of the quadratic functional equation on the pair $(G, \mathbb{R})$. Faiziev and Sahoo (2007b) proved that the quadratic functional equation is not stable on the pair $(G, \mathbb{R})$ when G is any arbitrary group. It is well known (see Skof (1983) and Cholewa (1984)) that the quadratic functional equation is stable on the pair $(G, \mathbb{R})$ when G is an abelian group. Thus it is interesting to know on which noncommutative groups the quadratic functional equation is stable in the sense of Hyers-Ulam. Faiziev and Sahoo (2007b) proved that the quadratic functional equation is stable on n-abelian groups and $T(2, \mathbb{K})$, where $\mathbb{K}$ is a commutative field. Further they also proved that every group can be embeded into a group in which the quadratic functional equation is stable. Yang (2004b) proved the stability of quadratic functional equation on amenable groups.

The following result has been proved by Czerwik (1992) and generalizes Theorem 21.5 due to Skof (1983) and Cholewa (1984).

Theorem 21.6. *Let X be a normed space and Y a Banach space and let $f : X \to Y$ be a function satisfying inequality*

$$\|f(x + y) + f(x - y) - 2f(x) - 2f(y)\| \leq \varphi(x, y)$$

with either

1. *$\varphi(x, y) = \eta + \theta(\|x\|^p + \|y\|^p)$, $p < 2$, $x, y \in X \setminus \{0\}$ or*

2. $\varphi(x, y) = \theta(||x||^p + ||y||^p), \quad p > 2, \ x, y \in X$

for some $\eta, \theta \geq 0$. Then there exists a unique quadratic function Q such that

$$||f(x) - Q(x)|| \leq \frac{1}{3}(\eta + ||f(0)||) + \frac{2\theta}{4 - 2^p}||x||^p, \quad x \in X \setminus \{0\}$$

in case 1 and

$$||f(x) - Q(x)|| \leq \frac{2\theta}{2^p - 4}||x||^p, \quad x \in X$$

in case 2.

In Theorem 21.6 the parameter p is assumed to take all values except 2. If $p = 2$, then Theorem 21.6 is no longer valid. Czerwik (1992) proved the following theorem.

Theorem 21.7. *Suppose the function $f : \mathbb{R} \to \mathbb{R}$ is defined by*

$$f(x) = \sum_{i=0}^{\infty} 4^{-n}\phi(2^n x),$$

where the function $\phi : \mathbb{R} \to \mathbb{R}$ is given by

$$\phi(x) = \begin{cases} \theta & \text{if } |x| \geq 1 \\ \theta x^2 & \text{if } |x| < 1 \end{cases} \tag{21.87}$$

with a constant $\theta > 1$. Then the function f satisfies the inequality

$$|f(x + y) + f(x - y) - 2f(x) - 2f(y)| \leq 32\,\theta\,(x^2 + y^2)$$

for all $x, y \in \mathbb{R}$. Moreover, there exists no quadratic function $q : \mathbb{R} \to \mathbb{R}$ such that the image set of $x^{-2}|f(x) - q(x)|$ (for $x \neq 0$) is bounded.

Ger (1992) observed that for $p = 2$, a stability result can be established if one takes $\varphi(x, y) = 2||x||^2 + 2||y||^2 - ||x + y||^2 - ||x - y||^2$. He proved the following more general theorem (see Ger (1992)).

Theorem 21.8. *Let $(G, +)$ be an abelian group and $(Y, || \cdot ||)$ a real n-dimensional normed space. Suppose $F : G \to \mathbb{R}$ is a non-negative function satisfying $2F(x) + 2F(y) - F(x + y) - F(x - y) \geq 0$ for all $x, y \in G$. Let $f : G \to Y$ be a function satisfying the inequality*

$$||f(x + y) + f(x - y) - 2f(x) - 2f(y)|| \leq \varphi(x, y),$$

where

$$\varphi(x, y) = 2F(x) + 2F(y) - F(x + y) - F(x - y).$$

Then there exists a quadratic function $Q : G \to Y$ such that

$$\|f(x) - Q(x)\| \le n\, F(x), \quad \forall\, x \in G.$$

Moreover, if

$$\lim_{k \to \infty} \inf \frac{F(kx)}{k^2} = 0$$

for all $x \in G$, then Q is uniquely determined.

Borelli and Forti (1995) have proven the following results which include some of the previous results concerning the stability of the quadratic functional equation.

Theorem 21.9. *Let $(G, +)$ be an abelian group, Y a Banach space and $f : G \to Y$ a function with $f(0) = 0$ and fulfilling*

$$\|f(x + y) + f(x - y) - 2f(x) - 2f(y)\| \le \varphi(x, y).$$

Assume that one of the series

$$\sum_{n=1}^{\infty} 2^{-2n} \varphi\left(2^{n-1}x, 2^{n-1}x\right) \quad \text{and} \quad \sum_{n=1}^{\infty} 2^{2(n-1)} \varphi\left(2^{-n}x, 2^{-n}x\right)$$

converges for every x and call $\Phi(x)$ its sum. If for every x, y as

$$\lim_{n \to \infty} 2^{-2n} \varphi\left(2^{n-1}x, 2^{n-1}y\right) = 0 \quad \lim_{n \to \infty} 2^{2(n-1)} \varphi\left(2^{-n}x, 2^{-n}y\right) = 0,$$

respectively, then there exists a unique quadratic function $Q : G \to Y$ such that

$$\|f(x) - Q(x)\| \le \Phi(x), \quad x \in X.$$

In the following theorem, Czerwik (1994) proved the generalized Hyers-Ulam stability for the "partially pexiderized" quadratic functional equation $f(x + y) + f(x - y) = 2\, g(x) + 2\, g(y)$.

Theorem 21.10. *Let G be an abelian group divisible by a natural number $k \ge 2$ and E a Banach space. Let $f, g : G \to E$ be mappings satisfying the inequality*

$$\|f(x + y) + f(x - y) - 2g(x) - 2g(y)\| \le \varphi(x, y), \quad x, y \in G,$$

where $\varphi : G \times G \to \mathbb{R}_+$ is a given mapping. Suppose that the series

$$\sum_{n=1}^{\infty} k^{2n} \varphi(mk^{-n}x, k^{-n}x) \quad \text{and} \quad \sum_{n=1}^{\infty} k^{2n} \varphi(k^{-n}x, 0)$$

converge for all $x \in G$ and for $m \in \{1, 2, ..., k-1\}$. Further, suppose

$$\lim_{n \to \infty} k^{2n} \varphi(k^{-n}x, k^{-n}y) = 0, \quad x, y \in G.$$

Then there exists a unique quadratic mapping $Q : G \to E$ such that

$$\|f(x) - f(0) - Q(x)\| \leq k^{-2} \sum_{m=1}^{k-1} \sum_{n=1}^{\infty} (k-m)k^{2n} \Phi(mk^{-n}x, k^{-n}x)$$

and

$$\|g(x) - g(0) - 2Q(x)\| \leq k^{-2} \sum_{m=1}^{k-1} \sum_{n=1}^{\infty} (k-m)k^{2n} \Psi(mk^{-n}x, k^{-n}x)$$

for all $x \in G$, where $\Phi(x, y) = \varphi(x, y) + \varphi(x, 0) + \varphi(y, 0) + \varphi(0, 0)$ and $\Psi(x, y) = 2\varphi(x, y) + \varphi(x + y, 0) + \varphi(x - y, 0)$.

Jung and Sahoo (2001b) studied the Hyers-Ulam stability of pexiderized quadratic functional equation $f_1(x + y) + f_2(x - y) = f_3(x) + f_4(y)$ and proved the following theorem

Theorem 21.11. *Suppose $(X, \|.\|)$ is a real normed space and $(Y, \|.\|)$ a Banach space. If functions $f_1, f_2, f_3, f_4 : X \to Y$ satisfy the inequality*

$$\|f_1(x + y) + f_2(x - y) - f_3(x) - f_4(y)\| \leq \varepsilon \tag{21.88}$$

for some $\varepsilon \geq 0$ and for all $x, y \in X$, then there exist a unique quadratic function $Q : X \to Y$ and exactly two additive functions $A_1, A_2 : X \to Y$ such that

$$\left. \begin{array}{r} \|f_1(x) - Q(x) - A_1(x) - A_2(x) - f_1(0)\| \leq \frac{137}{3}\varepsilon, \\[6pt] \|f_2(x) - Q(x) - A_1(x) + A_2(x) - f_2(0)\| \leq \frac{125}{3}\varepsilon, \\[6pt] \|f_3(x) - 2Q(x) - 2A_1(x) - f_3(0)\| \leq \frac{136}{3}\varepsilon, \\[6pt] \|f_4(x) - 2Q(x) - 2A_2(x) - f_4(0)\| \leq \frac{124}{3}\varepsilon \end{array} \right\} \tag{21.89}$$

for all $x \in X$. Moreover, if $f_3(tx)$ and $f_4(tx)$ are continuous in $t \in \mathbb{R}$ for each $x \in X$, then the function Q satisfies $Q(tx) = t^2 Q(x)$ for all $x \in X$ and the functions A_1, A_2 are linear.

In Theorem 21.11, Yang (2004a) has replaced the domain X of the functions f_1, f_2, f_3, f_4 by a 2-divisible abelian group G and has found sharper estimates. To the best of the knowledge of the authors no result has been established about the stability of pexiderized quadratic equation on noncommutative groups.

A functional equation that is closely related to quadratic functional equation is the Drygas functional equation

$$f(x + y) + f(x - y) = 2f(x) + f(y) + f(-y). \tag{21.90}$$

The Hyers-Ulam stability of this functional equation was studied by Jung and Sahoo (2001b). They proved the following theorem.

Theorem 21.12. *Let X be a real vector space and Y a Banach space. If a function $f : X \to Y$ satisfies the inequality*

$$\|f(x + y) + f(x - y) - 2f(x) - f(y) - f(-y)\| \leq \varepsilon \tag{21.91}$$

for some $\varepsilon > 0$ and for all $x, y \in X$, then there exist a unique additive function $A : X \to Y$ and a unique quadratic function $Q : X \to Y$ such that

$$\|f(x) - Q(x) - A(x)\| \leq \frac{25}{3}\varepsilon \tag{21.92}$$

for all $x \in X$.

Yang (2004a) improved the bound obtained by Jung and Sahoo (2001b) from $\frac{25}{3}\varepsilon$ to $\frac{3}{2}\varepsilon$. She also replaced the domain of f from a real vector space to an abelian group.

Let $(G, +)$ be an abelian group. The functional equation

$$f(x + y) + f(x - y) = 2f(x) + f(y) + f(-y) \quad \forall\, x, y \in G$$

is a variation of the quadratic equation. On an arbitrary group G, this equation takes the form

$$f(xy) + f(xy^{-1}) = 2f(x) + f(y) + f(y^{-1}) \quad \forall\, x, y \in G.$$

Consider the system of functional equations

$$f(xy) + f(xy^{-1}) = 2f(x) + f(y) + f(y^{-1})$$
$$f(yx) + f(y^{-1}x) = 2f(x) + f(y) + f(y^{-1}).$$

On groups, this system generalizes the functional equation introduced by Drygas (see Faiziev and Sahoo (2006b)). Faiziev and Sahoo (2006c) proved the following results: (1) the system, in general, is not stable on an arbitrary group; (2) the system is stable on Heisenberg group $UT(3, \mathbb{K})$, where $\mathbb{K}$ is a commutative field with characteristic different from two; (3) the system is stable on a certain class of n-abelian groups; (4) any group can be embeded into a group where this system is stable. In Faiziev and

Sahoo (2006d) the stability of this system in the noncommutative group $T(3, \mathbb{R})$ was established.

There are many variations of the quadratic functional equations. For instance,

$$f(x+y+z)+f(x)+f(y)+f(z) = f(x+y)+f(y+z)+f(z+x), \quad (21.93)$$

$$f(x+y+z)+f(x-y)+f(x-z) = f(x-y-z)+f(x+y)+f(x+z), \quad (21.94)$$

$$f(\lambda x + y) + f(x - \lambda y) = (1 + \lambda^2)[f(x) + f(y)], \quad (21.95)$$

$$f(\alpha x + \beta y) + f(\alpha x - \beta y) = (\alpha^2 + \beta^2)[f(x) + f(y)], \quad (21.96)$$

$$f(2x + y) + f(2x - y) = f(x + y) + f(x - y) + 6f(x), \quad (21.97)$$

$$f(2x + y) + f(x + 2y) = 4f(x + y) + f(x) + f(y) \quad (21.98)$$

are some variations of the quadratic functional equation.

The functional equation (21.93) was studied by Kannappan (1995). The Hyers-Ulam stability of the equation (21.93) was investigated by Jung (1998). He proved the following theorem.

Theorem 21.13. *Suppose X is a real norm space and Y a real Banach space. Let $f : X \to Y$ satisfy the inequalities*

$$\|f(x + y + z) + f(x) + f(y) + f(z)$$
$$- f(x + y) - f(y + z) - f(z + x)\| \leq \delta \quad (21.99)$$

and

$$\|f(x) - f(-x)\| \leq \theta \quad (21.100)$$

for some $\delta, \theta \geq 0$ and for all $x, y, z \in X$. Then there exists a unique quadratic mapping $Q : X \to Y$ which satisfies

$$\|f(x) - Q(x)\| \leq 3\delta \quad (21.101)$$

for all $x \in X$. If, moreover, f is measurable or $f(tx)$ is continuous in t for each fixed $x \in X$, then $Q(tx) = t^2 Q(x)$ for all $x \in X$ and $t \in \mathbb{R}$.

Jung (1998) also proved another theorem replacing the inequality $\|f(x) - f(-x)\| \leq \theta$ by $\|f(x) + f(-x)\| \leq \theta$. For this functional equation, Kim (2001) proved a generalized stability result in the spirit of Găvruta (1994). Chang and Kim (2002) generalized the theorem of Jung and proved the following theorem.

Theorem 21.14. *Suppose X is a real norm space and Y a real Banach space. Let $H : \mathbb{R}_+^3 \to \mathbb{R}_+$ be a function such that $H(tu, tv, tw) \leq t^p H(u, v, w)$ for all $t, u, v, w \in \mathbb{R}_+$ and for some $p \in \mathbb{R}$. Further, let $E : \mathbb{R}_+ \to \mathbb{R}_+$ satisfying $E(tx) \leq t^q E(x)$ for all $t, x \in \mathbb{R}_+$. Let $p.q < 1$ be real numbers and let $f : X \to Y$ satisfy the inequalities*

$$\|f(x + y + z) + f(x) + f(y) + f(z)$$
$$- f(x + y) - f(y + z) - f(z + x)\| \leq H(\|x\|, \|y\|, \|z\|) \quad (21.102)$$

and

$$\|f(x) - f(-x)\| \leq E(\|x\|) \quad (21.103)$$

for some $\delta, \theta \geq 0$ and for all $x, y, z \in X$. Then there exists a unique quadratic mapping $Q : X \to Y$ which satisfies

$$\|f(x) - Q(x)\| \leq \frac{H(\|x\|, \|x\|, \|x\|)}{2 - 2^p} + 2\|f(0)\| \quad (21.104)$$

for all $x \in X$. If, moreover, f is measurable or $f(tx)$ is continuous in t for each fixed $x \in X$, then $Q(tx) = t^2 Q(x)$ for all $x \in X$ and $t \in \mathbb{R}$.

Chang and Kim (2002a) also proved anathor similar theorem replacing the inequality $\|f(x) - f(-x)\| \leq E(\|x\|)$ by $\|f(x) + f(-x)\| \leq E(\|x\|)$.

The stability of the remaining five functional equations have been studied by many authors. The interested reader should refer to G.H. Kim (2001); Chang and Kim (2002a, 2002b); H.-M. Kim (2003); Bae, Jun and Jung (2003); Kang, Lee and Lee (2004) and Jun and Kim (2004). For more information on the stability of the quadratic functional equation and its generalizations the reader is referred to the books by Hyers, Isac and Rassias (1998), Jung (2001) and the survey papers by Rassias (1998a, 1998b, 2000a).

21.6 Exercises

1. If $f : \mathbb{R} \to \mathbb{R}$ satisfies the functional inequality

$$|f(2x + y) + f(2x - y) - f(x + y) - f(x - y) - 6f(x)| \leq \delta$$

for some positive δ and for all $x, y \in \mathbb{R}$, then show that there exists a unique function $q : \mathbb{R} \to \mathbb{R}$ satisfying

$$q(2x + y) + q(2x - y) = q(x + y) + q(x - y) + 6q(x)$$

for all $x, y \in \mathbb{R}$ such that

$$|f(x) - q(x)| \leq \frac{1}{6}\delta$$

for all $x \in \mathbb{R}$.

2. If $f : \mathbb{R} \to \mathbb{R}$ satisfies the functional inequality

$$|f(2x + y) + f(x + y) - 4f(x + y) - f(x) - f(y)| \leq \delta$$

for some positive δ and for all $x, y \in \mathbb{R}$, then show that there exists a unique function $q : \mathbb{R} \to \mathbb{R}$ satisfying

$$q(2x + y) + q(x + y) = q(x + y) + q(x) + q(y)$$

for all $x, y \in \mathbb{R}$ such that

$$|f(x) - q(x)| \leq \frac{5}{16}\delta$$

for all $x \in \mathbb{R}$.

3. If $f : \mathbb{R} \to \mathbb{R}$ satisfies the functional inequality

$$|f(x + y + z) + f(x + y - z) + f(y + z - x) \\ + f(z + x - y) - 4[f(x) + f(y) + f(z)]| \leq \delta$$

for some positive δ and for all $x, y, z \in \mathbb{R}$, then show that there exists a unique function $q : \mathbb{R} \to \mathbb{R}$ satisfying

$$q(x + y) + q(x - y) = 2q(x) + 2q(y)$$

for all $x, y \in \mathbb{R}$ such that

$$|f(x) - q(x)| \leq \frac{5}{8}\delta$$

for all $x \in \mathbb{R}$.

4. If $f : \mathbb{R} \to \mathbb{R}$ satisfies the functional inequality

$$|f(x + y + z) + f(x - y) + f(y - z) \\ + f(z - x) - 3[f(x) + f(y) + f(z)]| \leq \delta$$

for some positive δ and for all $x, y, z \in \mathbb{R}$, then show that there exists a unique function $q : \mathbb{R} \to \mathbb{R}$ satisfying

$$q(x + y + z) + q(x - y) + q(y - z) + q(z - x) = 3[q(x) + q(y) + q(z)]$$

for all $x, y \in \mathbb{R}$ such that

$$|f(x) - q(x)| \le \frac{1}{5}\delta$$

for all $x \in \mathbb{R}$.

5. If $f : \mathbb{R} \to \mathbb{R}$ satisfies the functional inequality

$$|f(x + y + z) + f(x - y) + f(x - z)$$
$$- f(x - y - z) - f(x + y) - f(x + z)| \le \delta$$

and

$$|f(x) + f(-x) - 2f(0)| \le \epsilon$$

for some positive δ and ϵ and for all $x, y, z \in \mathbb{R}$, then show that there exists a unique function $a : \mathbb{R} \to \mathbb{R}$ satisfying

$$a(x + y) = a(x) + a(y)$$

for all $x, y \in \mathbb{R}$ such that

$$|f(x) - f(0) - a(x)| \le 2\delta + \epsilon$$

for all $x \in \mathbb{R}$.

6. If $f : \mathbb{R} \to \mathbb{R}$ satisfies the functional inequality

$$|f(x + y) + f(x - y) - 2f(x) - 2f(y)| \le \delta \left(|x|^P + |y|^P\right)$$

for some positive δ and $p \in (2, \infty)$ and for all $x, y \in \mathbb{R}$, then show that there exists a unique function $q : \mathbb{R} \to \mathbb{R}$ satisfying

$$q(x + y) + q(x - y) = 2q(x) + 2q(y)$$

for all $x, y \in \mathbb{R}$ such that

$$|f(x) - q(x)| \le \frac{2}{2^P - 4}\delta\,|x|^P$$

for all $x \in \mathbb{R}$.

7. If $f : \mathbb{R} \to \mathbb{R}$ satisfies the functional inequality

$$|f(x + y + z) + f(x) + f(y) + f(z)$$
$$- f(x + y) - f(y + z) - f(z + x)| \le \delta$$

and

$$|f(x) - f(-x)| \le \theta$$

for some $\delta, \theta \ge 0$ and for all $x, y, z \in \mathbb{R}$, then show that there exists a unique quadratic mapping $q : \mathbb{R} \to \mathbb{R}$ which satisfies

$$|f(x) - q(x)| \le 3\,\delta$$

for all $x \in \mathbb{R}$.

8. Let $0 < p < 1$, $\delta \geq 0$ and $\theta \geq 0$ be given. If the function $f : \mathbb{R} \to \mathbb{R}$ satisfies the functional inequality

$$\left| 2f\left(\frac{x+y}{2}\right) - f(x) - f(y) \right| \leq \delta + \theta \left(|x|^p + |y|^p \right)$$

for all $x, y \in \mathbb{R}$, then show that there exists a unique additive mapping $A : \mathbb{R} \to \mathbb{R}$ such that

$$|f(x) - A(x)| \leq \delta + |f(0)| + \frac{1}{2^{1-p} - 1} \theta |x|^p$$

for all $x \in \mathbb{R}$.

9. Let $1 < p < \infty$ and $\theta \geq 0$ be given. If the function $f : \mathbb{R} \to \mathbb{R}$ satisfies the functional inequality

$$\left| 2f\left(\frac{x+y}{2}\right) - f(x) - f(y) \right| \leq \theta \left(|x|^p + |y|^p \right)$$

for all $x, y \in \mathbb{R}$, then show that there exists a unique additive mapping $A : \mathbb{R} \to \mathbb{R}$ such that

$$|f(x) - A(x)| \leq \delta + |f(0)| + \frac{2^{p-1}}{2^{1-p} - 1} \theta |x|^p$$

for all $x \in \mathbb{R}$.

Chapter 22

Stability of Davison Functional Equation

22.1 Introduction

In Chapter 13, among others we examined the general solutions of the Hosszú functional equation

$$f(x + y - xy) + f(xy) = f(x) + f(y).$$

In Chapter 14, we studied the Davison functional equation

$$f(x + y) + f(xy) = f(xy + x) + f(y)$$

and determined it is the general solution on the set of real numbers. In the Davison equation the variables enter in a bilinear way as they do in the Hosszú functional equation. So the Davison functional equation is a Hosszú-like functional equation (see Davison (2001)).

The aim of this chapter is to examine the Ulam-Hyers and the Ulam-Hyers-Rassias type stability of the Davison functional equation. The Hyers-Ulam stability of the Davison functional equation was first treated by Jung and Sahoo (1999).

22.2 Stability of Davison Functional Equation

In this section, we examine the Hyers-Ulam stability of the Davison functional equation following Jung and Sahoo (1999).

Theorem 22.1. *If the function* $f : \mathbb{R} \to \mathbb{R}$ *satisfies the inequality*

$$|f(xy) + f(x + y) - f(xy + x) - f(y)| \leq \delta \qquad (22.1)$$

for all $x, y \in \mathbb{R}$ and for some $\delta \geq 0$, then there exist an additive function $A : \mathbb{R} \to \mathbb{R}$ and a real constant b such that

$$|f(x) - A(x) - b| \leq 12\delta \qquad (22.2)$$

for all $x \in \mathbb{R}$.

Proof. Replacing y by $y + 1$ in (22.2), we have

$$|f(xy + x) + f(x + y + 1) - f(xy + 2x) - f(y + 1)| \leq \delta \qquad (22.3)$$

for any $x, y \in \mathbb{R}$. From (22.1) and (22.3), we obtain

$$
\begin{aligned}
|f(xy) &+ f(x + y) + f(x + y + 1) - f(y) - f(xy + 2x) - f(y + 1)| \\
&\leq |f(xy) + f(x + y) - f(xy + x)f(y)| \\
&\quad + |f(xy + x) + f(x + y + 1) - f(xy + 2x) - f(y + 1)| \\
&\leq 2\delta
\end{aligned} \qquad (22.4)
$$

for all $x, y \in \mathbb{R}$. Replacing x by $\frac{x}{2}$ and y by $2y$ in (22.4), we obtain

$$
\begin{aligned}
\Big| f(xy) &+ f\left(\frac{x}{2} + 2y\right) + f\left(\frac{x}{2} + 2y + 1\right) \\
&- f(2y) - f(xy + x) - f(2y + 1) \Big| \leq 2\delta.
\end{aligned}
$$

From (22.1) and the last inequality, we obtain

$$
\begin{aligned}
\Big| f&\left(\frac{x}{2} + 2y\right) + f\left(\frac{x}{2} + 2y + 1\right) \\
&\qquad\qquad - f(x + y) - f(2y) - f(2y + 1) + f(y) \Big| \\
&= \Big| f(xy) + f\left(\frac{x}{2} + 2y\right) + f\left(\frac{x}{2} + 2y + 1\right) \\
&\qquad - f(2y) - f(xy + x) - f(2y + 1) - \\
&\qquad [f(xy) + f(x + y) - f(xy + x) - f(y)] \Big| \\
&\leq 3\delta
\end{aligned} \qquad (22.5)
$$

for all $x, y \in \mathbb{R}$.

If we replace x by $x - y$ in (22.5), we get

$$
\begin{aligned}
\Big| f&\left(\frac{x}{2} + \frac{3y}{2}\right) + f\left(\frac{x}{2} + \frac{3y}{2} + 1\right) \\
&\qquad\qquad - f(x) - f(2y) - f(2y + 1) + f(y) \Big| \leq 3\delta
\end{aligned}
$$

for every $x, y \in \mathbb{R}$. Next substituting $\frac{y}{3}$ for y in the last inequality, we get

$$\left| f\left(\frac{x}{2} + \frac{y}{2}\right) + f\left(\frac{x}{2} + \frac{y}{2} + 1\right) - f(x) \right.$$
$$\left. - f\left(\frac{2y}{3}\right) - f\left(\frac{2y}{3} + 1\right) + f\left(\frac{y}{3}\right) \right| \le 3\delta. \tag{22.6}$$

Defining $g, h : \mathbb{R} \to \mathbb{R}$ as

$$\left. \begin{array}{l} g(x) = f\left(\dfrac{2x}{3}\right) + f\left(\dfrac{2x}{3} + 1\right) - f\left(\dfrac{x}{3}\right) \\[3mm] h(x) = f\left(\dfrac{x}{2}\right) + f\left(\dfrac{x}{2} + 1\right) \end{array} \right\} \tag{22.7}$$

for all $x, y \in \mathbb{R}$. From (22.6) and (22.7), we get

$$|h(x + y) - f(x) - g(y)| \le 3\delta \tag{22.8}$$

for all $x, y \in \mathbb{R}$. Putting $y = 0$ in (22.8), we get

$$|h(x) - f(x) - g(0)| \le 3\delta \tag{22.9}$$

for all $x \in \mathbb{R}$. Similarly letting $x = 0$ in (22.8), we have

$$|h(y) - f(0) - g(y)| \le 3\delta. \tag{22.10}$$

Next, we define

$$\left. \begin{array}{l} F(x) = f(x) - f(0) \\ G(x) = g(x) - g(0) \\ H(x) = h(x) - f(0) - g(0) \end{array} \right\} \tag{22.11}$$

for all $x \in \mathbb{R}$. Using (22.8), (22.9), (22.10) and (22.11), we get

$$\begin{aligned} |H(x + y) - H(x) - H(y)| &= |h(x + y) - h(x) - h(y) + f(0) + g(0)| \\ &\le |h(x + y) - f(x) - g(y)| \\ &\quad + |f(x) - h(x) + g(0)| \\ &\quad + |g(y) - h(y) + f(0)| \\ &\le 9\delta \end{aligned} \tag{22.12}$$

for all $x, y \in \mathbb{R}$.

Now using Hyers theorem (that is, Theorem 18.3), we get that

$$|H(x) - A(x)| \le 9\delta, \tag{22.13}$$

where $A : \mathbb{R} \to \mathbb{R}$ is a unique additive function. Now using (22.9), (22.11) and (22.12), we obtain

$$
\begin{aligned}
|f(x) - f(0) - A(x)| &\leq |f(x) + g(0) - h(x)| \\
&\quad + |h(x) - f(0) - g(0) - A(x)| \\
&\leq 12\delta
\end{aligned}
$$

for any $x \in \mathbb{R}$. Thus, we have the inequality (22.2) by putting $b = f(0)$ in the above inequality. $\square$

Remark 22.1. *Recently, Jun, Jung and Lee (2004) and Jung and Sahoo (2006) showed that the estimate 12δ in Theorem 22.1 can be improved to 9δ.*

22.3 Generalized Stability of Davison Equation

In this section, we prove a Hyers-Ulam-Rassias type stability result for the Davison functional equation. Theorem 22.2 of this section, which was initially proved by Jung and Sahoo (2006), is an improved version of the main results in Jung and Sahoo (1999, 2000a) because Theorem 22.2 gives improved bounds 9δ.

Let $\phi : \mathbb{R} \times \mathbb{R} \to [0, \infty)$ be a function with

$$
\Phi(x, y) := \sum_{i=0}^{\infty} 2^{-i} \varphi(2^i x, 2^i y) < \infty \tag{22.14}
$$

for all $x, y \in \mathbb{R}$, where we set

$$
\varphi(x, y) := \phi(6x - 2y, 2y) + \phi(3x - y, 4y) + \phi(3x - y, 4y + 1)
$$

for $x, y \in \mathbb{R}$.

Theorem 22.2. *If the function $f : \mathbb{R} \to \mathbb{R}$ satisfies the inequality*

$$
|f(xy) + f(x + y) - f(xy + x) - f(y)| \leq \phi(x, y) \tag{22.15}
$$

for all $x, y \in \mathbb{R}$, then there exists a unique additive function $A : \mathbb{R} \to \mathbb{R}$ such that

$$
|f(6x) - A(x) - f(0)| \leq \frac{1}{2} \Phi(x, -x) + \frac{1}{2} \Phi(x, 0) + \frac{1}{2} \Phi(2x, -x) \tag{22.16}
$$

for any $x \in \mathbb{R}$.

Proof. Let us define a function $g : \mathbb{R} \to \mathbb{R}$ by $g(x) = f(x) - f(0)$. Then, we see that $g(0) = 0$, and we get from (22.15) that

$$|g(xy) + g(x + y) - g(xy + x) - g(y)| \le \phi(x, y) \tag{22.17}$$

for any $x, y \in \mathbb{R}$. By (22.17), we have

$$\begin{aligned}
|g(xy) &+ g(x + y) + g(x + y + 1) - g(y) - g(xy + 2x) - g(y + 1)| \\
&\le |g(xy) + g(x + y) - g(xy + x) - g(y)| \\
&\quad + |g(xy + x) + g(x + y + 1) - g(xy + 2x) - g(y + 1)| \\
&\le \phi(x, y) + \phi(x, y + 1)
\end{aligned} \tag{22.18}$$

for all $x, y \in \mathbb{R}$. It follows from (22.17) and (22.18) that

$$\begin{aligned}
|g(x + 4y) &+ g(x + 4y + 1) - g(2x + 2y) - g(4y) - g(4y + 1) + g(2y)| \\
&\le |-g(4xy) - g(2x + 2y) + g(4xy + 2x) + g(2y)| \\
&\quad + |g(4xy) + g(x + 4y) + g(x + 4y + 1) \\
&\qquad - g(4y) - g(4xy + 2x) - g(4y + 1)| \\
&\le \phi(2x, 2y) + \{\phi(x, 4y) + \phi(x, 4y + 1)\}
\end{aligned}$$

for $x, y \in \mathbb{R}$. If we replace x by $x - y$ in the last inequality and if we replace x by $3x$ in the resulting one, then we have

$$\begin{aligned}
|g(3x + 3y) &+ g(3x + 3y + 1) - g(6x) - g(4y) - g(4y + 1) + g(2y)| \\
&\le \varphi(x, y)
\end{aligned} \tag{22.19}$$

for all $x, y \in \mathbb{R}$.

By (22.19), we obtain

$$\begin{aligned}
|g(1) - g(6x) - g(-4x) - g(-4x + 1) + g(-2x)| &\le \varphi(x, -x), \\
|g(3x) + g(3x + 1) - g(6x) - g(1)| &\le \varphi(x, 0), \\
|g(3x) + g(3x + 1) - g(12x) - g(-4x) \qquad\qquad & \\
- g(-4x + 1) + g(-2x)| &\le \varphi(2x, -x)
\end{aligned}$$

for $x, y \in \mathbb{R}$. From the above inequalities, we get

$$\begin{aligned}
|2g(6x) &- g(12x)| \\
&\le |-g(1) + g(6x) + g(-4x) + g(-4x + 1) - g(-2x)| \\
&\quad + |-g(3x) - g(3x + 1) + g(6x) + g(1)| \\
&\quad + |g(3x) + g(3x + 1) - g(12x) - g(-4x) - g(-4x + 1) + g(-2x)| \\
&\le \varphi(x, -x) + \varphi(x, 0) + \varphi(2x, -x)
\end{aligned} \tag{22.20}$$

for each $x \in \mathbb{R}$. Assume now that

$$|2^n g(6x) - g(2^n \cdot 6x)|$$
$$\leq \sum_{i=0}^{n-1} 2^{n-1-i} \left[\varphi(2^i x, -2^i x) + \varphi(2^i x, 0) + \varphi(2^{i+1} x, -2^i x) \right] \quad (22.21)$$

for some $n \in \mathbb{N}$ and for all $x \in \mathbb{R}$. By (22.20) and (22.21), we obtain

$$|2^{n+1} g(6x) - g(2^{n+1} 6x)|$$
$$\leq 2 |2^n g(6x) - g(2^n 6x)| + |2g(2^n 6x) - g(2^{n+1} 6x)|$$
$$\leq \sum_{i=0}^{n} 2^{n-i} \left[\varphi(2^i x, -2^i x) + \varphi(2^i x, 0) + \varphi(2^{i+1} x, -2^i x) \right],$$

which implies that the inequality (22.21) is true for all $n \in \mathbb{N}$.

In view of (22.14) and (22.21), we see

$$|2^{-m} g(2^m 6x) - 2^{-n} g(2^n 6x)|$$
$$\leq \sum_{i=m}^{n-1} 2^{-(i+1)} \left[\varphi(2^i x, -2^i x) + \varphi(2^i x, 0) + \varphi(2^{i+1} x, -2^i x) \right]$$
$$\to 0 \quad \text{as} \quad m \to \infty.$$

Hence, $\{2^{-n} g(2^n 6x)\}$ is a Cauchy sequence for each fixed $x \in R$. Therefore, we can define a function $A : \mathbb{R} \to \mathbb{R}$ by

$$A(x) = \lim_{n \to \infty} 2^{-n} g(2^n \cdot 6x). \quad (22.22)$$

By (22.19) again, we obtain the following inequalities

$$| - g(3y) - g(3y + 1) + g(6x + 6y) + g(-4x) + g(-4x + 1) - g(-2x)|$$
$$\leq \varphi(x + y, -x),$$
$$| - g(3x) - g(3x + 1) + g(6x + 6y) + g(-4y) + g(-4y + 1) - g(-2y)|$$
$$\leq \varphi(x + y, -y),$$
$$|g(3y) + g(3y + 1) - g(12y) - g(-4y) - g(-4y + 1) + g(-2y)|$$
$$\leq \varphi(2y, -y),$$
$$|g(3x) + g(3x + 1) - g(12x) - g(-4x) - g(-4x + 1) + g(-2x)|$$
$$\leq \varphi(2x, -x)$$

for all $x, y \in \mathbb{R}$. From the last four inequalities, we have

$$|2g(6x + 6y) - g(12x) - g(12y)|$$
$$\leq \varphi(x + y, -x) + \varphi(x + y, -y) + \varphi(2y, -y) + \varphi(2x, -x)$$

for any $x, y \in \mathbb{R}$. If we replace x and y by $2^n x$ and $2^n y$ in the last inequality and divide both sides by 2^n, then (22.14) and (22.22) yield that A is an additive function. By (22.14), (22.21) and (22.22), we conclude that the inequality (22.16) is true.

Now, let $A' : \mathbb{R} \to \mathbb{R}$ be another additive function that satisfies the inequality (22.16) for all $x \in \mathbb{R}$. Then, by (22.14) and (22.16), we have

$$
\begin{aligned}
|A(x) - A'(x)| \\
&= 2^{-n} |A(2^n x) - A'(2^n x)| \\
&\leq 2^{-n} |-f(2^n \cdot 6x) + A(2^n x) + f(0)| \\
&\quad + 2^{-n} |f(2^n \cdot 6x) - A'(2^n x) - f(0)| \\
&\leq \sum_{i=n}^{\infty} 2^{-i} \left[\varphi(2^i x, -2^i x) + \varphi(2^i x, 0) + \varphi(2^{i+1} x, -2^i x) \right] \\
&\to 0 \text{ as } n \to \infty,
\end{aligned}
$$

for all $x \in \mathbb{R}$, which implies the uniqueness of A. $\qquad \square$

Remark 22.2. If we let $\phi(x, y) = \delta \geq 0$ in our Theorem 22.2, then we obtain an improved version of Theorem 2 in Jung and Sahoo (2000a) (or we get the same result as that of Jun, Jung and Lee (2004) that is, there exists a unique additive function $A : \mathbb{R} \to \mathbb{R}$ such that

$$|f(x) - A(x) - f(0)| \leq 9\delta$$

for all $x \in \mathbb{R}$.

22.4 Concluding Remarks

The Hyers-Ulam stability of the Davison functional equation was first treated by Jung and Sahoo (1999). Let E_1 be a normed algebra with a unit element 1 and let E_2 be a Banach space. Let $\phi : E_1 \times E_1 \to [0, \infty)$ be a function with

$$\Phi(x, y) := \sum_{n=1}^{\infty} 2^{-n} [\varphi(2^{n-1} x, 2^{n-1} y) + \varphi(2^{n-1} x, 0) + \varphi(0, 2^{n-1} y)] < \infty$$

for all $x, y \in E_1$, where the function $\varphi : E_1 \times E_1 \to [0, \infty)$ is defined by

$$\varphi(x,y) := \phi\left(\frac{x}{2} - \frac{y}{6}, \frac{2}{3}y\right) + \phi\left(\frac{x}{2} - \frac{y}{6}, \frac{2}{3}y + 1\right) + \phi\left(x - \frac{y}{3}, \frac{y}{3}\right)$$

for any $x, y \in E_1$.

Jung and Sahoo (1999) proved the following theorem.

Theorem 22.3. *If a function $f : E_1 \to E_2$ satisfies the inequality*

$$\|f(xy) + f(x + y) - f(xy + x) - f(y)\| \leq \phi(x, y) \tag{22.23}$$

for all $x, y \in E_1$, then there exists a unique additive map $A : E_1 \to E_2$ such that

$$\|f(x) - A(x) - f(0)\| \leq \varphi(x, 0) + \Phi(x, x) \tag{22.24}$$

for all $x \in E_1$.

As a corollary, one can obtain the result:. If $f : E_1 \to E_2$ satisfies the inequality

$$\|f(xy) + f(x + y) - f(xy + x) - f(y)\| \leq \delta$$

for all $x, y \in E_1$, then there exists a unique additive map $A : E_1 \to E_2$ such that $\|f(x) - A(x) - f(0)\| \leq 12\delta$ for every $x \in E_1$. This result was established by Jung and Sahoo (2000) concerning the Hyers-Ulam stability of the Davison functional equation for a class of functions from a field (or a commutative algebra) of characteristics different from 2 and 3 into a Banach space. Jun, Jung and Lee (2004) have also studied the Hyers-Ulam stability of the Davison functional equation and improved the estimate obtained by Jung and Sahoo (1999). They showed that the estimate 12δ could be improved to 9δ.

Y.-H. Kim (2002) using mostly a similar argument of the proof of Theorem 22.3 due to Jung and Sahoo (1999) obtained the following result.

Theorem 22.4. *If a function $f : E_1 \to E_2$ satisfies the inequality*

$$\|f(xy) + f(x + y) - f(xy + x) - f(y)\| \leq \phi(x, y) \tag{22.25}$$

for all $x, y \in E_1$, then there exists a unique additive map $A : E_1 \to E_2$ such that

$$\|h(x) - A(x)\| \leq \|f(0)\| + \|g(0)\| + \Phi(x, x)$$
$$\|f(x) - A(x)\| \leq \|f(0)\| + 2\|g(0)\| + \varphi(x, 0) + \Phi(x, x)$$

$$\|g(x) - A(x)\| \leq 2\|f(0)\| + \|g(0)\| + \varphi(0, x) + \Phi(x, x)$$

for all $x \in E_1$, where the functions $g, h : E_1 \to E_2$ are defined by

$$g(x) = f(2x/3) + f(2x/3 + 1) - f(x/3),$$
$$h(x) = f(x/2) + f(x/2 + 1).$$

Jun, Jung and Lee (2004) proved the following result. Suppose F is a ring with the unit element. Let E be a Banach space. Further, let $\varphi : F \times F \to [0, \infty)$ satisfy the condition

$$\sum_{n=1}^{\infty} 2^{-n} \varphi(2^{n-1} x, 2^{n-1} y + z) < \infty \quad \text{for all } x, y, z \in F.$$

Theorem 22.5. *If a function $f : F \to E$ satisfies the inequality*

$$\|f(xy) + f(x + y) - f(xy + x) - f(y)\| \leq \varphi(x, y)$$

for all $x, y \in F$, then there exists a unique additive function $A : F \to E$ such that

$$\|f(6x) - A(x) - f(0)\| \leq \sum_{n=0}^{\infty} \frac{M(2^n x)}{2^n}$$

for all $x \in F$, where

$$M(x) = \frac{1}{2} \big[\varphi(4x, -4x) + \varphi(4x, -4x + 1) + \varphi(8x, -2x)$$
$$+ \varphi(3x, 0) + \varphi(3x, 1) + \varphi(6x, 0) + \varphi(7x, -x)$$
$$+ \varphi(7x, -4x + 1) + \varphi(14x, 2x) \big].$$

Wang (2002) studied the approximate remainder

$$\phi(x, y) = f(xy) + f(x + y) - f(xy + x) - f(y)$$

for function $f : G \to E$, where G is an algebra over the rational number field with a unit element e and E is a real or complex Hausdorff topological vector space. Let

$$\varphi(x, y) = \phi\left(\frac{x}{2} - \frac{y}{6}, \frac{2}{3}y\right) + \phi\left(\frac{x}{2} - \frac{y}{6}, \frac{2}{3}y + e\right) - \phi\left(x - \frac{y}{3}, \frac{y}{3}\right).$$

Wang found the conditions under which for some positive integer p and for all $x \in G$ there exists a limit $T(x) = \lim\limits_{n \to \infty} \frac{f(p^n x)}{p^n}$ and $T : G \to E$ is an additive function such that

$$T(x) - f(x) + f(0) = \lim_{n \to \infty} \frac{1}{p^n} \sum_{k=1}^{p^n - 1} \big[\varphi(x, kx) - \varphi(0, kx) \big].$$

Y.-H. Kim (2002) studied the Hyers-Ulam-Rassias stability of two Pexider-type Davison functional equation, namely,

$$f(xy) + f(x + y) = g(xy + x) + g(y)$$

and

$$f(xy) + g(x + y) = f(xy + x) + g(y)$$

for all $x, y \in E_1$ and $f, g : E_1 \to E_2$, where E_1 is a normed algebra with a unit element 1 and E_2 is a Banach space. Jun, Jung and Lee (2004) studied Hyers-Ulam-Rassias stability of the functional equation

$$f(xy) + g(x + y) = h(xy + x) + k(y)$$

for all $x, y \in E_1$ and $f, g, h, k : E_1 \to E_2$.

22.5   Exercises

1. Let $f : \mathbb{R} \to \mathbb{R}$ be a function. Prove or disprove that if the function

$$(x, y) \mapsto f(x + y + xy) - f(x + y) - f(xy) \qquad \forall\, x, y \in \mathbb{R}$$

is bounded on $\mathbb{R}^2$, then there exists an additive function $A : \mathbb{R} \to \mathbb{R}$ such that $f - A$ is bounded on $\mathbb{R}$.

2. Let $f : \mathbb{R} \to \mathbb{R}$ be a function. Prove or disprove that if the function

$$(x, y) \mapsto f(x + y - xy) + f(x + xy) - 2f(x) - f(y) \qquad \forall\, x, y \in \mathbb{R}$$

is bounded on $\mathbb{R}^2$, then there exists an additive function $A : \mathbb{R} \to \mathbb{R}$ such that $f - A$ is bounded on $\mathbb{R}$.

3. Let $f : \mathbb{R} \to \mathbb{R}$ be a function. Prove or disprove that if the function

$$(x, y) \mapsto f(x + y + xy) + f(x + y) - 2f(x) - 2f(y) - f(xy) \qquad \forall\, x, y \in \mathbb{R}$$

is bounded on $\mathbb{R}^2$, then there exists an additive function $A : \mathbb{R} \to \mathbb{R}$ such that $f - A$ is bounded on $\mathbb{R}$.

4. Let $f : \mathbb{R} \to \mathbb{R}$ be a function. Prove or disprove that if the function

$$(x, y) \mapsto f(x - xy) + f(xy) - f(x) \qquad \forall\, x, y \in \mathbb{R}$$

is bounded on $\mathbb{R}^2$, then there exists an additive function $A : \mathbb{R} \to \mathbb{R}$ such that $f - A$ is bounded on $\mathbb{R}$.

Chapter 23

Stability of Hosszú Functional Equation

23.1 Introduction

In this chapter, we prove the stability of the Hosszú functional equation and a functional equation that generalizes the Hosszú functional equation. In 1993, Kannappan and Sahoo determined the general solution $f, g, h, k : \mathbb{R} \to \mathbb{R}$ of the functional equation

$$f(x + y - \alpha xy) + g(xy) = h(x) + k(y) \tag{23.1}$$

for all $x, y \in \mathbb{R}$ (see also Ebanks, Kannappan and Sahoo (1992b)). Here α is an a priori chosen parameter. If $\alpha = 1$, then (23.1) is a pexiderized version of the Hosszú functional equation, namely,

$$f(x + y - xy) + f(xy) = f(x) + f(y).$$

If $\alpha = 0$, then (23.1) reduces to

$$f(x + y) + g(xy) = h(x) + k(y).$$

This functional equation was studied by Kannappan and Sahoo (1993) to characterize Cauchy differences that depend on the product of arguments.

In Chapter 18, we proved a result due to Hyers (1941). We reproduce that theorem for the sake of convenience.

Theorem 23.1. *If $f : \mathbb{R} \to \mathbb{R}$ is a real function satisfying*

$$|f(x + y) - f(x) - f(y)| \leq \delta$$

for some $\delta \geq 0$ and for all $x, y \in \mathbb{R}$, then there exists a unique additive function $A : \mathbb{R} \to \mathbb{R}$ such that

$$|f(x) - A(x)| \leq \delta$$

for any $x \in \mathbb{R}$.

We will use this theorem to establish a theorem concerning the Hyers-Ulam stability of the Hosszú functional equation in the next section.

23.2 Stability of Hosszú Functional Equation

The next result was established by Găvruta (2000) concerning the Hyers-Ulam stability of Hosszú's functional equation (cf. Losonczi (1996)).

Theorem 23.2. *Suppose that $f : \mathbb{R} \to \mathbb{R}$ satisfies the inequality*

$$|f(x + y - xy) + f(xy) - f(x) - f(y)| \le \delta \qquad (23.2)$$

for all $x, y \in \mathbb{R}$ with $\delta \ge 0$. Then there exists a unique additive function $A : \mathbb{R} \to \mathbb{R}$ such that

$$|f(x) - A(x) + A(1) - f(1)| \le 9\,\delta \qquad (23.3)$$

for all $x, y \in \mathbb{R}$.

Proof. We consider $x \in \mathbb{R} \setminus \{0\}$ and substitute $y = \frac{1}{x}$ in (23.2) to get

$$\left| f\left(x + \frac{1}{x} - 1\right) + f(1) - f(x) - f\left(\frac{1}{x}\right) \right| \le \delta. \qquad (23.4)$$

Next, replacing x by xy and y by $\frac{1}{x}$ in (23.2), we obtain

$$\left| f(y) - f(xy) - f\left(\frac{1}{x}\right) + f\left(xy + \frac{1}{x} - y\right) \right| \le \delta. \qquad (23.5)$$

From (23.2), (23.4) and (23.5), we get

$$2\left| f(x + y - xy) + f\left(xy + \frac{1}{x} - y\right) - f\left(x + \frac{1}{x} - 1\right) - f(1) \right|$$
$$\le |f(x + y - xy) - f(x) - f(y) + f(xy)|$$
$$+ \left| f(y) - f\left(xy + \frac{1}{x} - y\right) - f(xy) - f\left(\frac{1}{x}\right) \right|$$
$$+ \left| f\left(\frac{1}{x}\right) + f(x) - f(1) - f\left(x + \frac{1}{x} - 1\right) \right|$$
$$\le 3\,\delta.$$

If we set

$$x + y - xy = z, \qquad xy + \frac{1}{x} - y = t,$$

it follows that

$$x + \frac{1}{x} = z + t.$$

Hence, for $z + t > 2$, there exists a solution

$$x > 1 \quad \text{and} \quad y = \frac{z - x}{1 - x}.$$

Therefore

$$|f(z) + f(t) - f(z + t - 1) - f(1)| \leq 3\delta \qquad (23.6)$$

for $z + t > 2$. Let us set

$$g(z) = f(z + 1) - f(1)$$

for all $z \in \mathbb{R}$. From (23.6), we obtain

$$|g(z) + g(t) - g(z + t)| \leq 3\delta \qquad (23.7)$$

for all $z, t \in \mathbb{R}$ with $z + t > 0$.

If $z + t \leq 0$, we take $s \in \mathbb{R}$ such that

$$z + t + s > 0 \qquad \text{and} \qquad t + s > 0.$$

Thus

$$|g(z + t) + g(s) - g(z + t + s)| \leq 3\delta,$$
$$|g(t + s) - g(t) - g(s)| \leq 3\delta,$$
$$|g(z + t + s) - g(z) - g(t + s)| \leq 3\delta.$$

Hence

$$|g(z + t) - g(z) - g(t)| \leq 9\delta$$

for all $z, t \in \mathbb{R}$. From a result of Hyers, it follows that there exists a unique additive mapping $A : \mathbb{R} \to \mathbb{R}$ such that

$$|g(x) - A(x)| \leq 9\delta$$

for all $x \in \mathbb{R}$. Since $g(x) = f(x + 1) - f(1)$ for all $x \in \mathbb{R}$, therefore we have

$$\begin{aligned} |g(x) - A(x)| &= |f(x + 1) - f(1) - A(x)| \\ &= |f(x) - f(1) - A(x - 1)| \\ &= |f(x) - f(1) - A(x) + A(1)|, \end{aligned}$$

and hence

$$|f(x) - A(x) + A(1) - f(1)| \leq 9\delta$$

for all $x \in \mathbb{R}$. Now the proof of the theorem is complete. $\qquad \square$

Remark 23.1. *Volkmann (1998) has proved that the estimate given by Gávruta (2000) in Theorem 23.2 can be improved from 9δ to 4δ.*

Remark 23.2. *The proof of Theorem 23.2 works exactly the same way if one replaces the range of the function f by a Banach space.*

The following result was established by Jung and Sahoo (1999) (see also Jung and Sahoo (2000a)) and was proved in Chapter 22 (see the proof of Theorem 22.1). Here we reproduce the theorem for the benefit of the reader. Note that the estimate in Theorem 22.1 can be improved to $9\,\delta$ from 12δ.

Theorem 23.3. *If the function $f : \mathbb{R} \to \mathbb{R}$ satisfies the inequality*

$$|f(xy) + f(x + y) - f(xy + x) - f(y)| \leq \delta$$

for all $x, y \in \mathbb{R}$ and for some $\delta \geq 0$, then there exists a unique additive function $A : \mathbb{R} \to \mathbb{R}$ and a real constant $b := f(0)$ such that

$$|f(x) - A(x) - b| \leq 9\delta$$

for all $x \in \mathbb{R}$.

In the next section, we prove the Hyers-Ulam stability of the equation (23.1) by using Theorem 23.1 due to Hyers (1941), Theorem 23.2 due to Gávruta (2001), Remark 23.1 and Theorem 23.3 due to Jung and Sahoo (1999).

23.3 Stability of Pexiderized Hosszú Functional Equation

In the following theorem we show the Hyers-Ulam stability of the functional equation (23.1) for the case $\alpha = 0$. For the sake of convenience, we shall write the functional equation (23.1) as $f(x + y) - g(xy) = h(x) + k(y)$.

Theorem 23.4. *If functions $f, g, h, k : \mathbb{R} \to \mathbb{R}$ satisfy the functional inequality*

$$|f(x + y) - g(xy) - h(x) - k(y)| \leq \delta \tag{23.8}$$

for some $\delta \geq 0$ and for all $x, y \in \mathbb{R}$, then there exist unique additive functions $A_1, A_2 : \mathbb{R} \to \mathbb{R}$ such that for all $x \in \mathbb{R}$

$$|g(x) - 2\, A_1(x) - \delta_2| \leq 108\,\delta,$$

$$|f(x) - A_1(x^2) - A_2(x) - \delta_1| \le \frac{117}{2}\,\delta,$$

$$|h(x) - A_1(x^2) - A_2(x) - \delta_3| \le \frac{119}{2}\,\delta,$$

$$|k(x) - A_1(x^2) - A_2(x) - \delta_4| \le \frac{119}{2}\,\delta,$$

where $\delta_1, \delta_2, \delta_3, \delta_4$ *are constants in* $\mathbb{R}$ *satisfying* $|\delta_1 - \delta_2 - \delta_3 - \delta_4| \le \frac{\delta}{2}$.

Proof. Letting $x = 0$ in (23.8), we get

$$|f(y) - k(y) - b_1| \le \delta \tag{23.9}$$

where $b_1 = g(0) + h(0)$. Putting $y = 0$ in (23.8), we have

$$|f(x) - h(x) - b_2| \le \delta, \tag{23.10}$$

where $b_2 = g(0) + k(0)$. Finally, letting $x = 0$ and $y = 0$ in (23.8), we obtain

$$|f(0) - g(0) - h(0) - k(0)| \le \delta. \tag{23.11}$$

Using (23.8), (23.9) and (23.10), we see that

$$\begin{aligned}
|f(x+y) &- g(xy) - f(x) - f(y) + b_1 + b_2| \\
&= |\,f(x+y) - g(xy) - h(x) - k(y) \\
&\quad + h(x) - f(x) + b_2 + k(y) - f(y) + b_1\,| \\
&\le |\,f(x+y) - g(xy) - h(x) - k(y)\,| \\
&\quad + |\,h(x) - f(x) + b_2\,| + |\,k(y) - f(y) + b_1\,| \\
&\le 3\,\delta.
\end{aligned}$$

Hence, we have

$$|f(x+y) - g(xy) - f(x) - f(y) + b_1 + b_2| \le 3\,\delta \tag{23.12}$$

for all $x, y \in \mathbb{R}$. Defining

$$\phi(x) = f(x) - b_1 - b_2 \tag{23.13}$$

and using (23.13) in inequality (23.12), we have

$$|\phi(x+y) - g(xy) - \phi(x) - \phi(y)| \le 3\,\delta \tag{23.14}$$

for all $x, y \in \mathbb{R}$. From (23.14), we see that

$$|\phi(x+y+z) - g(xz + yz) - \phi(x+y) - \phi(z)| \le 3\,\delta, \tag{23.15}$$

$$|\phi(x+y+z) - g(xy+xz) - \phi(x) - \phi(y+z)| \leq 3\,\delta, \qquad (23.16)$$

$$|\phi(y+z) - g(yz) - \phi(y) - \phi(z)| \leq 3\,\delta. \qquad (23.17)$$

Now using (23.14), (23.15), (23.16) and (23.17), we obtain

$$
\begin{aligned}
&|g(xy+xz) + g(yz) - g(xy) - g(xz+yz)| \\
&\quad = |\phi(x+y) - g(xy) - \phi(x) - \phi(y)| \\
&\qquad + |\phi(x+y+z) - g(xz+yz) - \phi(x+y) - \phi(z)| \\
&\qquad + |\phi(x) + \phi(y+z) + g(xy+xz) - \phi(x+y+z)| \\
&\qquad + |\phi(y) + \phi(z) + g(yz) - \phi(y+z)| \\
&\quad \leq 12\,\delta
\end{aligned}
$$

which is

$$|g(xy+xz) + g(yz) - g(xy) - g(xz+yz)| \leq 12\,\delta \qquad (23.18)$$

for all $x, y, z \in \mathbb{R}$. Letting $z = 1$ in (23.18), we have

$$|g(xy+x) + g(y) - g(xy) - g(x+y)| \leq 12\,\delta$$

for all $x, y \in \mathbb{R}$. From Theorem 23.3, we see that

$$|g(x) - A(x) - \delta_2| \leq 108\,\delta, \qquad (23.19)$$

where $A : \mathbb{R} \to \mathbb{R}$ is a unique additive map and $\delta_2 = g(0)$. Now writing $A = 2A_1$ in (23.19), where A_1 is an additive map and uniquely determined by A, we have

$$|g(x) - 2A_1(x) - \delta_2| \leq 108\,\delta \qquad (23.20)$$

for all $x \in \mathbb{R}$.

Letting $y = -x$ in (23.14), we have

$$|\phi(0) - g(-x^2) - \phi(x) - \phi(-x)| \leq 3\,\delta \qquad (23.21)$$

for all $x \in \mathbb{R}$. Now using (23.20) and (23.21), we see that

$$
\begin{aligned}
&|\phi(x) + \phi(-x) - 2A_1(x^2) + g(0) - \phi(0)| \\
&\quad \leq |\phi(x) + \phi(-x) + g(-x^2) - \phi(0)| + |g(-x^2) - 2A_1(-x^2) - g(0)| \\
&\quad \leq 111\,\delta.
\end{aligned}
$$

Thus, we have

$$\left|\phi(x) + \phi(-x) - 2A_1(x^2) + g(0) - \phi(0)\right| \leq 111\,\delta \qquad (23.22)$$

for all $x \in \mathbb{R}$.

Replacing x by $-x$ and y by $-y$ in (23.14), we obtain

$$|\phi(-(x+y)) - g(xy) - \phi(-x) - \phi(-y)| \leq 3\,\delta \qquad (23.23)$$

for all $x, y \in \mathbb{R}$. From (23.14) and (23.23), we observe that

$$\begin{aligned}
|\phi(x+y) - \phi(-(x+y)) &- \phi(x) + \phi(-x) - \phi(y) + \phi(-y)| \\
&\leq |\phi(x+y) - g(xy) - \phi(x) - \phi(y)| \\
&\quad + |\phi(-x) + \phi(-y) + g(xy) - \phi(-(x+y))| \\
&\leq 6\,\delta
\end{aligned}$$

for all $x, y \in \mathbb{R}$. Defining $F : \mathbb{R} \to \mathbb{R}$ by

$$F(x) = \phi(x) - \phi(-x) \qquad \forall x \in \mathbb{R} \qquad (23.24)$$

and using this F in the last inequality, we have the functional inequality

$$|F(x+y) - F(x) - F(y)| \leq 6\,\delta$$

for all $x, y \in \mathbb{R}$. By Theorem 23.1, there is a unique additive function $A_0 : \mathbb{R} \to \mathbb{R}$ such that

$$|F(x) - A_0(x)| \leq 6\,\delta \qquad (23.25)$$

for all $x \in \mathbb{R}$. Writing $A_0 = 2A_2$ in (23.25), where $A_2 : \mathbb{R} \to \mathbb{R}$ is an additive map and then using (23.24), we have

$$|\phi(x) - \phi(-x) - 2\,A_2(x)| \leq 6\,\delta \qquad (23.26)$$

for all $x \in \mathbb{R}$.

Using (23.22) and (23.26), we see that

$$\begin{aligned}
|2\,\phi(x) - 2A_1(x^2) &- 2A_2(x) + g(0) - \phi(0)| \\
&\leq |\phi(x) + \phi(-x) - 2A_1(x^2) + g(0) - \phi(0)| \\
&\leq + |\phi(x) - \phi(-x) - 2\,A_2(x)| \\
&\leq 117\,\delta
\end{aligned}$$

for all $x \in \mathbb{R}$. Hence

$$\left|\phi(x) - A_1(x^2) - A_2(x) + \frac{1}{2}[g(0) - \phi(0)]\right| \leq \frac{117}{2}\,\delta. \qquad (23.27)$$

Since $\phi(x) = f(x) - b_1 - b_2 = f(x) - 2g(0) - h(0) - k(0)$, from the inequality (23.27), we have

$$\left|f(x) - A_1(x^2) - A_2(x) - \delta_1\right| \leq \frac{117}{2}\,\delta, \qquad (23.28)$$

where $\delta_1 = \frac{1}{2}[f(0) + g(0) + h(0) + k(0)]$. We can easily prove the uniqueness of A_2 satisfying the inequality (23.28).

Next, using (23.9) and (23.28), we see that

$$|k(x) - A_1(x^2) - A_2(x) + b_1 - \delta_1|$$
$$\leq |k(x) - f(x) + b_1| + |f(x) - A_1(x^2) - A_2(x) - \delta_1|$$
$$\leq \frac{119}{2}\delta$$

for all $x \in \mathbb{R}$. Hence

$$\left|k(x) - A_1(x^2) - A_2(x) - \delta_4\right| \leq \frac{119}{2}\delta,$$

where $\delta_4 = \frac{1}{2}[f(0) - g(0) - h(0) + k(0)]$.

Finally, using (23.10) and (23.28), we see that

$$|h(x) - A_1(x^2) - A_2(x) + b_2 - \delta_1|$$
$$\leq |h(x) - f(x) + b_2| + |f(x) - A_1(x^2) - A_2(x) - \delta_1|$$
$$\leq \frac{119}{2}\delta$$

for all $x \in \mathbb{R}$. Hence

$$\left|h(x) - A_1(x^2) - A_2(x) - \delta_3\right| \leq \frac{119}{2}\delta,$$

where $\delta_3 = \frac{1}{2}[f(0) - g(0) + h(0) - k(0)]$.

In view of the inequality (23.11), it is easy to check that the constants $\delta_1, \delta_2, \delta_3, \delta_4$ satisfy $|\delta_1 - \delta_2 - \delta_3 - \delta_4| = \frac{1}{2}|f(0) - g(0) - h(0) - k(0)| \leq \frac{\delta}{2}$. Now the proof of the theorem is complete. $\qquad\square$

In the next theorem we treat the stability of the functional equation (23.1) when the parameter $\alpha \neq 0$.

Theorem 23.5. *If functions $f, g, h, k : \mathbb{R} \to \mathbb{R}$ satisfy the functional inequality*

$$|f(x + y - \alpha xy) + g(xy) - h(x) - k(y)| \leq \delta \qquad (23.29)$$

for some $\delta \geq 0$ and for all $x, y \in \mathbb{R}$, then there exists unique additive function $A : \mathbb{R} \to \mathbb{R}$ such that for all $x \in \mathbb{R}$

$$|f(x) - A(\alpha x) - a| \leq 24\,\delta,$$

$$|h(x) - A(\alpha x) - a - b_1| \le 25\,\delta,$$
$$|k(x) - A(\alpha x) - a - b_2| \le 25\,\delta,$$
$$|g(x) - A(\alpha^2 x) - a - b_1 - b_2| \le 27\,\delta,$$

where $a = f\left(\frac{1}{\alpha}\right) - A(1)$, $b_1 = g(0) - k(0)$, *and* $b_2 = g(0) - h(0)$.

Proof. Letting $y = 0$ in (23.29), we get

$$|f(x) - h(x) + b_1| \le \delta \qquad (23.30)$$

where $b_1 = g(0) - k(0)$. Putting $x = 0$ in (23.29), we have

$$|f(y) - k(y) + b_2| \le \delta \qquad (23.31)$$

where $b_2 = g(0) - h(0)$. Using (23.29), (23.30) and (23.31), we see that

$$
\begin{aligned}
|f(x + y &- \alpha x y) + g(xy) - f(x) - f(y) - b_1 - b_2| \\
&= |\, f(x + y - \alpha x y) + g(xy) - h(x) - k(y) \\
&\quad\; +\ h(x) - f(x) - b_1 + k(y) - f(y) - b_2\,| \\
&\le |\, f(x + y - \alpha x y) + g(xy) - h(x) - k(y)\,| \\
&\quad\; +\ |\,h(x) - f(x) - b_1\,| +\ |\,k(y) - f(y) - b_2\,| \\
&\le 3\,\delta.
\end{aligned}
$$

Hence

$$|f(x + y - \alpha x y) + g(xy) - f(x) - f(y) - b_1 - b_2| \le 3\,\delta \qquad (23.32)$$

for all $x, y \in \mathbb{R}$. Since $\alpha \ne 0$, substituting $y = \frac{1}{\alpha}$ in (23.32), we obtain

$$\left| g\left(\frac{x}{\alpha}\right) - f(x) - b_1 - b_2 \right| \le 3\,\delta \qquad (23.33)$$

for all $x \in \mathbb{R}$. Now replacing x by αx in (23.33), we have

$$|g(x) - f(\alpha x) - b_1 - b_2| \le 3\,\delta \qquad (23.34)$$

for all $x \in \mathbb{R}$. From (23.32) and (23.34), we have

$$
\begin{aligned}
|f(x + y &- \alpha x y) + f(\alpha x y) - f(x) - f(y)| \\
&= |\, f(x + y - \alpha x y) + g(xy) - f(x) - f(y) - b_1 - b_2 \\
&\quad\; +\ f(\alpha x y) - g(xy) + b_1 + b_2\,| \\
&\le |\, f(x + y - \alpha x y) + g(xy) - f(x) - f(y) - b_1 - b_2\,| \\
&\quad\; +\ |\, f(\alpha x y) - g(xy) + b_1 + b_2\,| \\
&\le 6\,\delta.
\end{aligned}
$$

Thus we have

$$|f(x + y - \alpha xy) + f(\alpha xy) - f(x) - f(y)| \leq 6\,\delta \qquad (23.35)$$

for all $x, y \in \mathbb{R}$. Replacing x by $\frac{x}{\alpha}$ and y by $\frac{y}{\alpha}$ in (23.35), we obtain

$$\left| f\left(\frac{x + y - xy}{\alpha}\right) + f\left(\frac{xy}{\alpha}\right) - f\left(\frac{x}{\alpha}\right) - f\left(\frac{y}{\alpha}\right) \right| \leq 6\,\delta. \qquad (23.36)$$

Defining $\psi : \mathbb{R} \to \mathbb{R}$ by

$$\psi(x) = f\left(\frac{x}{\alpha}\right) \qquad \forall x \in \mathbb{R} \qquad (23.37)$$

and using it in (23.36), we see that

$$|\psi(x + y - xy) + \psi(xy) - \psi(x) - \psi(y)| \leq 6\,\delta$$

for all $x, y \in \mathbb{R}$. Hence by Theorem 23.2 and Remark 23.1, there exists a unique additive map $A : \mathbb{R} \to \mathbb{R}$ such that for all $x \in \mathbb{R}$

$$|\psi(x) - A(x) - a| \leq 24\,\delta, \qquad (23.38)$$

where $a = \psi(1) - A(1)$. Thus from (23.37) and (23.38), we obtain

$$|f(x) - A(\alpha x) - a| \leq 24\,\delta, \qquad (23.39)$$

where $a = f\left(\frac{1}{\alpha}\right) - A(1)$.

From (23.30) and (23.39), we get

$$\begin{aligned}
|h(x) - A(\alpha x) &- a - b_1| \\
&\leq |h(x) - f(x) - b_1| + |f(x) - A(\alpha x) - a| \\
&\leq 25\,\delta
\end{aligned}$$

for all $x \in \mathbb{R}$. Similarly, from (23.31) and (23.39), we have

$$\begin{aligned}
|k(x) - A(\alpha x) &- a - b_2| \\
&\leq |k(x) - f(x) - b_2| + |f(x) - A(\alpha x) - a| \\
&\leq 25\,\delta.
\end{aligned}$$

Finally, from (23.34) and (23.39), we get

$$\begin{aligned}
|g(x) - A(\alpha^2 x) &- a - b_1 - b_2| \\
&\leq |g(x) - f(\alpha x) - b_1 - b_2| + |f(\alpha x) - A(\alpha^2 x) - a| \\
&\leq 27\,\delta.
\end{aligned}$$

The proof of the theorem is now complete. $\qquad\qquad\qquad \square$

23.4   Concluding Remarks

Borelli (1994) was the first person to study the Hyers-Ulam stability of the Hosszú functional equation on a field of real numbers $\mathbb{R}$. He proved the following theorem.

Theorem 23.6. *Let $f : \mathbb{R} \to \mathbb{R}$ be a function satisfying*

$$|f(x + y - xy) + f(xy) - f(x) - f(y)| \leq \delta \tag{23.40}$$

for all $x, y \in \mathbb{R}$ and for some $\delta > 0$. There exists an additive function $A : \mathbb{R} \to \mathbb{R}$ such that the difference $f - A$ is bounded if and only if the even part h of f satisfies

$$|h(x + y - xy) + h(xy) - h(x) - h(y)| \leq \epsilon$$

for some positive ϵ.

Let Y be a Banach space. Losonczi (1996) improved Borelli's result by proving the following theorem.

Theorem 23.7. *Let $f : \mathbb{R} \to Y$ be a function satisfying*

$$\|f(x + y - xy) + f(xy) - f(x) - f(y)\| \leq \delta \tag{23.41}$$

for all $x, y \in \mathbb{R}$ and for some $\delta > 0$. Then there exists a unique additive function $A : \mathbb{R} \to Y$ and a unique constant b in Y such that

$$\|f(x) - A(x) - b\| \leq 20\delta \quad \text{for all } x \in \mathbb{R}.$$

The constant 20 appearing in the above theorem was improved to 4 by Volkmann (1998). Găvruta (2000a) also improved the constant 20 to 9. However, his bound was not sharper than Volkmann's.

Tabor (1996) investigated the stability of the Hosszú functional equation on the unit interval $[0, 1]$. He proved that the Hosszú functional equation on the unit interval $[0, 1]$ is not stable. For every $\delta > 0$, one can find a function $f_\delta : U \to \mathbb{R}$ such that

$$|f_\delta(x + y - xy) + f_\delta(xy) - f_\delta(x) - f_\delta(y)| \leq \delta \quad \text{for } x, y \in U,$$

but which cannot be "approximated" by any solution of the Hosszú functional equation on the unit interval.

Jung and Sahoo (2002a) investigated the Hyers-Ulam stability of a generalized Hosszú functional equation, namely,

$$f(x + y - \alpha xy) + g(xy) = h(x) + k(y),$$

where f, g, h, k are functions of a real variable with values in a Banach space.

Let $\mathbb{E} = \{ x^2 \mid x \in \mathbb{E} \} \cup \{ -x^2 \mid x \in \mathbb{E} \}$. It is easy to see that $\mathbb{R}$ is an example of such a space. Let Y be a real Banach space. Jung (1996b) considered the following functional equation

$$f(x^2 - y^2 + rxy) = f(x^2) - f(y^2) + rf(xy)$$

for all $x, y \in \mathbb{E}$, where r is a real number, and established the following stability result.

Theorem 23.8. *Let $p < 1$, $r > 0$, $r \neq 1$ and $\delta > 0$ be given. Suppose $f : \mathbb{E} \to Y$ is a mapping such that*

$$\|f(x^2 - y^2 + rxy) - f(x^2) + f(y^2) - rf(xy)\| \leq \delta \left(\|x\|^p + \|y\|^p \right)$$

for all $x, y \in \mathbb{E}$. Then there exists a unique mapping $T : \mathbb{E} \to Y$ satisfying

$$\|T(x) - f(x)\| \leq \frac{2\delta}{|r - r^p|} \|x\|^p \qquad \forall\, x \in \mathbb{E}$$

and

$$T(x^2 - y^2 + rxy) = T(x^2) - T(y^2) + r\, T(xy)$$

for all $x, y \in \mathbb{E}$.

The above theorem of Jung (1996b) was generalized by G.H. Kim (1997).

Theorem 23.9. *Let $p < 1$, $r > 0$, $r \neq 1$ and $\delta > 0$ be given. Let $\varphi : \mathbb{E} \times \mathbb{E} \to [0, \infty)$ such that*

$$\Phi(x, y) := \sum_{k=0}^{\infty} r^{-k} \varphi \left(r^{\frac{k}{2}} x, r^{\frac{k}{2}} y \right) < \infty$$

for all $x, y \in \mathbb{E}$. Suppose $f : \mathbb{E} \to Y$ is a mapping such that

$$\|f(x^2 - y^2 + rxy) - f(x^2) + f(y^2) - rf(xy)\| \leq \varphi(x, y)$$

for all $x, y \in \mathbb{E}$. Then there exists a unique mapping $T : \mathbb{E} \to Y$ satisfying

$$\|T(x) - f(x)\| \leq \begin{cases} \frac{1}{r}\, \Phi(z, z), & \text{if } x = z^2, \\[2mm] \frac{1}{r}\, \Phi(z, -z), & \text{if } x = -z^2. \end{cases}$$

and

$$T(x^2 - y^2 + rxy) = T(x^2) - T(y^2) + r\,T(xy)$$

for all $x, y \in \mathbb{E}$.

At this time we are not aware of any result related to the stability of the Hosszú functional equation when the domain of the unknown function is other than $\mathbb{R}$. It would be nice to have a Hyers-Ulam type result on an arbitrary field $\mathbb{F}$. Also, there is no stability result when one replaces the inequality

$$|f(x + y - xy) + f(xy) - f(x) - f(y)| \leq \delta$$

by

$$|f(x + y - xy) + f(xy) - f(x) - f(y)| \leq \delta(|x|^p + |y|^p),$$

where $p \in \mathbb{R} \setminus \{1\}$.

23.5   Exercises

1. Let $\alpha \geq 0$ be a real number. Show that if $f : \mathbb{R} \to \mathbb{R}$ satisfies the functional inequality

$$|f(x + y) - f(x) - f(y)| \leq \delta \qquad (x + y \geq \alpha)$$

for some nonnegative δ, then there exists a unique solution $a : \mathbb{R} \to \mathbb{R}$ of the functional equation $a(x + y) = a(x) + a(y)$ for all $x, y \in \mathbb{R}$ such that

$$|f(x) - a(x)| \leq \delta$$

for all $x \in \mathbb{R}$.

2. Show that if $f : \mathbb{R} \to \mathbb{R}$ satisfies the functional inequality

$$|f(x + y - xy) + f(xy) - f(x) - f(y)| \leq \delta$$

for some nonnegative δ and for all $x, y \in \mathbb{R}$, then there exists a unique solution $a : \mathbb{R} \to \mathbb{R}$ of the functional equation $a(x + y) = a(x) + a(y)$ for all $x, y \in \mathbb{R}$ and real constant b such that

$$|f(x) - a(x) - b| \leq 4\,\delta$$

for all $x \in \mathbb{R}$.

3. Let $r \in (0, 1) \cup (1, \infty)$ be a real number. Show that if $f : \mathbb{R} \to \mathbb{R}$ satisfies the functional inequality

$$|f(x^2 - y^2 + rxy) - f(x^2) + f(y^2) - r\,f(xy)| \leq \delta$$

for some nonnegative δ and for all $x, y \in \mathbb{R}$, then there exists a unique solution $g : \mathbb{R} \to \mathbb{R}$ of the functional equation $g(x^2 - y^2 + rxy) = g(x^2) - g(y^2) + r\,g(xy)$ for all $x, y \in \mathbb{R}$ such that

$$|f(x) - g(x)| \leq \frac{\delta}{|r - r^p|}$$

for all $x \in \mathbb{R}$.

4. Prove or disprove, if $f : \mathbb{R} \to \mathbb{R}$ satisfies the functional inequality

$$|f(x + y + xy) - f(xy) - f(x) - f(y)| \leq \delta$$

for some nonnegative δ and for all $x, y \in \mathbb{R}$, then there exists a unique solution $a : \mathbb{R} \to \mathbb{R}$ of the functional equation $a(x + y) = a(x) + a(y)$ for all $x, y \in \mathbb{R}$ and real constant b such that $|f(x) - a(x) - b|$ is bounded on the set of reals $\mathbb{R}$.

5. If the function $f : \mathbb{R} \to \mathbb{R}$ satisfies the functional inequality

$$|f(xy + x) + f(y) - f(xy + y) - f(x)| \leq \delta$$

for some nonnegative δ and for all $x, y \in \mathbb{R}$, then show that there exists a unique additive function $A : \mathbb{R} \to \mathbb{R}$ such that

$$|f(x) - A(x) - b)| \leq 12\,\delta$$

for some constant $b \in \mathbb{R}$ and for all $x \in \mathbb{R}$.

Chapter 24

Stability of Abel Functional Equation

24.1   Introduction

In his 1823 manuscript, Abel had considered, among others, the functional equation

$$f(x + y) = g(xy) + h(x - y), \qquad x, y \in \mathbb{R}, \qquad (24.1)$$

where $f, g, h : \mathbb{R} \to \mathbb{R}$. In the same manuscript he gave the differentiable solutions of (24.1). Hilbert suggested in connection with his fifth problem, that, while the theory of differential equations provides elegant and powerful techniques for solving functional equations, the differentiability assumptions are not inherently required (see Aczél (1989)). Motivated by Hilbert's suggestion many researchers in functional equations have treated various functional equations without any regularity assumptions. The general solution of (24.1) was given by Aczél (1989) and also independently by Lajkó (1994) (see also Lajkó (1987)) without any regularity assumption. Chung, Ebanks, Ng and Sahoo (1994) determined the general solution of the Abel functional equation for $f, g, h : \mathbb{F} \to \mathbb{G}$, where $\mathbb{F}$ is a field belonging to a certain class, and $\mathbb{G}$ is an abelian group. When the general solution of a functional equation is known without any regularity assumptions, it is possible to obtain the Hyers-Ulam type stability result. In this chapter, we investigate the Hyers-Ulam stability of this Abel functional equation.

24.2   Stability Theorem

In the following theorem we present the Hyers-Ulam stability of the functional equation (24.1).

Theorem 24.1. *If functions $f, g, h : \mathbb{R} \to \mathbb{R}$ satisfy the functional in-equality*

$$|f(x+y) - g(xy) - h(x-y)| \leq \varepsilon \qquad (24.2)$$

for some $\varepsilon \geq 0$ and for all $x, y \in \mathbb{R}$, then there exists a unique additive function $A : \mathbb{R} \to \mathbb{R}$ such that

$$\left| f(x) - A\left(\frac{x^2}{4}\right) - f(0) \right| \leq 22\,\varepsilon,$$

$$|g(x) - A(x) - f(0) + h(0)| \leq 21\,\varepsilon,$$

$$\left| h(x) - A\left(\frac{x^2}{4}\right) - h(0) \right| \leq 22\,\varepsilon$$

for all $x \in \mathbb{R}$.

Proof. Letting $y = 0$ in (24.2), we get

$$|f(x) - g(0) - h(x)| \leq \varepsilon \qquad (24.3)$$

for all $x \in \mathbb{R}$. Putting $x = 0 = y$ in (24.2), we have

$$|f(0) - g(0) - h(0)| \leq \varepsilon. \qquad (24.4)$$

Next letting $x = y = \frac{t}{2}$ in (24.2), we obtain

$$\left| f(t) - g\left(\frac{t^2}{4}\right) - h(0) \right| \leq \varepsilon \qquad (24.5)$$

for all $t \in \mathbb{R}$. Finally substituting $x = -y = \frac{t}{2}$ in (24.2), we have

$$\left| f(0) - g\left(-\frac{t^2}{4}\right) - h(t) \right| \leq \varepsilon \qquad (24.6)$$

for all $t \in \mathbb{R}$. Using (24.2), (24.5) and (24.6), we see that

$$\left| g\left(\frac{(x+y)^2}{4}\right) - g(xy) + h(0) - f(0) + g\left(-\frac{(x-y)^2}{4}\right) \right|$$

$$= \left| f(x+y) - g(xy) - h(x-y) \right.$$

$$+ \; g\left(\frac{(x+y)^2}{4}\right) - f(x+y) + h(0)$$

$$+ \; g\left(-\frac{(x-y)^2}{4}\right) + h(x-y) - f(0) \Bigg|$$

$$\leq |f(x+y) - g(xy) - h(x-y)|$$

$$+ \left| g\left(\frac{(x+y)^2}{4}\right) - f(x+y) + h(0) \right|$$

$$+ \left| g\left(-\frac{(x-y)^2}{4}\right) + h(x-y) - f(0) \right|$$

$$\leq 3\,\varepsilon.$$

Hence, we have

$$\left| g\left(\frac{(x+y)^2}{4}\right) - g(xy) + h(0) - f(0) + g\left(-\frac{(x-y)^2}{4}\right) \right| \leq 3\,\varepsilon \quad (24.7)$$

for all $x, y \in \mathbb{R}$. If we substitute

$$xy = u \qquad \text{and} \qquad \frac{(x-y)^2}{4} = v$$

then

$$u + v = \frac{(x+y)^2}{4}$$

and hence $u + v \geq 0$.

Now substituting $xy = u$ and $\frac{(x-y)^2}{4} = v$ in (24.7), we obtain

$$|g\,(u+v) - g(u) + g(-v) + h(0) - f(0)| \leq 3\,\varepsilon \qquad (24.8)$$

for all $u, v \in \mathbb{R}$ with $u + v \geq 0$. Defining

$$G(u) = g(u) + h(0) - f(0) \qquad (24.9)$$

for all $u \in \mathbb{R}$, we see that the inequality (24.8) becomes

$$|G\,(u+v) - G(u) + G(-v)| \leq 3\,\varepsilon \qquad (24.10)$$

for all $u, v \in \mathbb{R}$ with $u + v \geq 0$. Letting $u = 0$ in (24.10), we obtain

$$|G\,(v) - G(0) + G(-v)| \leq 3\,\varepsilon \qquad (24.11)$$

for all $v \in \mathbb{R}$. Hence using (24.10), (24.11), (24.9) and (24.4), we see that

$$|G(u+v) - G(u) - G(v)|$$
$$= |\,G(u+v) - G(u) + G(-v) - G(-v) - G(v) + G(0) - G(0)|$$
$$\leq |G(u+v) - G(u) + G(-v)| + |G(-v) + G(v) - G(0)| + |G(0)|$$
$$\leq 7\,\varepsilon.$$

Thus, we have

$$|G(u+v) - G(u) - G(v)| \leq 7\,\varepsilon \qquad (24.12)$$

for all $u, v \in \mathbb{R}$ with $u + v \geq 0$.

Choose $u, v \in \mathbb{R}$ such that $u + v < 0$. Then there exists a real number $x \geq 0$ such that $x + u \geq 0$ and $x + u + v \geq 0$. By (24.12), we have

$$|G(x) + G(u + v) - G(x + u + v)| \leq 7\varepsilon, \qquad (24.13)$$

$$|G(x + u + v) - G(x + u) - G(v)| \leq 7\varepsilon \qquad (24.14)$$

and

$$|G(x + u) - G(x) - G(u)| \leq 7\varepsilon. \qquad (24.15)$$

Therefore, we have

$$
\begin{aligned}
|G(u + v) &+ G(u) - G(v)| \\
&= |G(x) + G(u + v) - G(x + u + v) + G(x + u + v) - G(x + u) \\
&\quad - G(v) + G(x + u) - G(x) - G(u)| \\
&\leq |G(x) + G(u + v) - G(x + u + v)| \\
&\quad + |G(x + u + v) - G(x + u) - G(v)| \\
&\quad + |G(x + u) - G(x) - G(u)| \\
&\leq 21\varepsilon.
\end{aligned}
$$

Hence

$$|G(u + v) - G(u) - G(v)| \leq 21\varepsilon \qquad (24.16)$$

for all $u, v \in \mathbb{R}$ with $u + v < 0$. From (24.12) and (24.16), we see that

$$|G(u + v) - G(u) - G(v)| \leq 21\varepsilon \qquad (24.17)$$

holds for all $u, v \in \mathbb{R}$. Using Theorem 18.3 and Remark 18.1, we obtain

$$|G(u) - A(u)| \leq 21\varepsilon \qquad (24.18)$$

for all $u \in \mathbb{R}$ and for some unique additive map $A : \mathbb{R} \to \mathbb{R}$. From (24.18) and (24.9), we have

$$|g(x) - A(x) - f(0) + h(0)| \leq 21\varepsilon \qquad (24.19)$$

for all $x \in \mathbb{R}$.

From (24.19) and (24.5), we get

$$
\begin{aligned}
\left| f(x) - A\left(\frac{x^2}{4}\right) - f(0) \right| \\
\leq \left| f(x) - g\left(\frac{x^2}{4}\right) - h(0) \right| + \left| g\left(\frac{x^2}{4}\right) - A\left(\frac{x^2}{4}\right) - f(0) + h(0) \right| \\
\leq 22\varepsilon.
\end{aligned}
$$

Hence

$$\left| f(x) - A\left(\frac{x^2}{4}\right) - f(0) \right| \le 22\,\varepsilon. \tag{24.20}$$

Similarly, from (24.19) and (24.6), we get

$$\left| h(x) - A\left(\frac{x^2}{4}\right) - h(0) \right|$$

$$\le \left| h(x) + g\left(-\frac{x^2}{4}\right) - f(0) \right|$$

$$\quad + \left| g\left(-\frac{x^2}{4}\right) - A\left(-\frac{x^2}{4}\right) - f(0) + h(0) \right|$$

$$\le 22\,\varepsilon.$$

Hence

$$\left| h(x) - A\left(\frac{x^2}{4}\right) - h(0) \right| \le 22\,\varepsilon \tag{24.21}$$

for all $x \in \mathbb{R}$. The proof of the theorem is now complete. □

24.3 Concluding Remarks

Smajdor (1999) studied the stability of the Abel functional equation

$$f(x + y) = g(xy) + h(x - y), \qquad \text{for all } 0 \le y \le x$$

when the unknown functions f, g, h are defined on the interval $[0, \infty)$ with range on an abelian semigroup with zero and satisfying cancellation law. Sahoo (2003) studied the Hyers-Ulam stability of this functional equation when the unknown functions f, g, h are defined on the set of real numbers $\mathbb{R}$ with range on an a Banach space.

Although the Abel functional equation has been studied on certain type of fields (see Chung et. al (1994)), to our best knowledge, the Hyers-Ulam stability of this equation has not been studied on any field other than the field of real numbers. Also, there is no Hyers-Ulam-Rassias type stability result regarding this functional equation.

Since this is the last occurrence of Concluding Remarks for this book, we point out some results on the stability of few functional equations that we have treated in this book but whose stability we have not considered. In Chapter 16, we studied several functional equations that originated from the Lagrange mean value theorem. There are only three papers related to the stability of the mean value type functional equations. Jung and Sahoo (2000b) studied the stability of the functional equation $f(x) - g(y) = (x - y)h(x + y)$. Let $\varphi : \mathbb{F} \times \mathbb{F} \to [0, \infty)$ be a symmetric function with the property

$$\varphi(-x, y) = \varphi(x, y)$$

for all x and y in $\mathbb{F}$, where $\mathbb{F}$ is a normed algebra with a unit element 1 (or a normed field of characteristic different from 2). We will use the following notation

$$\Phi(x, y) = 2\varphi\left(\frac{x + y}{2}, 0\right) + 4\varphi\left(0, \frac{x - y}{2}\right)$$
$$+ 2\varphi\left(\frac{x + y}{2}, \frac{x - y}{2}\right) + \varphi\left(\frac{x - y}{2}, \frac{x - y}{2}\right) + 3\varphi(0, 0)$$

for all $x, y \in \mathbb{F}$. Jung and Sahoo (2000b) proved the following theorem.

Theorem 24.2. *Let $\mathbb{F}$ be a normed algebra with a unit element 1 (or a normed field of characteristic different from 2). If maps $f, g, h : \mathbb{F} \to \mathbb{F}$ satisfy the functional inequality*

$$\|f(x) - g(y) - (x - y)h(x + y)\| \le \varphi(x, y) \qquad (24.22)$$

for all $x, y \in \mathbb{F}$, then there exist constants $a, b, c, d \in \mathbb{F}$ satifying the inequality $\|c - d\| \le \varphi(0, 0)$ such that

$$\|f(x) - ax^2 - bx - c\| \le \varphi(x, 0) + \|x\|\Phi(x, 1),$$
$$\|g(x) - ax^2 - bx - d\| \le \varphi(x, 0) + \|x\|\Phi(x, 1),$$
$$\|h(x) - ax - b\| \le \Phi(x, 1)$$

for all $x \in \mathbb{F}$.

As a corollary one obtains the following Hyers-Ulams type stability results for the mean value type functional equation

$$f(x) - g(y) = (x - y)h(x + y).$$

Corollary 24.1. *Let $\mathbb{F}$ be a normed algebra with a unit element 1 (or a normed field of characteristic different from 2). If maps $f, g, h : \mathbb{F} \to \mathbb{F}$ satisfy the functional inequality*

$$\|f(x) - g(y) - (x - y)h(x + y)\| \le \varepsilon$$

for all $x, y \in \mathbb{F}$, *then there exist constants* $a, b, c, d \in \mathbb{F}$ *with* $\|c - d\| \leq \varepsilon$ *such that*

$$\|f(x) - ax^2 - bx - c\| \leq \varepsilon + 12\varepsilon \|x\|,$$
$$\|g(x) - ax^2 - bx - d\| \leq \varepsilon + 12\varepsilon \|x\|,$$
$$\|h(x) - ax - b\| \leq 12\varepsilon$$

for all $x \in \mathbb{F}$

Jung and Sahoo (2001a) also studied the stability of the functional equation

$$f[x, y, z] = h(x + y + z) \tag{24.23}$$

for all $x, y, z \in \mathbb{R}$ with $x \neq y$, $y \neq z$ and $z \neq x$. They proved the following results.

Let G be an additive subgroup of $\mathbb{C}$ and let $\varphi : G^3 \rightarrow [0, \infty)$ be a control function. In the following theorem, Jung and Sahoo (2001a) investigated the stability of the functional equation (24.23) for cubic polynomials.

Theorem 24.3. *Let* $\alpha \in G \setminus \{0\}$ *and* $\beta \in G \setminus \{-\alpha, 0, \alpha\}$ *be fixed. If functions* $f, h : G \rightarrow \mathbb{C}$ *satisfy the inequality*

$$|(y - z)f(x) + (z - x)f(y) + (x - y)f(z)$$
$$- (x - z)(x - y)(y - z)h(x + y + z)| \leq \varphi(x, y, z) \tag{24.24}$$

for all $x, y, z \in G$, *then there exist constants* a, b, c, d *such that*

$$|f(x) - ax^3 - bx^2 - cx - d|$$
$$\leq \frac{|x^2 - \alpha^2|}{2|\beta||\beta^2 - \alpha^2|}\varphi(x, \beta, -\beta) + \frac{|x^2 - \beta^2|}{2|\alpha||\beta^2 - \alpha^2|}\varphi(x, \alpha, -\alpha)$$

for all $x \in G$, *and*

$$|h(x) - ax - b|$$
$$\leq \frac{|x^2 - \beta^2| + |\beta^2 - \alpha^2|}{2|\alpha||\beta^2 - \alpha^2||x^2 - \alpha^2|}\varphi(x, \alpha, -\alpha) + \frac{1}{2|\beta||\beta^2 - \alpha^2|}\varphi(x, \beta, -\beta)$$

for all $x \in G \setminus \{-\alpha, \alpha\}$. *Moreover, the constants* a, b, c, d *are explicitly given by*

$$a = \frac{f(\beta) - f(-\beta)}{2\beta(\beta^2 - \alpha^2)} - \frac{f(\alpha) - f(-\alpha)}{2\alpha(\beta^2 - \alpha^2)},$$
$$b = \frac{f(\beta) + f(-\beta)}{2(\beta^2 - \alpha^2)} - \frac{f(\alpha) + f(-\alpha)}{2(\beta^2 - \alpha^2)},$$

$$c = \frac{f(\alpha) - f(-\alpha)}{2\alpha(\beta^2 - \alpha^2)}\beta^2 - \frac{f(\beta) - f(-\beta)}{2\beta(\beta^2 - \alpha^2)}\alpha^2,$$

$$d = \frac{f(\alpha) + f(-\alpha)}{2(\beta^2 - \alpha^2)}\beta^2 - \frac{f(\beta) + f(-\beta)}{2(\beta^2 - \alpha^2)}\alpha^2.$$

Corollary 24.2. *Suppose that the control function* $\varphi : G^3 \to [0, \infty)$ *is given by*

$$\varphi(x, y, z) = \varepsilon |x - y|\,|y - z|\,|z - x|$$

for some given $\varepsilon > 0$. *If functions* $f, h : G \to \mathbb{C}$ *satisfy the inequality* (24.24) *for any* $x, y, z \in G$, *then there exist constants* a, b, c, d *such that*

$$|f(x) - ax^3 - bx^2 - cx - d| \leq \frac{2\varepsilon}{|\beta^2 - \alpha^2|}\,|x^2 - \alpha^2|\,|x^2 - \beta^2|$$

and

$$|h(x) - ax - b| \leq \varepsilon + \frac{2\varepsilon}{|\beta^2 - \alpha^2|}\,|x^2 - \beta^2|$$

for any x *of* G.

Given a control function $\psi : G^3 \to [0, \infty)$, we can also prove the Hyers-Ulam-Rassias stability of the functional equation (24.23) in the original setting:

Theorem 24.4. *Let* $\alpha \in G \backslash \{0\}$ *and* $\beta \in G \backslash \{-\alpha, 0, \alpha\}$ *be given. If functions* $f, h : G \to \mathbb{C}$ *satisfy the inequality*

$$|f[x, y, z] - h(x + y + z)| \leq \psi(x, y, z)$$

for all $x, y, z \in G$ *with* $x \neq y$, $y \neq z$ *and* $z \neq x$, *then there exist constants* a, b, c, d *such that*

$$|f(x) - ax^3 - bx^2 - cx - d|$$
$$\leq \frac{|x^2 - \alpha^2|\,|x^2 - \beta^2|}{|\beta^2 - \alpha^2|}\,(\psi(x, \alpha, -\alpha) + \psi(x, \beta, -\beta))$$

and

$$|h(x) - ax - b|$$
$$\leq \psi(x, \alpha, -\alpha) + \frac{|x^2 - \beta^2|}{|\beta^2 - \alpha^2|}\,(\psi(x, \alpha, -\alpha) + \psi(x, \beta, -\beta))$$

for all $x \in G$, *where* a, b, c, d *are explicitly given in Theorem 24.3.*

Another mean value type functional equation is

$$\frac{x\,f(y) - y\,f(x)}{x - y} = h(x + y) \quad \forall\, x, y \in \mathbb{K} \text{ with } x \neq y, \qquad (24.25)$$

where $\mathbb{K} \in \{\mathbb{R}, \mathbb{C}\}$. Jabłoński and Pekala (2003) proved the following Hyers-Ulam type stability result for the functional equation (24.25).

Theorem 24.5. *Suppose X is a vector space over $\mathbb{K}$. If the functions $f, h : \mathbb{K} \to X$ satisfy the inequality*

$$\left\| \frac{x\,f(y) - y\,f(x)}{x - y} - h(x + y) \right\| \leq \epsilon$$

for all $x, y \in \mathbb{K}$ with $x \neq y$, then there are constants $a, b \in X$ such that

$$\|f(x) - (ax + b)\| \leq 2\epsilon \quad \text{and} \quad \|h(x) - b\| \leq 3\epsilon.$$

In Chapter 17, we studied several functional equations that are used in the characterization of distance measures between discrete probability distributions. Recently, Kim and Sahoo (2010, 2011) have studied the stability of some of these functional equations on groups G. These equations are:

$$f(pr, qs) + f(ps, qr) = f(p, q)f(r, s),$$
$$f(pr, qs) + f(ps, qr) = f(p, q)g(r, s),$$
$$f(pr, qs) + f(ps, qr) = g(p, q)f(r, s),$$
$$f(pr, qs) + f(ps, qr) = g(p, q)g(r, s),$$
$$f(pr, qs) + f(ps, qr) = g(p, q)h(r, s),$$

where $p, q, r, s \in G$. Let S be a semigroup written multiplicatively and let $\mathbb{F}$ be either the field of real numbers $\mathbb{R}$ or the field of complex numbers $\mathbb{C}$. Kim and Sahoo (2010, 2011) proved the following theorems:

Theorem 24.6. *Let $f, g : S^2 \to \mathbb{F}$ be functions satisfying*

$$|f(pr,\, qs) + f(ps,\, qr) - f(p, q)\, g(r, s)| \leq \phi(r, s)$$

for all $p, q, r, s \in S$ and for some $\phi : S^2 \to \mathbb{R}_+$. Then either f is bounded or g satisfies

$$g(pr,\, qs) + g(ps,\, qr) = g(p, q)\, g(r, s)$$

for all $p, q, r, s \in S$.

Theorem 24.7. *Let* $f, g : S^2 \to \mathbb{F}$ *be functions satisfying*

$$|f(pr, qs) + f(ps, qr) - g(p,q)f(r,s)| \le \phi(p,q) \quad and \quad \phi(r,s)$$

for all $p, q, r, s \in S$ *and for some* $\phi : S^2 \to \mathbb{R}_+$. *Then either* f *(or* g*) is bounded or* g *satisfies*

$$g(pr, qs) + g(ps, qr) = g(p,q)\, g(r,s)$$

for all $p, q, r, s \in S$.

Theorem 24.8. *Let* $f, g : S^2 \to \mathbb{F}$ *be functions satisfying*

$$|f(pr, qs) + f(ps, qr) - g(p,q)f(r,s)| \le \phi(p,q) \quad and \quad \phi(r,s)$$

for all $p, q, r, s \in S$ *and for some* $\phi : S^2 \to \mathbb{R}_+$. *Then either* f *and* g *are bounded or* f *and* g *satisfy the equation*

$$f(pr, qs) + f(ps, qr) = g(p,q)f(r,s).$$

Moreover, if $f \neq 0$, *then* g *also satisfies*

$$g(pr, qs) + g(ps, qr) = g(p,q)\, g(r,s)$$

for all $p, q, r, s \in S$.

In the next two theorems, G is a group written multiplicatively.

Theorem 24.9. *Let* $f, g : G^2 \to \mathbb{R}$ *and* $\phi : G^2 \to \mathbb{R}$ *be a nonzero function satisfying*

$$|f(pr, qs) + f(ps, qr) - g(p,q)g(r,s)| \le \phi(p,q) \qquad \forall\, p, q, r, s \in G$$

and $|f(p,q) - g(p,q)| \le M$, *and* $|f(p,q) + f(q,p)| \le M'$ *for all* $p, q \in G$ *and some constants* M, M'. *Then either* g *is bounded or* g *satisfies the equation*

$$g(pr, qs) + g(ps, qr) = g(p,q)g(r,s).$$

Theorem 24.10. *Let* $f, g, h : G^2 \to \mathbb{R}$ *and* $\phi : G^2 \to \mathbb{R}$ *be a nonzero function satisfying*

$$|f(pr, qs) + f(ps, qr) - g(p,q)h(r,s)| \le \phi(r,s)$$

for all $p, q, r, s \in G$. *If* $|f(p,q) - g(p,q)| \le M$ *for all* $p, q \in G$ *and some constant* M, *then* g *is bounded or* h *satisfies*

$$g(pr, qs) + g(ps, qr) = g(p,q)g(r,s)$$

for all $p, q, r, s \in G$.

24.4 Exercises

1. Let $f, g : \mathbb{R} \to \mathbb{R}$. If the map

$$(x, y) \mapsto f(x + y) - f(x - y) - g(xy)$$

is bounded on $\mathbb{R}^2$, then prove or disprove that there exists a unique additive function $A : \mathbb{R} \to \mathbb{R}$ such that the map $x \mapsto f(x) - A\left(\frac{x^2}{4}\right) - f(0)$ is bounded on $\mathbb{R}$.

2. Let $f, g : \mathbb{R} \to \mathbb{R}$. If the map

$$(x, y) \mapsto f(x + y + xy) - g(xy) - f(x + y - xy)$$

is bounded on $\mathbb{R}^2$, then prove or disprove that there exists a unique additive function $A : \mathbb{R} \to \mathbb{R}$ such that the maps $x \mapsto f(x) - A(x)$ and $x \mapsto g(x) - 2A(x)$ are bounded on $\mathbb{R}$.

Bibliography

[1] N. H. Abel. Methode generale pour trouver des fonctions d'une seule quantite variable lorsqu'une propriete des fonctions est exprimee par une equation entre deux variables (Norwegian). *Mag. Naturvidenskab.*, 1:1–10, 1823.

[2] J. Aczél. *Bemerkungen 11-12, Funktionalgleichungen 7-11 Okt 1963 (2. Tagung)*. Mathematisches Forschungsinstitut, Oberwolfach, Germany, p. 14, 1963.

[3] J. Aczél. On a generalization of the functional equations of Pexider. *Publ. Inst. Math. Beogard*, 18:77–80, 1964.

[4] J. Aczél. The general solution of two functional equations by reduction to functions additive in two variables and with aid of Hamel-bases. *Glasnik Mat.-Fiz. Astronom. Drustvo Mat. Fiz. Hrvatske*, 20:65–73, 1965.

[5] J. Aczél. *Lectures on Functional Equations and Their Applications*. Academic Press, New York, London, 1966.

[6] J. Aczél. A mean value property of the derivative of quadratic polynomials–without mean values and derivatives. *Math. Mag.*, 58:42–45, 1985.

[7] J. Aczél. *A Short Course on Functional Equations*. D. Reidel Publishing Company, Dordrecht, Holland, 1987.

[8] J. Aczél. The state of the second part of Hilbert's fifth problem. *Bull. Amer. Math. Soc.*, 20:153–163, 1989.

[9] J. Aczél, J. A. Baker, D. Ž. Djoković, Pl. Kannappan, and F. Radó. Extensions of certain homomorphisms of subsemigroups to homomorphisms of groups. *Aequationes Math.*, 6:263–271, 1971.

[10] J. Aczél and Z. Daróczy. *On Measures of Information and Their Characterizations*. Academic Press, New York, 1975.

[11] J. Aczél and J. Dhombres. *Functional Equations in Several Variables*. Cambridge University Press, Cambridge, 1989.

[12] J. Aczél and P. Erdős. The non-existence of a Hamel-basis and the general solution of Cauchy's functional equation for non-negative numbers. *Publ. Math. Debrecen*, 12:259–263, 1965.

[13] J. Aczél and M. Kuczma. On two mean value properties and functional equations associated with them. *Aequationes Math.*, 38:216–235, 1989.

[14] J. Aczél and E. Vincze. Uber eine gemeinsame Verallgemeinerung zweier Funktionalgleichungen von Jensen. *Publ. Math. Debrecen*, 10:326–344, 1963.

[15] K. M. Andersen. A characterization of polynomials. *Math. Mag.*, 69:137–142, 1996.

[16] Th. Angheluta. Sur deux systemes d'equations fonctionnelles. *Mathematica (Cluj)*, 19:19–22, 1943.

[17] H. Anton. *Calculus with Analytic Geometry*. John Wiley & Sons, Inc., New York, 1992.

[18] T. Aoki. On the stability of the linear transformation in Banach spaces. *J. Math. Soc. Japan*, 2:64–66, 1950.

[19] L. M. Arriola and W. A. Beyer. Stability of the Cauchy functional equation over p-adic fields. *Real Anal. Exchange*, 31:125–132, 2005.

[20] T. A. Azlarov and N. A. Volodin. *Characterization Problems Associated with the Exponential Distribution*. Springer-Verlag, New York, 1989.

[21] A. Azzalini and M.G. Genton. On Gauss's characterization of the normal distribution. *Bernoulli*, 13:169–174, 2007.

[22] R. Badora. On the stability of the cosine functional equation. *Rocznik Nauk.–Dydakt. Prace Mat.*, 15:5–14, 1998.

[23] R. Badora and R. Ger. On some trigonometric functional inequalities. In Z. Daróczy and Z. Páles, editors, *Functional Equations–Results and Advances*, volume 3, pages 3–15, Kluwer Academic, Dordrecht, 2002.

[24] J.-H. Bae, K.-W. Jun, and S.-M. Jung. On the stability of a quadratic functional equation. *Kyungpook Math. J.*, 43:415–423, 2003.

[25] J. Bagyinszki. The solution of the Hosszú equation over finite fields. *Közl.–MTA Számitástech. Automat. Kutató Int. Budapest No.*, 25:25–33, 1982.

[26] D. F. Bailey. A mean–value property of cubic polynomials-without mean value. *Mathematics Magazine*, 65:123–124, 1992.

[27] J. Baker. A sine functional equation. *Aequationes Math.*, 4:56–62, 1970.

[28] J. Baker. The stability of the cosine equation. *Proc. Amer. Math. Soc.*, 80:411–416, 1980.

[29] J. Baker, J. Lawrence, and F. Zorzitto. The stability of the equation $f(x + y) = f(x)f(y)$. *Proc. Amer. Math. Soc.*, 74:242–246, 1979.

[30] S. Banach. Sur l'équation fonctionnelle $f(x + y) = f(x) + f(y)$. *Fund. Math.*, 1:123–124, 1920.

[31] W. Benz. Remark on problem 191. *Aequationes Math.*, 20:307, 1980.

[32] G. Birkhoff. *Lattice Theory, 3rd ed.* American Mathematical Society, Providence, RI, 1967.

[33] D. Blanusa. The functional equation $f(x + y - xy) + f(xy) = f(x) + f(y)$. *Aequationes Math.*, 5:63–67, 1970.

[34] C. Borelli. On Hyers-Ulam stability of Hosszú's functional equation. *Results Math.*, 26:221–224, 1994.

[35] C. Borelli and G. L. Forti. On a general Hyers–Ulam stability result. *Internat. J. Math. Math. Sci.*, 18:229–236, 1995.

[36] D. G. Bourgin. Classes of transformations and bordering transformations. *Bull. Math. Soc.*, 57:223–237, 1951.

[37] M. Bousquet-Mélou. On (some) functional equations arising in enumerative combinatorics. In *Proceedings of FPSAC'01*, pages 83–89, Arizona State University, Arizona, USA, 2001.

[38] M. Bousquet-Mélou. On (some) functional equations arising in enumerative combinatorics. In *Proceedings of FPSAC'02*, pages 1–12, University of Melbourne, Melbourne, Australia, 2002.

[39] S. Butler. Problem no. 11030. *Amer. Math. Monthly*, 110:637, 2003.

[40] R. D. Carmichael. On certain functional equations. *Amer. Math. Monthly*, 16:180–183, 1909.

[41] E. Castillo, A. Cobo, J. M. Gutiérrez, and E. Pruneda. *Functional Networks with Applications: A Neural-Based Paradim.* Kluwer Academic Publishers, Boston, 1999.

[42] E. Castillo and M. R. Ruiz-Cobo. *Functional Equations and Modelling in Science and Engineering.* Marcel Dekker, Inc., New York, 1992.

[43] A. L. Cauchy. *Cours d'Analyse de l'Ecole Polytechnique, Vol. 1.* Analyse algebrique, V., Paris, 1821.

[44] I.-S. Chang and H.-M. Kim. Hyers-Ulam-Rassias stability of a quadratic equation. *Kyungpook Math. J.*, 42:71–86, 2002.

[45] I.-S. Chang and H.-M. Kim. On the Hyers-Ulam stability of quadratic functional equations. *J. Inequal. Pure and Appl. Math.*, 3:1–12, 2002.

[46] B. Choczewski and M. Kuczma. Sur certaines equationes fonctionnelles considerees par I. Stamate. *Mathematica (Cluj)*, 27:225–233, 1962.

[47] P. W. Cholewa. The stability of the sine equation. *Proc. Amer. Math. Soc.*, 88:631–634, 1983.

[48] P. W. Cholewa. Remarks on the stability of functional equations. *Aequationes Math.*, 27:76–86, 1984.

[49] J. Chung, S.-Y. Chung, and D. Kim. Generalized Pompeiu equation in distributions. *Appl. Math. Lett.*, 19:485–490, 2006.

[50] J. K. Chung, B. R. Ebanks, C. T. Ng, and P. K. Sahoo. On a functional equation connected with Rao's quadratic entropy. *Proc. Amer. Math. Soc.*, 120:843–848, 1994.

[51] J. K. Chung, B. R. Ebanks, C. T. Ng, and P. K. Sahoo. On a quadratic-trigonometric functional equation and some applications. *Trans. Amer. Math. Soc.*, 347:1131–1161, 1995.

[52] J. K. Chung, B. R. Ebanks, C. T. Ng, P. K. Sahoo, and W. B. Zeng. On a functional equation of Abel. *Results in Math.*, 26:241–252, 1994.

[53] J. K. Chung, Pl. Kannappan, and C. T. Ng. A generalization of the cosine-sine functional equation on groups. *Linear Algebra Appl.*, 66:259–277, 1985.

[54] J. K. Chung, Pl. Kannappan, C. T. Ng, and P. K. Sahoo. Measures of distance between probability distributions. *J. Math. Anal. Appl.*, 139:280–292, 1989.

[55] J. K. Chung and P. K. Sahoo. Characterizations of polynomials with mean value properties. *Appl. Math. Lett.*, 6:97–100, 1993.

[56] S.-Y. Chung. Reformulation of some functional equations in the space of Gevrey distributions and regularity of solutions. *Aequationes Math.*, 59:108–123, 2001.

[57] I. Corovei. The cosine functional equation for nilpotent groups. *Aequationes Math.*, 15:99–106, 1977.

[58] I. Corovei. The sine functional equation for groups. *Mathematica (Cluj)*, 25:11–19, 1983.

[59] I. Corovei. On semi-homomorphisms and Jensen's equation. *Mathematica (Cluj)*, 37:59–64, 1995.

[60] I. Corovei. The d'Alembert functional equation on metabelian groups. *Aequationes Math.*, 57:201–205, 1999.

[61] I. Corovei. The sine functional equation on 2-divisible groups. *Mathematica (Cluj)*, 47:49–52, 2005.

[62] S. Czerwik. The stability of the quadratic functional equation. *Abh. Math. Sem. Univ. Hamburg*, 62:59–64, 1992.

[63] S. Czerwik. The stability of the quadratic functional equation. In *Stability of Mappings of Hyers-Ulam Type* (Th. M. Rassias and J. Tabor, editors), 1:81–91, 1994.

[64] S. Czerwik. *Functional Equations and Inequalities in Several Variables*. World Scientific, Singapore, 2002.

[65] J. d'Alembert. Recherches sur la courbe que forme une corde tendue mise en vibration, I. *Mem. Acad. Sci. Berlin*, 3:214–219, 1747.

[66] J. d'Alembert. Recherches sur la courbe que forme une corde tendue mise en vibration, II. *Mem. Acad. Sci. Berlin*, 3:220–249, 1747.

[67] J. d'Alembert. Addition au Mémoire sur la courbe que forme une corde tendue mise en vibration. *Hist. Acad. Berlin*, pages 355–360, 1750.

[68] J. d'Alembert. Mémoire sur les principes de mécanique. *Hist. Acad. Sci. Paris*, pages 278–286, 1769.

[69] G. Darboux. Sur la composition des forces en statique. *Bull. Sci. Math.*, 9:281–288, 1875.

[70] Z. Daróczy. Elementare Lösung einer mehrere unbekannte Funktionen enthaltenden Funktionalgleichung. *Publ. Math. Debrecen*, 8:160–168, 1961.

[71] Z. Daróczy. Uber die funktionalgleichung $f(x + y - xy) + f(xy) = f(x) + f(y)$. *Publ. Math. Debrecen*, 16:123–132, 1969.

[72] Z. Daróczy. On the general solution of the functional equation $f(x + y - xy) + f(xy) = f(x) + f(y)$. *Aequationes Math.*, 6:130–132, 1971.

[73] Z. Daróczy. On a functional equation of Hosszú type. *Math. Pannon.*, 10:77–82, 1999.

[74] Z. Daróczy and A. Jarai. On the measurable solution of a functional equation of the information theory. *Acta Math. Acad. Sci. Hungaricae*, 34:105–116, 1979.

[75] Z. Daróczy and L. Losonczi. Über die Erweiterung der auf einer Punktmenge additiven Funktionen. *Publ. Math. Debrecen*, 14:239–245, 1967.

[76] R. Dasgupta. Cauchy equation on discrete domain and some characterization theorems. *Theory Probab. Appl.*, 38:520–524, 1993.

[77] R. O. Davies and G. Rousseau. A divided-difference characterization of polynomials over a general field. *Aequationes Math.*, 55:73–78, 1998.

[78] T. M. K. Davison. On the functional equation $f(m + n - mn) + f(mn) = f(m) + f(n)$. *Aequationes Math.*, 10:206–211, 1974.

[79] T. M. K. Davison. The complete solution of Hosszu's functional equation over a field. *Aequationes Math.*, 11:273–276, 1974.

[80] T. M. K. Davison. Problem 191. *Aequationes Math.*, 20:306, 1980.

[81] T. M. K. Davison. A Hosszú-like functional equation. *Publ. Math. Debrecen*, 58:505–513, 2001.

[82] T. M. K. Davison and L. Redlin. Hosszu's functional equation over rings generated by their units. *Aequationes Math.*, 21:121–128, 1980.

[83] R. M. Davitt, R. C. Powers, T. Riedel, and P. K. Sahoo. Flett's mean value theorem for holomorphic functions. *Math. Magazine*, 72:304–307, 1999.

[84] N. G. de Bruijn. On almost additive functions. *Colloq. Math.*, 15:59–63, 1967.

[85] B. de Finetti. Sul concetto di media. *Giorn. Ist. Ital. Attuari*, 2:369–396, 1931.

[86] P. de Place Friis and H. Stetkaer. On the cosine-sine functional equation on groups. *Aequationes Math.*, 64:145–164, 2002.

[87] P. de Place Friis and H. Stetkaer. On the quadratic functional equation on groups. *Publ. Math. Debrecen*, 69:65–93, 2006.

[88] E. Y. Deeba and E. L. Koh. The Pexider functional equations in distributions. *Canad. J. Math.*, 42:304–314, 1990.

[89] S. W. Dharmadhikari. On the functional equation $f(x + y) = f(x) \cdot f(y)$. *Amer. Math. Monthly*, 72:847–851, 1965.

[90] J. Dhombres. *Some Aspects of Functional Equations*. Chulalongkorn University, Department of Mathematics, Bankok, 1979.

[91] H. Drygas. Quasi-inner products and their applications. In *Advances in Multivariate Statistical Analysis,* (A. Gupta, editor), D. Reidel Publishing Co., Dordrecht, 1:13–30, 1987.

[92] E3338. A characterizatic of low degree polynomials. *Amer. Math. Monthly*, 98:268–269, 1991.

[93] B. R. Ebanks, Pl. Kannappan, and P. K. Sahoo. A common generalization of functional equations characterizing normed and quasi-inner-product spaces. *Canad. Math. Bull.*, 35:321–327, 1992.

[94] B. R. Ebanks, Pl. Kannappan, and P. K. Sahoo. Cauchy differences that depend on the product of arguments. *Glasnik Matematicki*, 27:251–261, 1992.

[95] B. R. Ebanks, P. K. Sahoo, and W. Sander. *Characterizations of Information Measures*. World Scientific, Singapore, 1998.

[96] P. Erdős. Problem 310. *Colloq. Math.*, 7:311, 1960.

[97] F. Ergün, S. R. Kumar, and R. Rubinfeld. Approximate checking of polynomials and functional equations. *SIAM J. on Comput.*, 31:550–576, 2001.

[98] L. Euler. *Institutiones calculi integralis, I.* Petropol, Berlin, 1768.

[99] V. Faber, M. Kuczma, and J. Mycielski. Some models of plane geometries and a functional equation. *Colloq. Math.*, 62:279–281, 1991.

[100] V. A. Faiziev. The stability of a functional equation on groups. *Russian Math. Surveys*, 48:165–166, 1993.

[101] V. A. Faiziev, Th. M. Rassias, and P. K. Sahoo. The space of (ψ, γ)-additive mappings on semigroups. *Trans. Amer. Math. Soc.*, 354:4455–4472, 2002.

[102] V. A. Faiziev and P. K. Sahoo. On Drygas functional equation on groups. *Internat. J. Appl. Math. Stat.*, 7:59–69, 2007.

[103] V. A. Faiziev and P. K. Sahoo. On the stability of Drygas functional equation on groups. *Banach J. Math. Analysis*, 1:43–55, 2007.

[104] V. A. Faiziev and P. K. Sahoo. On the stability of the quadratic equation on groups. *Bull. Belgian Math. Soc.*, 1:1–15, 2007.

[105] V. A. Faiziev and P. K. Sahoo. Stability of Drygas functional equation on $T(3, \mathbb{R})$. *Internat. J. Appl. Math. Stat.*, 7:70–81, 2007.

[106] I. Fenyö. Über eine Lösungsmethode gewisser Funktionalgleichungen. *Acta. Math. Acad. Sci. Hung.*, 7:383–396, 1956.

[107] I. Fenyö. On the general solution of a functional equation in the domain of distributions. *Aequationes Math.*, 3:236–246, 1969.

[108] I. Fenyö. On an inequality of P. W. Cholewa. In *General Inequalities* Birkhäuser, Basel-Boston, MA, 5:277–280, 1987.

[109] A. St. Filipescu. Asupra unui sistem de ecuatii functionale. *St. Cerc. Mat.*, 5:739–746, 1969.

[110] T. M. Flett. A mean value theorem. *Math. Gazette*, 42:38–39, 1958.

[111] T. M. Flett. Continuous solutions of the functional equation $f(x+y)+f(x-y) = 2f(x)f(y)$. *Amer. Math. Monthly*, 70:392–397, 1963.

[112] G. L. Forti. An existence and stability theorem for a class of functional equations. *Stochastica*, 4:23–30, 1980.

[113] G. L. Forti. The stability of homomorphisms and amenability, with applications to functional equations. *Abh. Math. Sem. Univ. Hamburg*, 57:215–226, 1987.

[114] G. L. Forti. Hyers-Ulam stability of functional equations in several variables. *Aequationes Math.*, 50:143–190, 1995.

[115] Z. Gajda. On stability of additive mappings. *Internat. J. Math. Math. Sci.*, 14:431–434, 1991.

[116] C. F. Gauss. *Theoria Motus Corporum Coelestium in Sectionibus Conicis Solem Ambientium.* Perthes et Besser, Hamburg, 1809.

[117] J. Ger. On Sahoo-Riedel equations on a real interval. *Aequationes Math.*, 63:168–179, 2002.

[118] R. Ger. Functional inequalities stemming from stability questions. In W. Walter, editor, *General Inequalities*, volume 6, pages 227–240, Birkhaäuser, Basel, 1992.

[119] R. Ger. Superstability is not natural. *Rocznik Naukowo-Dydaktyczny WSP w Krakowie, Prace Mat.*, 159:109–123, 1993.

[120] R. Ger and P. Šemrl. The stability of the exponential equation. *Proc. Amer. Math. Soc.*, 124:779–787, 1996.

[121] J. C. H. Gerretsen. De karakteriseering van de goniometrische functies door middel van een funktionaalbetrekking. *Euclides (Groningen)*, 16:92–99, 1939.

[122] M. Ghermanescu. Caracterisation fonctionnelle des fonctions trigonometriques. *Bull. Inst. Polyt. Jassy*, 4:362–368, 1949.

[123] M. Ghermănescu. *Ecuatii functionale (Romanian).* Editura Academiei Republicii Populare Romine, Bucharest, 1960.

[124] R. Girgensohn and K. Lajkó. A functional equation of Davison and its generalization. *Aequationes Math.*, 60:219–224, 2000.

[125] E. Glowacki and M. Kuczma. Some results on Hosszú's functional equation on integers. *Uniw. Śpohlkaski w Katowicach Prace Nauk-Prace Mat.*, 9:53–63, 1979.

[126] P. M. Gruber. Stability of Isometries. *Trans. Amer. Math. Soc.*, 245:263–277, 1978.

[127] A. Grzaślewicz. Some remarks to additive functions. *Math. Japon*, 23:573–578, 1978.

[128] P. Găvruta. A generalization of the Hyers-Ulam-Rassias stability of approximately additive mappings. *J. Math. Anal. Appl.*, 184:431–436, 1994.

[129] P. Găvruta. Hyers-Ulam stability of Hosszú's equation. In Z. Daróczy and Z. Páles, editors, *Functional Equations and Inequalities*, volume 518, pages 105–110, Kluwer Academic, Dordrecht, 2000.

[130] J. Hadamard. Résolution d'une question relative aux déterminants. *Bull. des Sciences Math.*, 17:240–248, 1893.

[131] G. Hamel. Eine Basis aller Zahlen und die unstetigen Lösungen der Funktionalgleichung $f(x + y) = f(x) + f(y)$. *Math. Ann.*, 60:459–462, 1905.

[132] G. Hardy, J. E. Littlewood, and G. Pólya. *Inequalities*. Cambridge University Press, Cambridge, 1934.

[133] H. Haruki and Th. M. Rassias. New generalizations of Jensen's functional equation. *Proc. Amer. Math. Soc.*, 123:495–503, 1995.

[134] Sh. Haruki. A property of quadratic polynomials. *Amer. Math. Monthly*, 86:577–579, 1979.

[135] K. J. Heuvers. Another logarithmic functional equation. *Aequations Math.*, 58:260–264, 1999.

[136] K. J. Heuvers and Pl. Kannappan. A third logarithmic functional equation and Pexider generalizations. *Aequations Math.*, 70:117–121, 2005.

[137] E. Hewitt and S. Zuckerman. Remarks on the functional equation $f(x + y) = f(x) + f(y)$. *Math. Mag.*, 42:121–123, 1969.

[138] D. Hilbert. Mathematical problems, Lecture delivered before the International Congress of Mathematicians at Paris in 1900. *Bull. Amer. Math. Soc.*, 8:437–479, 1902.

[139] O. Hölder. Über einen Mittelwertsatz. *Göttinger Nachrichten*, pages 38–47, 1889.

[140] M. Hosszú. Remarks on the Pexider's functional equation. *Studia Univ. Babes-Bolyai Ser. Math.-Phys.*, 7:99–102, 1962.

[141] M. Hosszú. A remark on the dependence of functions. *Zeszyty Nauk. Uniw. Jagiellonskiego Prace Mat. Zeszyt.*, 14:127–129, 1969.

[142] D. H. Hyers. On the stability of the linear functional equation. *Proc. Nat. Acad. Sci. U.S.A.*, 27:222–224, 1941.

[143] D. H. Hyers, G. Isac, and Th. M. Rassias. *Stability of Functional Equations in Several Variables*. Birkhauser, Boston, 1998.

[144] W. Jabłoński and A. Pekala. Stability of generalized mean value type functional equation. *Mat. Methody i Fiz.-Mekh. Polya*, 46:110–113, 2003.

[145] A. Jarai. *Regularity properties of functional equations in several variables*. Springer-Verlag, New York, 2005.

[146] J. L. W. V. Jensen. Om Fundamentalligningers Oplösning ven elementäre Midler. *Tidsskr. Mat.*, 4:149–155, 1878.

[147] J. L. W. V. Jensen. Om Lösning at Funktionalligninger med det mindste Maal af Forudsätninger. *Tidsskr. Mat. B*, 8:25–28, 1897.

[148] J. L. W. V. Jensen. Om konvekse funktioner og uligheder imellem middelvaerdier. *Mat. Tidsskr.*, 1905:49–68, 1905.

[149] B. Jessen. Bemaerkinger om konvekse Funktioner og Uligheder imellem Middelvaerdier, I. *Mat. Tidsskr.*, 1931:17–28, 1931.

[150] F. B. Jones. Connected and disconnected plane sets and the functional equation $f(x + y) = f(x) + f(y)$. *Bull. Amer. Math. Soc.*, 48:115–120, 1942.

[151] P. Jordan and J. von Neumann. On the inner products in linear metric spaces. *Ann. Math.*, 36:719–823, 1935.

[152] K.-W. Jun, S.-M. Jung, and Y.-H. Lee. A generalization of the Hyers-Ulam stability of a functional equation of Davison. *J. Korean Math. Soc.*, 41:501–511, 2004.

[153] K.-W. Jun and H.-M. Kim. On the stability of the quadratic equation in Banach modules over a Banach algebra. *Kyungpook Math. J.*, 44:385–393, 2004.

[154] S.-M. Jung. On the Hyers-Ulam-Rassias stability of approximately additive mappings. *J. Math. Anal. Appl.*, 204:221–226, 1996.

[155] S.-M. Jung. On the Hyers-Ulam-Rassias stability of the equation $f(x^2 - y^2 + rxy) = f(x^2) - f(y^2) + rf(xy)$. *Bull. Korean Math. Soc.*, 33:513–519, 1996.

[156] S.-M. Jung. Hyers-Ulam-Rassias stability of functional equations. *Dynamic Syst. Appl.*, 6:541–566, 1997.

[157] S.-M. Jung. On the superstability of some functional inequalities with the unbounded Cauchy difference $f(x+y) - f(x)f(y)$. *Commun. Korean Math. Soc.*, 12:287–291, 1997.

[158] S.-M. Jung. Hyers–Ulam–Rassias stability of Jensen's equation and its application. *Proc. Amer. Math. Soc.*, 126:3137–3143, 1998.

[159] S.-M. Jung. Quadratic functional equations of Pexider type. *Internat. J. Math. Math. Sci.*, 24:351–359, 2000.

[160] S.-M. Jung. *Hyers-Ulam-Rassias Stability of Functional Equations in Mathematical Analysis.* Hadronic Press, Palm Harbor, FL, 2001.

[161] S.-M. Jung. Hyers-Ulam stability of Butler-Rassias functional equation. *J. Inequal. Appl.*, 2005:41–47, 2005.

[162] S.-M. Jung. On a functional equation of Butler-Rassias type and its Hyers-Ulam stability. *Nonlinear Func. Anal. Appl.*, 5:833–840, 2006.

[163] S.-M. Jung and B. Chung. Remarks on Hyers-Ulam stability of Butler-Rassias functional equation. *Dynam. Cont., Dis. Impul. Syst. Ser. A*, 13:193–197, 2006.

[164] S.-M. Jung and P. K. Sahoo. Hyers–Ulam–Rassias stability of an equation of Davison. *J. Math. Anal. Appl.*, 238:297–304, 1999.

[165] S.-M. Jung and P. K. Sahoo. On the Hyers–Ulam stability of a functional equation of Davison. *Kyungpook Math. J.*, 40:87–92, 2000.

[166] S.-M. Jung and P. K. Sahoo. On the stability of a mean value type functional equation. *Demonstr. Math.*, 33:793–796, 2000.

[167] S.-M. Jung and P. K. Sahoo. A functional equation characterizing cubic polynomials and its stability. *Internat. J. Math. & Math. Sci.*, 27:301–307, 2001.

[168] S.-M. Jung and P. K. Sahoo. Hyers-Ulam stability of the quadratic equation of Pexider type. *J. Korean Math. Soc.*, 38:645–665, 2001.

[169] S.-M. Jung and P. K. Sahoo. Hyers-Ulam stability of a generalized Hosszú's functional equation. *Glas. Mat. Ser. III*, 37:283–292, 2002.

[170] S.-M. Jung and P. K. Sahoo. Stability of a functional equation of Drygas. *Aequationes Math.*, 64:263–273, 2002.

[171] S.-M. Jung and P. K. Sahoo. Hyers-Ulam-Rassias stability of a functional equation of Davison in rings. *Nonlin. Funct. Anal. and Appl.*, 11:891–896, 2006.

[172] W. B. Jurkat. On Cauchy's functional equation. *Proc. Amer. Math. Soc.*, 16:683–686, 1965.

[173] S. Kaczmarz. Sur l'équation fonctionnelle $f(x) + f(x + y) = \phi(y)f(x + \frac{1}{2}y)$. *Fund. Math.*, 6:122–129, 1924.

[174] E. Kamke. Zur definition der affinen abbildung. *Jahresbericht der Deutschen Mathematischer-Vereinigung*, 36:145–156, 1927.

[175] J.-H. Kang, C.-J. Lee, and Y.-H. Lee. A note on the Hyers-Ulam-Rassias stability of a quadratic equation. *Bull. Korean Math. Soc.*, 41:541–557, 2004.

[176] Pl. Kannappan. The functional equation $f(xy) + f(xy^{-1}) = 2f(x)f(y)$ for groups. *Proc. Amer. Math. Soc*, 19:69–74, 1968.

[177] Pl. Kannappan. On cosine and sine functional equations. *Ann. Polon. Math.*, 20:245–249, 1968.

[178] Pl. Kannappan. On sine functional equation. *Studia Sci. Math. Hungar.*, 4:331–333, 1969.

[179] Pl. Kannappan. Quadratic functional equation and inner product spaces. *Results Math.*, 27:368–732, 1995.

[180] Pl. Kannappan. Quadratic functional equation and inner product spaces. *Internal. J. Math & Statist. Sci.*, 9:10–35, 2000.

[181] Pl. Kannappan. Rudin's problem on groups and a generalization of mean value theorem. *Aequationes Math.*, 65:82–92, 2003.

[182] Pl. Kannappan. *Functional Equations and Inequalities with Applications*. Springer, New York, 2009.

[183] Pl. Kannappan and G. H. Kim. The stability of the generalized cosine functional equations. *Annales Academiae Paedagogicae Cracoviensis; Studia Mathematica*, 1:49–58, 2001.

[184] Pl. Kannappan, T. Riedel, and P. K. Sahoo. On a generalization of a functional equation associated with Simpson's rule. *Results Math.*, 31:115–126, 1997.

[185] Pl. Kannappan, T. Riedel, and P. K. Sahoo. On a functional equation associated with Simpson's rule. *Wyz. Szkola Ped. Krakow Rocznik Nauk.-Dydakt. Prace Matematyczne*, 196:95–101, 1998.

[186] Pl. Kannappan and P. K. Sahoo. Rotation invariant separable functions are Gaussian. *SAIM J. Math. Anal.*, 23:1341–1351, 1992.

[187] Pl. Kannappan and P. K. Sahoo. Cauchy difference–a generalization of Hosszú functional equation. *Proc. Nat. Acad. Sci. India*, 63:541–550, 1993.

[188] Pl. Kannappan and P. K. Sahoo. Characterization of polynomials and the divided difference. *Proc. Indian Acad. Sci. (Math. Sci.)*, 105:287–290, 1995.

[189] Pl. Kannappan and P. K. Sahoo. Sum form distance measures between probability distributions and functional equations. *Int. J. of Math. Stat. Sci.*, 6:91–105, 1997.

[190] Pl. Kannappan and P. K. Sahoo. A property of quadratic polynomials in two variables. *Jour. Math. Phy. Sci. (Madras)*, 31:65–74, 1997.

[191] Pl. Kannappan and P. K. Sahoo. On generalizations of the Pompeiu functional equation. *Internat. J. Math. Math. Sci.*, 21:117–124, 1998.

[192] Pl. Kannappan, P. K. Sahoo, and J. K. Chung. On a functional equation associated with the symmetric divergence measures. *Utilitas Math.*, 44:75–83, 1993.

[193] Pl. Kannappan, P. K. Sahoo, and J. K. Chung. An equation associated with the distance between probability distributions. *Ann. Math. Silesianal*, 8:39–58, 1994.

[194] Pl. Kannappan, P. K. Sahoo, and M. S. Jacobson. A characterization of low degree polynomials. *Demonstratio Math.*, 28:87–96, 1995.

[195] H. Kestelman. On the functional equation $f(x+y) = f(x)+f(y)$. *Fund. Math.*, 34:144–147, 1948.

[196] G. H. Kim. On the modified Hyers-Ulam-Rassias Stability of the equation $f(x^2-y^2+rxy) = f(x^2)-f(y^2)+rf(xy)$. *J. Chungcheong Math. Soc.*, 10:109–116, 1997.

[197] G. H. Kim. On the stability of the quadratic mapping in normed spaces. *Internat. J. Math. Math Sci.*, 25:217–229, 2001.

[198] G. H. Kim. On the stability of the quadratic equation in Banach modules over Banach algebra. *Kyungpook Math. J.*, 44:385–393, 2004.

[199] G. H. Kim. The stability of the generalized sine functional equations, II. *J. Inequal. Pure Appl. Math.*, 7:1–10, 2006.

[200] G. H. Kim and S. S. Dragomir. The stability of the generalized d'Alembert and Jensen functional equations. *Internat. J. Math. Math. Sci.*, 6:1–12, 2006.

[201] G. H. Kim and P. K. Sahoo. Stability of some functional equations related to distance measure–II. *Annal Funct. Anal.*, 1, 2010.

[202] G. H. Kim and P. K. Sahoo. Stability of some functional equations related to distance measure–I. *To appear*, 2011.

[203] H.-M. Kim. Hyers-Ulam stability of a general quadratic functional equation. *Publ. de L'Institut Math.*, 73:129–137, 2003.

[204] Y.-H. Kim. On the Hyers-Ulam-Rassias stability of an equation of Davison. *Indian J. Pure and Appl. Math.*, 33:713–726, 2002.

[205] M. Kiwi, F. Magniez, and M. Santha. Approximate testing with errors relative to input size. *J. Comp. and Syst. Sci.*, 66(2):371–392, 2003.

[206] E. L. Koh. On Hosszú's functional equation in distributions. *Proc. Amer. Math. Soc.*, 120:1123–1129, 1994.

[207] A. N. Kolmogorov. Sur la notion de la moyenne. *Atti Accad. Nazl. Lincei, Rend.*, 12:388–391, 1930.

[208] M. Kuczma. A survey on the theory of functional equations. *Univ. Beogard. Publ. Elektrotehn. Fak. Ser. Mat. Fiz.*, 130:1–64, 1964.

[209] M. Kuczma. *Functional Equations in a Single Variable*. Monografie Mat. 46. Polish Scientific Publishers, Warsaw, 1968.

[210] M. Kuczma. Note on additive functions of several variables. *Prace naukowe Uniwersytetu Śląskiego Nr 30, Prace matematyczne*, 3:75–78, 1973.

[211] M. Kuczma. *An Introduction to the Theory of Functional Equations and Inequalities*. Uniwersytet Slaski, Warszawa-Krakow-Katowice, 1985.

[212] M. Kuczma. On a Stamate-type functional equation. *Publ. Math. Debrecen*, 39:325–338, 1991.

[213] M. Kuczma. On the quasiarithmetic mean in a mean value property and the associated functional equation. *Aequationes Mathematicae*, 41:33–54, 1991.

[214] M. Kuczma, B. Choczewski, and R. Ger. *Iterative Functional Equations*. Cambridge University Press, Cambridge, 1990.

[215] S. Kurepa. On the quadratic functional. *Publ. Inst. Math. Acad. Serbe Sci.*, 13:58–78, 1959.

[216] S. Kurepa. Functional equation $F(x+y) \times F(x-y) = F^2(x) - F^2(y)$ in n-dimensional vector space. *Monatsh. Math.*, 64:321–329, 1960.

[217] S. Kurepa. On the functional equation $T_1(t + s) T_2(t - s) = T_3(t) T_4(s)$. *Publ. Inst. Math. Beogard*, 2:99–108, 1962.

[218] S. Kurepa. Remarks on the Cauchy functional equation. *Publ. Inst. Math. (Beograd) (N.S.)*, 5:85–88, 1965.

[219] K. Lajkó. Über die allgemeinen Lösungen der Funktionalgleichung $F(x) + F(y) - F(xy) = H(x + y - xy)$. *Publ. Math. Debrecen*, 19:219–223, 1973.

[220] K. Lajkó. Remark on the Hosszu functional equation. *Wyz. Szkola Ped. Krakow, Rocznik Nauk-Dydakt. Prace Mat.*, 12:192–193, 1987.

[221] K. Lajkó. The general solution of Abel-type functional equations. *Results Math.*, 26:336–341, 1994.

[222] K. S. Lau. Characterization of Rao's quadratic entropies. *Sankhya A*, 47:295–309, 1985.

[223] J. Lawrence. The stability of multiplicative semigroup homomorphisms to real normed algebra. *Aequationes Math.*, 28:94–101, 1985.

[224] S. H. Lee and K.-W. Jun. On a generalized Pompeiu functional equation. *Aequationes Math.*, 62:201–210, 2001.

[225] A. M. Legendre. *Elements de géométrie*. Note II, Paris, 1791.

[226] T. Levi-Civita. Sulle funzioni che ammetono una formula d'addizione del tipo $f(x + y) = \sum_{i=1}^{n} X_i(x) Y_i(y)$. *R. C. Accad. Lincei*, 22:181–183, 1913.

[227] N. I. Lobačevskii. Pangeometrie (Russian). *Kasaner gelehrte Schriften*, 1:1–56, 1855.

[228] L. Losonczi. On the stability of Hosszú's functional equation. *Results Math.*, 29:305–310, 1996.

[229] F. Magniez. Multi-linearity self-testing with relative error. *J. Comp. Syst. (TOCS)*, 38(5):573–591, 2005.

[230] B. M. Makarov, M. G. Goluzina, A. A. Lodkin, and A. N. Podkory-tov. *Selected Problems in Real Analysis*. American Mathematical Society, Providence, RI, 1991.

[231] Gy. Maksa. Problem. *Aequationes Math.*, 46:301, 1993.

[232] A. W. Marshall and I. Olkin. Maximum likelihood characterizations of distributions. *Statist. Sinica*, 3:157–171, 1993.

[233] S. C. Milne. New infinite families of exact sums of squares formulas, Jacobi elliptic functions, and Ramanujan's tau function. *Proc. Nat. Acad. Sci. U.S.A.*, 93:15004–15008, 1996.

[234] A. Monreal and M. S. Tomás. On some functional equations arising in computer graphics. *Aequationes Math.*, 55:61–72, 1998.

[235] M. S. Moslehian and Th. M. Rassias. Stability of functional equations in non-Archimedean spaces. *Appl. Anal. Dis. Math.*, 1:1–10, 2007.

[236] M. Nagumo. Über eine Klasse der Mittlewerte. *Japan J. Math.*, 7:71–79, 1930.

[237] A. Najdecki. The stability of a functional equation connected with the Reynolds operator. *J. Inequal. Appl.*, 2007:1–3, 2007.

[238] P. Nakmahachalasint. The stability of a cosine functional equation. *KMITL Sci. J.*, 7:49–53, 2007.

[239] P. Nath and V. D. Madaan. On some extension theorems concerning generalized Cauchy functional equations. *Istanbul niv. Fen Fak. Mecm. Ser. A*, 41:39–52, 1976.

[240] M. Neagu. About the Pompeiu equation in distributions. *Inst. Politehn. Traian Vuia Timisoara Lucrar. Sem. Mat. Fiz.*, 4:62–66, 1984.

[241] M. Neagu. About the symmetric distributions and some extensions of the Pompeiu equation in distributions. *Inst. Politehn. Traian Vuia Timisoara Lucrar. Sem. Mat. Fiz.*, 5:38–40, 1985.

[242] C. T. Ng. Jensen's functional equation on groups. *Aequationes Math.*, 39:85–99, 1990.

[243] C. T. Ng. Jensen's functional equation on groups, II. *Aequationes Math.*, 58:311–320, 1999.

[244] C. T. Ng. Jensen's functional equation on groups, III. *Aequationes Math.*, 62:143–159, 2001.

434   *Bibliography*

[245] A. Ostrowski.   Mathematische Miszellen, XIV. Über die funck-
      tionalgleichung der exponentialfunktion und verwandte funktion-
      algleichungen. *Jber. Deutsch. Math.-Verein.*, 38:54–62, 1929.

[246] R. C. Penney and A. L. Rukhin. D'Alembert's functional equation
      on groups. *Proc. Amer. Math. Soc.*, 77:73–80, 1979.

[247] J. V. Pexider. Notiz über funktionaltheoreme. *Monatsh. Math.
      Phys.*, 14:293–301, 1903.

[248] E. Picard. Deux lecons sur certaines equations fonctionnelles et la
      geometrie non-euclidienne, II. *Bull. Sci. Math.*, 46:425–443, 1922.

[249] E. Picard. *Lecons sur quelques equations fonctionneles avec des
      applications a divers problemes d'analyse et de physique mathe-
      matique.* Gauthier-Villars, Paris, 1928.

[250] S. Pincherle. *Funktionaloperationen und Gleichungen.* In *Encyk-
      lopadie der mathematischen Wissenschaften mit Einschluss ihrer
      Anwendungen, Vol. II.1.2, 761-781.* Teubner, Leipzig, 1906.

[251] S. Pincherle. *Equations et operations fonctionnelles.* In *Encyclo-
      pedie des sciences mathematiques pures et appliquees, Vol. II.5.1,
      II, 26.* Gauthier-Villars, Paris, 1912.

[252] S. D. Poisson. Du paralellogramme des forces. *Correspondance
      sur l'Ecole Polytechnique*, 1:356–360, 1804.

[253] D. Pompeiu. Sur une equation fonctionnelle qui s'introduit dans un
      probleme de moyenne. *Comptes rendus hebdomadaires des seances
      de l'Acade'mie des sciences*, 190:1107–1109, 1930.

[254] D. Pompeiu. Sur une proposition analogue au theoreme des ac-
      croissements finis. *Mathematica (Cluj)*, 22:143–146, 1946.

[255] T. Popoviciu. Sur certaines inégalités qui caractérisent les fonc-
      tions convexes. *An. Stiint. Univ. Al. I. Cuza Iasi Sect. Ia Mat.*,
      11:155–164, 1965.

[256] R. C. Powers, T. Riedel, and P. K. Sahoo. Some models of ge-
      ometries and a functional equation. *Colloq. Math.*, 66:165–173,
      1993.

[257] R. C. Powers, T. Riedel, and P. K. Sahoo. On a model of plane
      geometry. *Elemente der Mathematik*, 51:171–175, 1996.

[258] B. Ramachandran and K. S. Lau. *Functional Equations in Proba-
      bility Theory.* Academic Press, San Diego, 1991.

[259] C. R. Rao and D. N. Shanbhag. *Choquet-Deny Type Functional Equations with Applications to Stochastic Models*. John Wiley & Sons, New York, 1994.

[260] M. Th. Rassias. Solution of a functional equation problem of Steven Butler. *Octogon Math. Mag.*, 12:152–153, 2004.

[261] Th. M. Rassias. On the stability of the linear mapping in Banach spaces. *Proc. Amer. Math. Soc.*, 72:297–300, 1978.

[262] Th. M. Rassias. On the stability of the quadratic functional equation and its applications. *Studia Univ. Babes–Bolyai Math.*, 43:89–124, 1998.

[263] Th. M. Rassias. On the stability of the quadratic functional equation and its applications. *Studia Univ. Babes – Bolyai Math.*, 43:89–124, 1998.

[264] Th. M. Rassias. On the stability of functional equations in Banach spaces. *J. Math. Anal. Appl.*, 251:264–284, 2000.

[265] Th. M. Rassias. On the stability of the quadratic functional equation. *Studia Univ. Babes-Bolyai Math.*, 45:77–114, 2000.

[266] Th. M. Rassias. The problem of S. M. Ulam for approximately multiplicative mappings. *J. Math. Anal. Appl.*, 246:352–378, 2000.

[267] Th. M. Rassias and P. Šemrl. On the behavior of mappings which do not satisfy Hyers-Ulam stability. *Proc. Amer. Math. Soc.*, 114:989–993, 1992.

[268] Th. M. Rassias and P. Šemrl. On the Hyers-Ulam stability of linear mappings. *J. Math. Anal. Appl.*, 173:325–338, 1993.

[269] T. Riedel and M. Sablik. On a functional equation related to a generalization of Flett's mean value theorem. *Internat. J. Math. Math. Sci*, 23:103–107, 2000.

[270] T. Riedel and P. K. Sahoo. On a generalization of a functional equation associated with the distance between the probability distributions. *Publ. Math. Debrecen*, 46:125–135, 1995.

[271] T. Riedel and P. K. Sahoo. A functional equation characterizing low degree polynomials in two variables. *Demonstratio Math.*, 30:85–94, 1997.

[272] T. Riedel and P. K. Sahoo. On two functional equations connected with the characterizations of the distance measures. *Aequationes Math.*, 54:242–263, 1998.

[273] I. Risteski and V. Covachev. *Complex Vector Functional Equations*. World Scientific, Singapore, 2002.

[274] H. Roşcău. Sur l'équation fonctionnelle $\psi(x+y) = f(xy)+\varphi(x-y)$. *Inst. Politehn. Cluj Lucrări Şti.*, 1960:43–46, 1960.

[275] R. A. Rosenbaum and S. L. Segal. A functional equation characterising the sine. *Math. Gaz.*, 44:97–105, 1960.

[276] R. Rubinfeld. On the robustness of functional equations. *SIAM J. Comput.*, 28:1972–1997, 1999.

[277] W. Rudin. Problem E3338. *Amer. Math. Monthly*, 96:641, 1989.

[278] M. Sablik. A remark on a mean value property. *C.R. Math. Rep. Acad. Sci. Canada*, 24:207–212, 1992.

[279] P. K. Sahoo. Circularly symmetric separable functions are Gaussian. *Appl. Math. Letters*, 3:111–113, 1990.

[280] P. K. Sahoo. Three open problems in functional equations. *Amer. Math. Monthly*, 102:741–742, 1995.

[281] P. K. Sahoo. On a functional equation associated with stochastic distance measures. *Bull. Korean Math. Soc.*, 36:287–303, 1999.

[282] P. K. Sahoo. Hyers-Ulam stability of an Abel functional equation. In *Stability of Functional Equations of Ulam-Hyers-Rassias Type*, (Stefan Czerwik, editor), 1:143–149, 2003.

[283] P. K. Sahoo. On a sine-cosine functional equation on commutative semigroups. Technical Report, Department of Mathematics, University of Louisville, Louisville, Kentucky, USA, pages 1-9, 2007.

[284] P. K. Sahoo and T. Riedel. *Mean Value Theorems and Functional Equations*. World Scientific, Singapore, 1998.

[285] P. K. Sahoo and L. Székelyhidi. On a functional equation related to digital filtering. *Aequationes Math.*, 62:280–285, 2001.

[286] P. K. Sahoo and L. Székelyhidi. A functional equation on $\mathbb{Z}_n \oplus \mathbb{Z}_n$. *Acta Math. Hungar.*, 94:93–98, 2002.

[287] P. K. Sahoo and L. Székelyhidi. On the general solution of a functional equation on $\mathbb{Z} \oplus \mathbb{Z}$. *Archives of Mathematics*, 81:233–239, 2003.

[288] A. N. Sarkovskii and G. P. Reljuch. *Introduction to the Theory of Functional Equations (Russian)*. Naukova Dumka, Kiev, 1974.

[289] J. Schwaiger. On a characterization of polynomials by divided differences. *Aequationes Math*, 48:317–323, 1994.

[290] S. L. Segal. On a sine functional equation. *Amer. Math. Monthly*, 70:306–308, 1963.

[291] P. Šemrl. The stability of approximately additive functions. In Th. M. Rassias and J. Tabor, editors, *Stability of Mappings of Hyers-Ulam Type*, pages 135–140, Hadronic Press, Palm Harbor, FL, 1994.

[292] H. N. Shapiro. A micronote on a functional equation. *Amer. Math. Monthly*, 80:1041, 1973.

[293] J. Shore and R. Johnson. Axiomatic derivation of the principle of maximum entropy and the principle of minimum cross-entropy. *IEEE Trans. Inform. Theory*, 26:26–37, 1980.

[294] W. Sierpiński. Sur l'équation fonctionnelle $f(x+y) = f(x)+f(y)$. *Fund. Math.*, 1:116–122, 1920.

[295] P. Sinopoulos. Functional equations on semigroups. *Aequationes Math.*, 59:255–261, 2000.

[296] F. Skof. Proprieta locali e approssimazione di operatori. *Rend. Sem. Mat. Fis. Milano*, 53:113–129, 1983.

[297] F. Skof. Approssimazione di funzioni δ-quadratiche su dominio restretto. *Atti Accad. Sci. Torino Cl. Sci. Fis. Mat. Natur.*, 118:58–70, 1984.

[298] W. Smajdor. On a functional equation of Abel. *Rocznik Nauk.-Dydakt. Prace Mat.*, 16:85–93, 1999.

[299] J. Smital. *On Functions and Functional Equations*. Adam Hilger, Bristol and Philadelphia, 1988.

[300] D. R. Snow. Formulas for sums of powers of integers by functional equations. *Aequationes Math.*, 18:269–285, 1978.

[301] W. Stadje. ML characterization of the multivariate normal distributions. *J. Multivariate Anal.*, 46:131–138, 1993.

[302] I. Stamate. Ecuatii functionale de tip Pexider, nota a II-a. *Buletinul Stiintific al Inst. Politechnic din Cluj*, 7:60–61, 1964.

[303] I. Stamate. Equations fonctionnelles contenant plusieurs functions inconnues (Roumanian). *Univ. Beogard. Publ. Elektrotehn. Fak. Ser. Mat. Fiz.*, 356:123–156, 1971.

[304] H. Stetkaer. d'Alembert's equation and spherical functions. *Aequationes Math.*, 48:220–227, 1994.

[305] G. Stokes. On the intensity of light reflected from or transmitted through a pile of plates. *Proc. Roy. Soc. (London)*, 11:545–557, 1860.

[306] H. Swiatak. On the functional equation $f(x + y - xy) + f(xy) = f(x) + f(y)$. *Aequationes Math.*, 1:239–241, 1968.

[307] H. Swiatak. On the functional equation $f(x + y - xy) + f(xy) = f(x) + f(y)$. *Mat. Vesnik*, 5:177–182, 1968.

[308] H. Swiatak. A proof of the equivalence of the equation $f(x + y - xy) + f(xy) = f(x) + f(y)$ and Jensen's functional equation. *Aequationes Math.*, 6:24–29, 1971.

[309] L. Székelyhidi. On a theorem of Baker, Lawrence and Zorzitto. *Proc. Amer. Math. Soc.*, 84:95–96, 1982.

[310] L. Székelyhidi. The stability of d'Alembert-type functional equations. *Acta Sci. Math. (Szeged)*, 44:313–320, 1983.

[311] L. Székelyhidi. *Convolution Type Functional Equations on Topological Abelian Groups.* World Scientific, Singapore, 1991.

[312] L. Székelyhidi. Fréchet's equation and Hyers theorem on noncommutative semigroups. *Ann. Polon. Math.*, 48:183–189, 1998.

[313] J. Tabor. Hosszú's functional equation on the unit interval is not stable. *Publ. Math. Debrecen*, 49:335–340, 1996.

[314] S.-E. Takahasi, T. Miura, and H. Takagi. Exponential type functional equation and its Hyers-Ulam stability. *J. Math. Anal. Appl.*, 329:1191–1203, 2007.

[315] H. Teicher. Maximum likelihood characterization of distributions. *Ann. Math. Statist.*, 32:1214–1222, 1961.

[316] H. P. Thielman. On generalized Cauchy functional equations. *Amer. Math. Monthly*, 56:452–457, 1949.

[317] T. Trif. Hyers-Ulam-Rassias stability of a Jensen type functional equation. *J. Math. Anal. Appl.*, 250:579–588, 2000.

[318] T. Trif. On the stability of a functional equation deriving from an inequality of Popoviciu for convex function. *J. Math. Anal. Appl.*, 272:604–616, 2002.

[319] I. Tyrala. The stability of d'Alembert functional equation. *Aequationes Math.*, 69:250–256, 2005.

[320] S. M. Ulam. *A Collection of Mathematical Problems*. Chapter VI, Interscience Publishers, New York, 1960.

[321] S. M. Ulam. *Problems in Modern Mathematics*. Science Editions, John Wiley & Sons, New York, 1964.

[322] R. Vakil. *A Mathematical Mosaic: Patterns and Problem Solving*. Brendan Kelly Pub., Burlington, Ont., 1996.

[323] J. G. van der Corput. A remarkable family. *Euclides (Groningen)*, 18:50–64, 1941.

[324] E. Van Vleck. A functional equation for the sine. *Ann. of Math.*, 11:161–165, 1910.

[325] H. E. Vaughan. Characterization of the sine and cosine. *Amer. Math. Monthly*, 62:707–713, 1955.

[326] L. Vietoris. Zur Kennzeichnung des Sinus und verwandter Funktionen durch Funktionalgleichungen. *J. Reine Angew. Math.*, 186:1–15, 1944.

[327] E. Vincze. Über die Verallgemeinerung der trigometrischen und verwandten Funktionalgleichungen. *Ann. Univ. Sci. Budapestensis, Sect. Math.*, 4:389–404, 1960.

[328] E. Vincze. Eine Allgemeinere Methode in der Theorie der Funktionalgleichungen, II. *Publ. Math. Debrecen*, 9:314–323, 1962.

[329] E. Vincze. *Über eine Verallgemeinerung der Pexiderschen Funktionalgleichungen*. Studia Univ., Babes-Bolyai, 1962.

[330] E. Vincze. Eine Allgemeinere Methode in der Theorie der Funktionalgleichungen, III. *Publ. Math. Debrecen*, 10:283–318, 1963.

[331] P. Volkmann. Zur stabilität der Cauchyschen und der Hosszúschen funktionalgleichung. *Seminar*, 5:1–5, 1998.

[332] J. Wang. The stability of approximately additive solutions on Davison equations. *Nonlinear Funct. Anal. Appl.*, 7:101–114, 2002.

[333] K. Weierstrass. *Zur Determinantentheorie*. Berlin, 1886.

[334] J. H. C. Whitehead. A certain exact sequence. *Ann. of Math.*, 52:51–110, 1950.

[335] A. Wilansky. Additive functions. In Kenneth O. May, editor, *Lectures on Calculus*, pages 97–124, Holden-Day, Inc., Toronto, 1967.

[336] D. Yang. Remarks on the stability of Drygas' equation and the Pexider-quadratic equation. *Aequationes Math.*, 68:108–116, 2004.

[337] D. Yang. The stability of the quadratic functional equation on amenable groups. *J. Math. Anal. Appl.*, 291:666–672, 2004.

[338] D. Yang. The quadratic functional equation on groups. *Publ. Math. Debrecen*, 66:327–348, 2005.

[339] G. S. Young. The linear functional equation. *Amer. Math. Monthly*, 65:37–38, 1958.

Index